A pair of WHARFEDALES

A pair of WHARFEDALES

The Story of Gilbert Briggs and his Loudspeakers

DAVID BRIGGS

Published by:
IM Publications LLP, 6 Charlton Mill, Charlton, Chichester, West Sussex PO18 0HY, UK
Tel: +44-1243-811334, Fax: +44-1243-811711, E-mail: info@impublications.co.uk
Web: www.impublications.com
Book website: www.apairofwharfedales.com

© David Briggs 2012

David Briggs asserts his moral rights to be identified as the author of this work.

ISBN: 978-1-906715-14-4
British Library Cataloguing-in-Publication Data
A catalogue record for this book is available from the British Library

All rights reserved. No part of this publication may be reproduced, stored or transmitted, in any form or by any means, except with the prior permission in writing of the publisher.

Designed and typeset by Edge Creative (www.edgecreative.com)
Printed in the UK by Latimer Trend & Company Ltd, Plymouth

Contents

Foreword .... Page viii

Preface .... Page x

Introduction .... Page xiii

Chapter 1: Origins .... Page 3

Chapter 2: 1894–1905 .... Page 9
School in Clayton, King's Lynn and Halifax

Chapter 3: 1906–1933 .... Page 17
Textiles Export Merchant in Bradford,
Manchester and the Far East

Chapter 4: 1932–1937 .... Page 39
Part 1. Wharfedale Wireless Works in Bradford
Part 2. Family Life and Music

Chapter 5: 1937–1946 .... Page 67
Part 1. Wharfedale Wireless Works in Brighouse
Part 2. Home Life During the War

Chapter 6: 1946–1952 .... Page 97
Part 1. Wharfedale Wireless Works in Idle
Part 2. Away from Wharfedale

Chapter 7: 1953–1958 .... Page 137
Part 1. Wharfedale Wireless Works Ltd in Idle
Part 2. Away from Wharfedale

Chapter 8: 1959–1965 .... Page 171
Part 1. Wharfedale Wireless Works Ltd (within Rank
Organisation Ltd) in Idle
Part 2. Away from Wharfedale

Chapter 9: 1965–1978 .... Page 215
The Evolution of Wharfedale under Rank

Chapter 10: 1965–1978 .... Page 237
        The Wharfedale Book Department, Ilkley

Chapter 11: The Concert Hall Lecture-Demonstrations .... Page 253

Chapter 12: The Audio Books .... Page 293

Epilogue .... Page 320

Appendices: .... Page 322
        1. Reproduced letters (BBC Written Archive)
        2. Concert-demonstration programme details
        3. G.A. Briggs publications
        4. Wharfedale products from 1932 to 1978 (with prices)

Notes and References .... Page 344

Index .... Page 352

For Ninetta and Valerie

Foreword

This biography is important both because it records the life of one of Britain's most successful 20th century engineers, who many have referred to as the father of hi-fi sound, and because it puts on record the leading role played by British engineers in perfecting the reproduction of sound. Gilbert Briggs was a self-made engineer who recognised that the loudspeaker was the weakest link in an audio system and set out to overcome this limitation, and to a large extent he succeeded. David Briggs tells his story in a clear and captivating style that is sympathetic, sensitive and highly readable, and yet, through his comprehensive research, it is detailed and complete. The book is greatly enhanced by fine photographs and illustrations that immediately transport the reader back to the decades before and after the Second World War, when Gilbert was founding and growing his company. He also takes time to set the technological scene, explaining how audio technologies evolved before and during the early days of domestic electronics, and reminding us of the important role played by British engineers.

I learned of this book as a result of the latter. I was in conversation with Mark Tully on *Something Understood*, a BBC radio programme that gently challenges its audience early on Sunday mornings to think in some depth, and with the aid of well chosen music, about a subject of current interest. We were discussing, in the words of the programme's creators, 'the history and contribution, the rise and fall and potential new rise of the professional engineer' and were looking back at how the role of engineers had changed over the 20th century, especially how it had declined over the last fifty years. Mark asked me what had brought me from Australia to Britain in 1959. I grew up in Melbourne and had just completed my undergraduate studies in physics and electronics at Melbourne University, and in my spare time had been making hi-fi sets for rich people, especially graziers (they did not call themselves farmers in those heady days of high wool prices), and was intent upon pursuing a career in electronics. I told Mark that I came to England because it led the world in high fidelity sound and in many other areas of consumer electronics, such as colour television, and I cited Wharfedale loudspeakers as the archetypal example of world-leading British components. I had used them in the sets I built, together with Wharfedale designs for cabinets. The last of these was a stereo set with column speakers which were highly efficient. The grazier for whom it was built placed the set on his Australian-style veranda, with the speakers about 20 feet apart, and sat on the lawn 40 feet away to listen to it. He was thrilled when his neighbour about half a mile away called him to ask where the brass band had come from. I had bought him a microgroove recording of Souza marches as a test record!

Shortly after the programme was broadcast I received a letter from David, who told me that he was working on this book and asked me, amongst other things, if I had any material relating to Wharfedale speakers in Australia in the 1950s. I did, and I also had information about the speakers that they sold in the USA, as I had purchased a pair of large Wharfedale speakers in sand-filled enclosures when I moved to the USA to work for IBM in the 1960s. I built my own electronics to drive these speakers, with 60 watt valve amplifiers that I still use today, as their bandwidth and distortion characteristics remain unsurpassed. Today's sound systems employ many channels to create surround sound effects that were not available in those days, but the fidelity of the sound produced has changed little since that time, largely because the performance of loudspeakers was already so advanced.

Building a technology business takes a broad set of skills; Gilbert Briggs possessed almost all of them and was a good judge of others whom he could attract to complete the set. In addition to his engineering abilities, which he acquired by seeking knowledge from others and self study, he was a natural business man, who knew that high quality was essential but cost control of equal importance. At the same time he was a brilliant salesman who understood his market and, indeed, through inspired outreach he created and nurtured the market. He was a talented writer and speaker, who knew and loved music, and he used his knowledge and talents to educate his customers and attract new ones. The book captures skillfully the risk and excitement that surrounded the staging of the remarkable concert-demonstrations in the Festival Hall in London and in Carnegie Hall in New York. There are few engineers or businessmen with the chutzpah to mount such high profile events.

As an engineer he realised that it was not just the loudspeaker driver that was important but also the ways in which the electronic signal was coupled to the driver and the configuration, shape, and materials of the enclosure. Gilbert was truly the renaissance audio expert and his company manufactured a complete set of components to enable customers to produce sound of world-leading fidelity.

Enthusiasts of hi-fi sound in its early days will find this book a compelling read as they, at last, learn who was behind the Wharfedale speakers that were their pride and joy, and how their creator took on international competition and produced the best high fidelity speakers of his time. For those born in the last fifty years, most of whom take high fidelity sound as given, it presents the chance to learn of the creative advances that were needed to achieve the clear and transparent sound they enjoy and take for granted.

Professor the Lord Broers FREng, FRS

Preface

Anyone who has driven along the M62 between Manchester and Leeds will know the Pennine moors which stretch across Lancashire and Yorkshire between Oldham and Halifax. I was born in 1948 and brought up on the very edge of these moors just on the Lancashire side of the border, not far from Oldham. My father, on the other hand, had been born in Queensbury, near Bradford, which from my childhood perspective was across the moors and just the other side of Halifax. His sister lived there all her life. We were lucky to have a car and some of my earliest memories are of challenging car journeys to see my older cousins in Queensbury. At one or more of these gatherings, I recall the grown-ups talking about 'Cousin Gilbert'. He was clearly a person of some note—practically famous—who was an inventor, an authority on loudspeakers, and author of a book or books on 'sound reproduction'. Fascinating stuff. If there was also mention of his running a company, or its name, it did not register.

In 1971 when I was earning a living and had a decent-sized space to fill with music I abandoned my record player and bought my first hi-fi system. This included an amplifier and speakers made by Wharfedale (I still possess the speakers, fortunately). When my father heard of this he exclaimed 'that was Cousin Gilbert's Company'. He later produced a copy of *Sound Reproduction* signed by G.A. Briggs, which I eventually inherited. At that point I knew I would some day find out more about Gilbert and Wharfedale. Retirement, and the ability to work on family trees via the internet, finally provided the opportunity. Only then did I actually find out what kind of 'cousin' he was and, inevitably, become drawn into his remarkable life-story.

I discovered that my grandfather was Gilbert's first cousin (their fathers were brothers). Therefore, Gilbert's grandchildren would be my first cousins, three times removed. By trawling through post-1910 marriages and births records and BT residential numbers on-line I eventually traced one of them and this led to contact with both of Gilbert's daughters, Ninetta and Valerie. Amazingly, the latter lives less than half a mile from me and this turned out to be a very good omen. Without their help and support this book would have been impossible. In a way, this research has been guided by Gilbert himself who provided plenty of signposts in his own books. His brief reminiscences written up in two chapters of the fifth edition of *Loudspeakers* and his entry in *Audio Biographies* were the milestones but the other books are littered with facts and clues about his life in and out of Wharfedale, often in the most unlikely places, with his introductions to the contributors in *Audio Biographies* a particularly rich source. Material from his early life and nearly 28 years in the textile export business has miraculously survived, only coming to light as a result of this research. As to the history of Wharfedale, I have been very fortunate to trace several key employees, or their family members, and, through them, gain invaluable personal insights, photographs and technical information. These are acknowledged below.

Acknowledgements

I owe a particular debt to the Escott family. Bill Escott worked at Wharfedale from 1946 to 1973 and was Gilbert's deputy for much of that time. His younger brother, Phil, was there from 1971 to 1998 and for many years he ran the service department. I was unable to interview Bill because of his health, but I am grateful to his wife, Jacqueline, for allowing Phil to search through his files. Between them they have provided me with important facts and photographs, many of Gilbert's books and a treasure-trove of technical literature. Dorothy Stevens joined Wharfedale as 'office junior' in 1943 and left in 1968; she was my link to the early days. Dorothy Dawson became Gilbert's secretary in 1961 and she stayed with him, following his 'retirement' to run the 'Wharfedale Book Department' in Ilkley, until his death in 1978. Ken Russell was technical manager from 1962 to 1982. These three all knew Gilbert well and between them covered most of the period of this Wharfedale story. As well as providing pictures and documents they patiently put up with my probing of events which were at least 40 years ago, and to which I returned many times over the course of the project, as the jigsaw came together.

Raymond Cooke, one of the founders of KEF Electronics, was Gilbert's technical manager from 1955 to 1961. I was fortunate to discover his first wife, Marjorie, and two children, Ann Crayford and Martin Cooke, late in the project. Through them I found John Ball, co-founder of KEF, who was briefly Gilbert's understudy from 1959 to 1961. They provided recollections, photographs and documents which answered several key remaining questions. John Collinson, whilst working for Quad, was very involved in the 'live versus recorded' concert-demonstrations and he also worked at Wharfedale from 1966 to 1973. He too, provided me with unique memories and material. Alex Garner, worked in R&D with both Collinson and Russell; he was my source of detailed information about that activity during his time from 1969 to 1976.

Away from the direct connections with Wharfedale there have been many contributors. My cousin, Susan Smith, looked after me during my trips to the Bradford area and also carried out local research. John Patchett made available his extensive research into the origins of Queensbury and the early Briggses. My other cousin, Mansell Jagger, loaned me Phineas Briggs' cuttings book and material about Black Dyke Mills and early Clayton, whilst Stuart Downey sent me old pictures of the village. Rose Taylor explored the archives of Crossley and Porter's School, David Pyett provided information about musical societies in Ilkley and Michael Callaghan found original negatives in the C.H. Wood collection, held in the Bradford Industrial Museum. Paddy O'Connell (presenter of *Broadcasting House* on BBC Radio 4) was my intro. into the BBC Written Archive, where Erin O'Neill and Trish Hayes uncovered facts and documents. Roger Beardsley professionally digitised tape recordings of Gilbert and Ian Thompson carried out photography and difficult image restoration. I obtained invaluable help from many librarians, in particular John Shepherd of the University of California at Berkley and John Hillsden of the Radcliffe Science Library, Oxford University. John Grant scanned material held by the University of Dayton, Ohio. Staff at Southampton and Birmingham University libraries, the Bodleian Library (Oxford) and at the libraries in Bradford, Brighouse, King's Lynn and Malvern all found information for me, as did staff of the West Yorkshire Archive in Wakefield. Heather Lane of the Audio Engineering Society and Anne Locker of the Institute of Engineering and Technology did the same. The following all contributed to the project in one way or another: Julian Alderton, Sue Bee, John Borwick, Fr Anselm Cramer, Michael Fountain, Malcolm Green, Jennie Goossens, Mervyn Grimshaw, John Handley, Tim Harris, Clare Lee, Stuart McLaughlan, David Patching, John and Chris Pitchford, Christine Randall, Stefan Sergent, Stephen Spicer, Judy Smith, Felicity Stubbs, Leslie, Ruth and Jonathan Theobald, Simon Waddington and Andrew Watson. My grateful thanks to them all.

The current owners of 'Wharfedale' are the International Audio Group (IAG) which continues to manufacture award-winning loudspeakers. I should like to thank the President, Michael Chang, for all the help I have received from IAG during this project and for the permission to reproduce

xi

images relating to Wharfedale products. The source of all other images is provided in the captions, with reproduction permission details where the original photographer could be identified.

Finally, my thanks to Lord Broers for his Foreword and other contributions. It was my hope, once the project got underway, to find someone who had built loudspeakers to Wharfedale designs and using their drive units during the late 1950s, when Gilbert Briggs' influence was at its peak. A sound-bite from a radio programme put me on the trail of Alec Broers, distinguished engineer/scientist and Vice-Chancellor of the University of Cambridge from 1996 to 2003. I am confident that had Gilbert heard the story about the Australian grazier, related in the Foreword, he would have dined out on it for years.

Introduction

During the four years of researching and writing this book, there were many occasions on which I asked complete strangers whether they had heard of Wharfedale loudspeakers. Almost invariably in the case of men, and also for a substantial number of women, aged over 50, the answer would not only be 'yes', but would be followed by either: I have a pair (in use, in the garage or in the loft); I once had a pair; my father had a pair; I wished I could have afforded a pair. Out of their mouths came the title for the book. Not many, though, knew anything about the remarkable man behind the loudspeakers and the firm that made them.

Gilbert Arthur Briggs, G.A.B. to those who knew him well, was born into a humble family in a Yorkshire textile village in 1890. His father died when he was nine and he went to an orphans' school. Despite this inauspicious start in life he rose to become a director of a firm of textile export merchants in Bradford, but was virtually bankrupted in the Depression. A passion for music and a love affair with the piano had led to an interest in loudspeakers and, as the textile industry collapsed, he started a sideline to make them, called Wharfedale Wireless Works. He was forced to turn this into his full-time occupation in 1933. He had no relevant, theoretical or practical training, yet both he and Wharfedale became internationally famous, and when he died Gilbert's obituaries referred to him as 'the father of hi-fi'. His contribution to the development and popularity of 'hi-fi' was unique and profound.

As well as designing loudspeakers and growing his company, he wrote twenty-one books on various aspects of sound reproduction and audio for amateurs, published through Wharfedale, with total sales worldwide of well over a quarter of a million copies. He also staged a series of over twenty audacious lecture-demonstrations in major concert halls during the 1950s (including the Royal Festival Hall, London and Carnegie Hall, New York) which featured 'live versus recorded' performances.

A straight talking Yorkshireman, with an impish sense of humour, Gilbert possessed many human qualities which drew people to him. Very few recruits to Wharfedale during his 30 years at the helm left the firm and, through a combination of sincerity, integrity, charm and prodigious letter writing, he built up, and maintained, a huge network of friends throughout the audio world. It was their willingness to help him, when asked, which made his writing and concert activities both possible and successful.

Once he left the 'rag trade', Gilbert rarely referred to this first career of nearly 28 years. Unlike most of his peers amongst the 'audio pioneers' he had not gone into the industry as a young man and his formative experiences and influences were quite different. However, the foundations for the things that he did with Wharfedale, which so surprised his colleagues at the time, were all laid during this period—indeed by the time he was 30. One of the fascinations of my research was to uncover these traits and see how they played out in later life. This, then, is the story of Wharfedale, the company and its products, and the remarkable G.A. Briggs who started it all.

For the benefit of readers with different interests, I have tried to keep these two narrative streams separate, in as much as this is possible, whilst writing chronologically. Chapters covering the period when Gilbert was running Wharfedale (1933–64) are in two parts, the first dealing with Wharfedale and the second mainly biographical. Once he retired, but continued to run the Wharfedale 'Book Department' from his Ilkley office, separate chapters cover this period in his life and events at Wharfedale until his death in 1978. Although the sagas of his books and concert-demonstrations are covered briefly in the chronology, they are described fully in two final chapters. A short Epilogue brings the Wharfedale story up-to-date. The history of the company's products is quite detailed and a full list, with prices, from during Gilbert's lifetime, can be found in Appendix 4.

I have avoided excessive referencing. Where possible and appropriate I have quoted from Gilbert's own writing; the book extracts are referenced. I have also referenced specific information from books, journal/magazine articles, tape recordings and the occasional website, as well as books which were major sources of background material. Other background information which is readily available from the internet, has been corroborated, but not referenced. The remaining information came from interviews with, and documents provided by, the Briggs family and the many others listed in the Preface.

Finally, some comments about units. During the period covered, dimensions were in feet and inches and these have been retained, the latter being abbreviated, e.g. 6" for 6 inches. Prices were in pounds, shillings and pence (£, s, d) with 20 shillings to the pound and 12 pence to the shilling. Often prices would be expressed in just shillings and pence using a shorthand form, e.g. 39/6 for 39s 6d. In Appendix 4 the prices have been converted into the current decimalised system, so 39/6 becomes £1 98p (rounded up). From time to time, in the text, an approximate equivalent price today is given, for which I used the inflation calculator at http://safalra.com/other/historical-uk-inflation-price-conversion/.

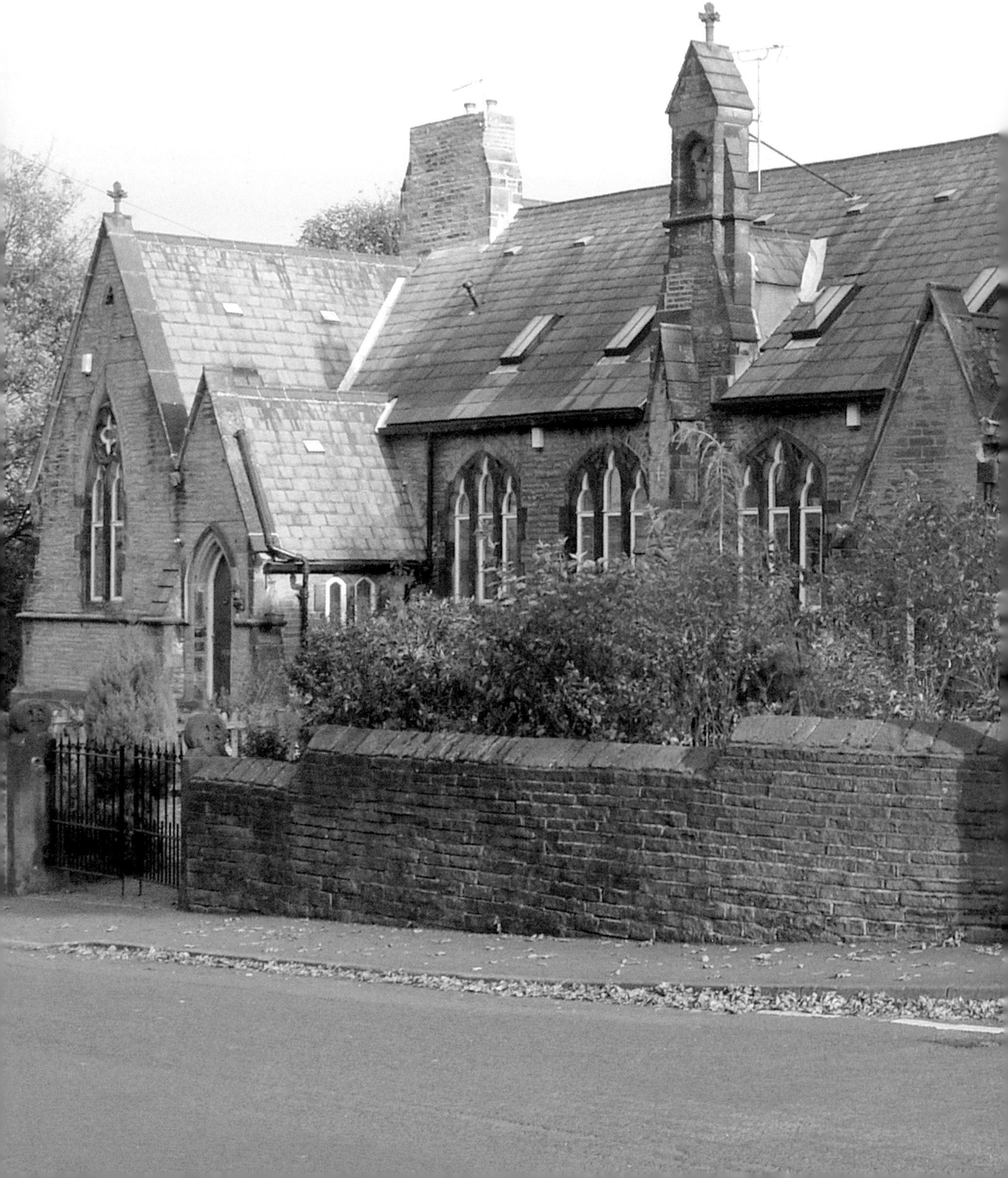

CHAPTER ONE: ORIGINS

Gilbert Arthur Briggs was born on 29 December 1890 in the village of Clayton, on the western edge of Bradford, just off today's main Bradford to Halifax road, the A647. Originally a Saxon farming settlement, the village, which retains most of its 19th century character (1.1–1.3), was incorporated into Bradford in 1930. When Gilbert was born, the population of Clayton (the village and surrounding farms) was under 4000, but only 10 years later, at the turn of the century, it had risen to nearly 5000. At that time the local economy was dominated by the mainly wool-based textiles industry, with quarrying and associated masonry trades also providing significant employment. A fireclay works, established in 1880, exploited the local clay and was known for its glazed bricks.

In 1961, when Gilbert was asked indirectly by 'Free Grid', the anonymous columnist of *Wireless World,* whether he was ' …descended from Henry Briggs, also a Yorkshireman,* who collaborated with Napier on the production of Logarithms.' Gilbert replied that he ' …could only trace [his] ancestry back two generations (in spinning and weaving—not science)'.[2] In fact, spinning and weaving had been an integral part of his family history for far longer than that and his ancestors can be traced back at least six generations, to the time when they were known, not as Briggs, but as Brigg.

Ancestry and the Importance of Textiles

Gilbert's earliest ancestor, for certain, is Jonathan Brigg (~1713–1788) who, from 1742, was tenant of an estate consisting of a farmhouse with several fields in the 'township' of Clayton, which was to become known as New House. (A 'township' was a civil administration district covering a rather large area of land which in this case included, and was named after, Clayton village.) Jonathan Brigg was a 'manufacturer' as well as a farmer. He would buy wool in bulk and distribute it to local cottages, where it would go through a sequence of processes, including washing, combing, spinning and weaving, to produce worsted cloth—locally known as 'stuff'—in standard lengths or pieces. Workers would be paid 'piece rate' and the finished cloth would be taken to the Piece-Halls at Halifax and Bradford for sale. This area, and the West Riding of Yorkshire in general, was especially suitable for this activity, because the coarse grass growing on the local peat provided excellent grazing for sheep and the soft water was ideal for wool washing. Farming and textile work within the home, involving most family members, went hand-in-hand.

Jonathan's only son Abraham (1738–1809) took over the New House tenancy in 1789, whilst one of his grandsons, John (1764–1842), was tenant of the nearby Harrowins estate and was also described as a manufacturer. With the coming of mechanisation and the harnessing of water power, this domestic system of textile manufacturing was gradually replaced by the factory system (in

---

*A genealogical link to the mathematician Henry Briggs (1561–1630) is not such a fanciful notion, because he was born at Warleywood, a parish of Halifax.[1]

A Pair of Wharfedales: The Story of Gilbert Briggs and his Loudspeakers

Figure 1.1 Green End, Clayton in 1912. The village centre is out of view, beyond the left hand bend, see Figure 1.2, and Clayton Lane begins further up the street, behind the camera, see Figure 1.3. The building by the gas lamp was an infants' school, built by the Baptists in 1873 and taken over by the School Board in 1891, which Gilbert probably attended. (Reproduced from their collection, by permission of the Clayton History Group.)

Figure 1.2 Wells, Clayton in 1930. Looking from Clayton village centre towards Clayton Green, 1930. (Reproduced, from their collection, by permission of the Clayton History Group.)

Chapter One: Origins

Figure 1.3 Greenside, Clayton in 1910. Looking from Green End up Clayton Lane. (Reproduced, from their collection, by permission of the Clayton History Group.)

'mills') from about 1810. Around 1750 the surname Brigg started to transform into Briggs, with the two forms becoming interchangeable in records and documents; by about 1830 the form Briggs had become dominant, although census entries can be ambiguous as late as 1861.

John Brigg(s) of Harrowins was Gilbert's great-great grandfather. He married Mary Firth and had two sons, the younger, Joseph Firth Briggs (1790–1857), being Gilbert's great grandfather. Joseph was a textiles worker, recorded as a weaver in 1841 and a woolcomber in 1851 whilst living at Swamp/Causeway End, now in Queensbury. Joseph and his wife Hannah Field had seven children and the youngest, George (1830–1896), was Gilbert's grandfather. George married Mary Sutcliffe in 1855 and they lived at Hill Top/Moor End, again now in Queensbury, before moving to Back Lane in Clayton village. Having followed his father's trades as a young man he became a warehouseman, possibly at the only textile mill in Clayton at the time, Beck Mill, which was opened in 1845 by John Milner and Company. George and Mary had six children, the fourth of whom was Phineas, Gilbert's father, born in 1864.

Phineas Briggs

Phineas would have started work part-time aged 10–11 and in the 1881 census, aged 17, he is described as a 'mill hand'. He was clearly very capable because in 1891 he was already overlooker, or foreman, in a worsted spinning mill and by 1900 he was a mill manager. Progress beyond overlooker was unusual. Phineas married Mary Ann Emsley, a farmer's daughter, in 1889 and they set up home in a two-bedroom, back-to-back, terraced house at 7 Cranbrook Street, off Clayton Lane, in Clayton. Claris, Gilbert Arthur and Mabel were born there in 1889, 1890 and 1892 respectively (1.4). A fourth child, Bernard Joseph, was born in 1899.

We know something of Phineas from a 'cuttings' book dated 6 February 1892. This includes letters which he wrote to the *Bradford Observer* during 1890–1893, and related correspondence, a paper written by him and read to the Airdale District Co-operative Conference of 1893, held at Clayton, and a large number of cuttings from newspapers and other publications covering a bewildering range of subjects. Marginal notes are in Pitman shorthand and the book is indexed in beautiful copper-plate script. Evidently he was keenly interested in politics, an active member of the Liberal party and heavily involved in the Co-operative movement. His writings of 1891–1893, when he was still in his late 20's, show him to be well-informed, passionate and confident and well able to argue his case in print. He was also interested in natural history as well as medical progress. He must have enjoyed poetry, owning several volumes, and was partial to introducing verses, sometimes his own, into his writings. In a cutting which reported a Liberal party candidature meeting over which he presided, Phineas describes himself as 'a religious man and a teetotaller and a Liberal'.

Phineas almost certainly worked at Oak Mills on Station Road in Clayton, for J. Benn and Co. Joseph Benn had started textile manufacturing by leasing Beck Mill in 1860 and expanded by building Oak Mills in 1870 with partners Asa Briggs and Alfred Wallis. By 1881 the company had nearly 900 employees. The three partners, all local men, became very wealthy and were major benefactors of Clayton and its inhabitants. Asa Briggs (1832–1912) was a distant cousin of Phineas; both were descendants of Abraham Brigg (1738–1809), the only son of Jonathan Brigg described earlier.

The very first item in the cuttings book has overtones of premonition. It is an 'in memoriam' to one Brigg Briggs (his first cousin, in fact) written by Phineas, aged 26, to the *Bradford Observer* on 6 January 1890. Brigg was a mechanic working at Black Dyke Mills in Queensbury* who had been responsible for several improvements in weaving looms. Despite a lack of formal training, he was a skilled maker of fine barometers and clocks and had made an impressive model of the Black Dyke Mills complex. He had a weak constitution and died of pulmonary tuberculosis at the age of 34. Phineas ended his piece:

'*the following epitaph is by one who knew him:*

At noon-time there is darkness. One has gone
In whom the light of genius brightly shone
The skilful and deft hand can move no more
Sweet peace be his upon the other shore.'

Nine years after Phineas penned these lines, his elder brother, Arthur, died of the same disease, also aged 34. Then, just over a year later, Phineas went down with pleuro-pneumonia. After nine days illness he died of cardiac failure on 12 April 1900, aged 36. He was buried four days later at the Methodist chapel in Clayton Lane.

*Black Dyke was one of several estates tenanted by Briggses during the eighteenth century, most, if not all, of whom were related. Abraham Briggs (1766-1844) eventually acquired the estate and his daughter, Ruth, married John Foster who owned a small textile mill. Wanting to expand, Foster obtained some estate land from his father-in-law and built the Black Dyke Mill. The company, John Foster and Son, became famous for the quality and range of its cloths and for its brass bands. The huge expansion of Black Dyke Mills during the mid-19th century, around an old 'packhorse pub' called the Queens Head, created the town of Queensbury, so named in 1863.

Chapter One: Origins

Figure 1.4 Family photo of about 1893, Gilbert in the centre. (Photo courtesy of the Briggs family.)

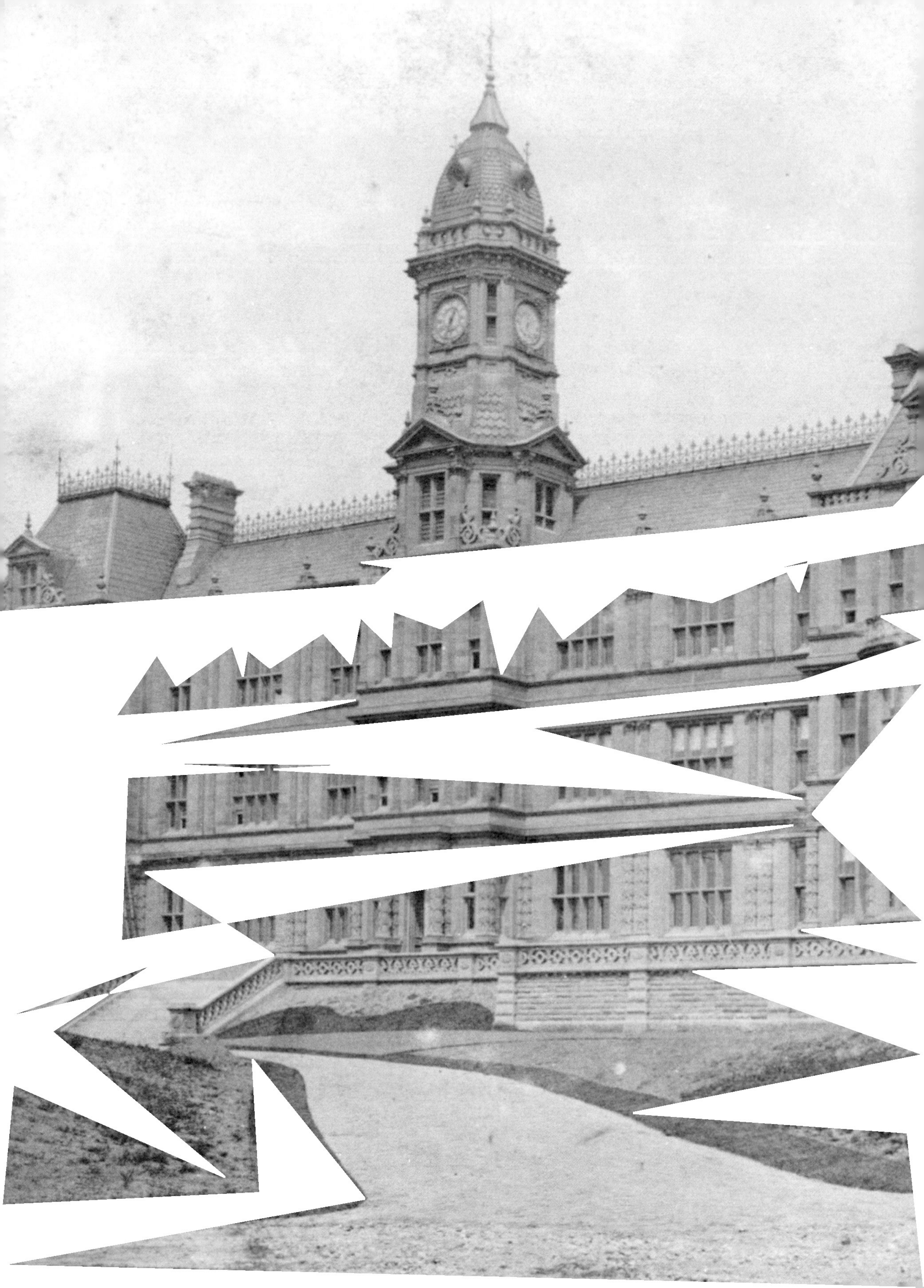

CHAPTER TWO: 1894–1905

At School in Clayton, King's Lynn and Halifax

Commenting on his fractured childhood nearly 60 years on, Gilbert says:[1]

'My father died when I was nine, and my mother brought up a family of four by her own efforts. We were poor but very happy.'

This conjures up a rather misleading picture and what actually happened to him is much more interesting.

When Phineas died he had become a mill manager and was already an overlooker when Gilbert was born. Even with three children, Bernard having only just been born, the family must have enjoyed a fairly good standard of living. Gilbert claimed that he started going to school aged three and this would have been at one of the two village infants' schools. One of these was a Church of England school, the other a Baptists' school, which had been taken over by the School Board. The latter is perhaps the most likely, given that the Briggs family were staunch Methodists and it was also closer to home, in which case it appears in the old Clayton photograph of the previous chapter (1.1). The school did take children aged three into a 'baby class', although the usual age range was four to seven. Phineas believed passionately in the value of education, not simply as the means to individual advancement, but as the pre-eminent means by which society could be rid of the many ills that were so depressingly evident all around him. Therefore, if Gilbert had shown signs of being bright, it would not be surprising for him to have started his schooling at the earliest possible opportunity, despite the modest cost of 2 d per week (roughly 76 p in 2010).[2]

Phineas' sudden death was devastating for the family. With no income and with four children under 10, including a one year-old infant, his widow, Mary Ann, was in a desperate situation. The Cranbrook Street home was abandoned and the family moved in with Mary Ann's mother at 5 Clayton Lane, originally 5 Clayton Green, just down the road (2.1). Emma Emsley was 72, widowed nine years previously, and still had two sons living with her: Mary Ann's younger brothers Arthur (33) and Phineas (30). It must have been a very cramped arrangement, although this terraced house was rather larger than the Cranbrook Street one, not being back-to-back. However, for Gilbert it did not last very long, because he was soon nearly 150 miles away in King's Lynn, Norfolk at a technical school in the town.

Figure 2.1 5–9 Clayton Lane, Clayton. © D. Briggs 2008.

King's Lynn Technical School

How this came to pass requires some conjecture. On census night, 31 March 1901, Gilbert was a 'visitor' at the house of John and Mary Haigh in South Lynn, a district of King's Lynn, where John was Headmaster of a technical school. The King's Lynn Technical School on London Road was very new, having begun in 1893 with 350 students. It was officially opened by the Duke of York on 2 February 1894 and the local newspaper report[3] included a partial sketch of the building (2.2).

John H. Haigh, previously of Leeds Higher Grade Science School, was its first Headmaster.

Figure 2.2 King's Lynn Technical School. From the *Lynn News* of 1894, courtesy of King's Lynn Library.

Chapter Two: 1894–1905

Although born in Shelley, Huddersfield in 1866, John had moved to Clayton in about 1875, when his father left the weaving trade to become clerk/cashier at the gasworks on Low Lane. He was a pupil-teacher and then elementary school teacher in the village until 1891. John's father, Henry, was a lay-Methodist preacher and must have known Phineas Briggs through the chapel.

Somehow, through these connections, it was arranged that Gilbert should attend Haigh's school. In about 1960, when Gilbert was staying with his daughter Valerie in nearby Cambridge, he asked to be taken to King's Lynn to revisit the school. He recognised the building, said he had been a boarder there and recalled the matron. In 1901, Easter day was on 7 April, just a week after the census, so he was probably staying with the Haighs, who had no children, during the Easter break. The earliest of the few surviving letters to Gilbert is from John Haigh; dated 16 February 1911, it shows that he had followed Gilbert's fortunes with interest and that they had stayed in touch. The Technical School would have been fee-paying, which raises the question 'who paid'?

It is family lore that Gilbert had a benefactor, a philanthropist, who recognised his potential and was willing to make sure he had a good education. It is most likely that this person was, directly or indirectly, Alfred Wallis (1829–1912), one of the partners in Oak Mills, mentioned in Chapter 1. He had established the 'Alfred Wallis Trust' to help further the education of worthy Clayton inhabitants. By 1900 Oak Mills was owned by Alfred and his sons, Frank and Joah, so, if only through his job as a manager at the Mills, Phineas would be well known to the Wallis family.

To be spirited so far away from his family under these circumstances was probably expedient in the short-term, but unsatisfactory in the longer-term. Gilbert probably only spent a year during 1900–1901 in King's Lynn; we have it from Gilbert himself[4] that he was at Crossley and Porter's Orphans School in Halifax between 1900 and 1905, which ignores the period in Norfolk. Exactly when he started there is not known, but it was probably autumn 1901 following the summer holiday. Again this was a boarding school, but only a few miles from Clayton.

Figure 2.3 Crossley and Porter's School, Halifax, newly built in 1864. (Photo courtesy of Crossley Heath School.)

Crossley and Porter's School[5]

Originally the Crossley's Orphan School and Home, this interesting institution was the result of a major act of philanthropy by the Crossley brothers—Francis, John and Joseph—who had made their fortunes in the Halifax textile business (Crossley's later being renowned for carpets). The 'Orphan School for Boys and Girls' took its first few pupils before it was fully completed in 1864 (2.3).

ADDENDA

The following information is supplied in response to frequent enquiries:

The Crossley and Porter Orphan Home and School is situate in the Borough of Halifax, and is about a mile and a half from the centre of the town. The site is all that can be desired, so far as pure air, pleasant prospect, and open space are concerned. The structure is capable of accommodating three hundred children. It contains Reception Room, Museum, Board Room, Spacious Dining Hall, House for the Principal, School Rooms, Class Rooms, Play Rooms, Apartments for the Head Mistress and Matron, Sick Ward, Lavatories, large and well-ventilated Dormitories; besides Kitchen Premises with every appliance for Baking, Washing and Cooking; a detached Sanatorium; also a playshed and Gymnasium. There are also two spacious Swimming baths, available for every season of the year, and ten Slipper baths, for the use of the children. As respects Warming, Ventilation, Water Supply and Drain-age, the arrangements are as perfect as modern improvements can make them.

The object of the Institution is to receive orphan children of both sexes, between the ages of two and ten (in **SPECIAL CASES ONLY**, up to twelve years of age), and to provide them with board, clothing and education until the Boys have reached fifteen years of age and the Girls fifteen or seventeen years of age at the discretion of the Governors. In addition to the usual branches of an English Education, those children who show a capacity for such studies are taught some of the following subjects:- Latin, French, Book-keeping, Composition, Human Physiology, Physical Geography, Chemistry, Hygiene, Shorthand and Typewriting, Drawing, English Literature, Vocal and Instrumental Music, Mensuration, Algebra, Euclid, Trigonometry and Mechanics. The Girls are also taught needlework and Cutting-out, and such departments of household service as are likely to prove useful to them in after life.

Applications for admission are considered at Meetings of the Board of Governors, with whom alone is vested the election of children.

A preference is given to orphans born in the county of York; to children who have lost both father and mother; to orphans of parents who have been in full communion with a Nonconformist Church, or who have been regular communicants in the Church of England; and **especially to children of families whose temporal condition has been reduced.**

Except in a limited number of cases, an annual premium of Ten Pounds is required with each child admitted. When the premiums are commuted by a payment of one sum in advance, an allowance is made at the rate of three per cent. compound interest; and in the event of the child's death, or removal with the consent of the Governors before the expiration of the term, a return will be made for the period paid for by anticipation and during which the benefits of the Institution have not been enjoyed.

Persons wishing to see the Institution may do so by obtaining an order from one of the Governors or the Secretary, available any Tuesday Afternoon from two to four.

The friends or relations of the orphans are permitted to visit them on the first Tuesday in every month, between the hours of two and four.

Further information, or forms of application, may be obtained of MR. D. LORD, Secretary, Orphan Home and School, Halifax.

Chapter Two: 1894-1905

In 1877, when there were some 250 pupils, the provision for girls was enhanced and they got their own headmistress, effectively creating two single-sex schools in the same building. The number of pupils was limited by the size of the endowment fund for running the school, an important contributor being Titus Salt, the Bradford industrialist, who gave £100 per annum. In 1887 a Manchester yarn merchant, Thomas Porter, gifted over £50,000 on condition that that school be renamed to recognise his contribution. The information supplied to potential applicants to the school in 1904 is reproduced, in facsimile, above[6] and the criteria to be met by the child are unchanged from those laid down by the Crossleys in 1861.

Since Gilbert turned 10 just before the start of 1901, he was borderline on age grounds, unless his application had been made when he was nine and he could not be admitted because the school was full. In all other respects he satisfied the criteria. The term 'orphan' is somewhat ambiguous as it seems also to have been used to mean fatherless, and there were many pupils to whom this applied. The cost was £10 per year (the equivalent of over £900 today).

Gilbert was admitted as 'boy 1126'. Although only a few miles from home, he still did not see that much of his family; relatives and friends were only permitted to visit for two hours on the first Tuesday of each month. There were initially two terms in the school year, which started in January, with a summer vacation of four weeks and a Christmas break of two. In Gilbert's time there may also have been a short Easter break. However, at the beginning of the 1902 school year his younger sister Mabel joined him ('girl 604' supported, financially, in the same way). Although the boys and girls were taught separately and lived in different buildings, there were opportunities for them to meet. The school had an excellent Principal, William Cambridge Barber, FRGS, who had been at the helm since 1872. He was religious and scholarly, with a commanding personality which generated a certain veneration despite the fact that discipline was strict. Gilbert was a red-head (nicknamed 'carrot-head') and this distinctiveness could have led to him being 'picked on'. He countered this by teaming up with a muscular boy who became his 'minder' in return for help with his Maths prep. Mr Barber the disciplinarian was also the Maths master!

As shown in the Application Criteria, the range of subjects which was available to pupils with aptitude, in addition to the standard curriculum, was extensive and there was good provision for sporting activities. Although the number of full-time teachers was small, there were also resident assistant staff and visiting teachers. Mr Barber, being a cellist and Vice-Principal of the Halifax Choral Society, ensured that music featured in school life. Pupils were taken to concerts in the new Victoria Hall in Halifax, opened in 1901. Crossley and Porter's School definitely provided a good education and, moreover, the boys stayed until the age of 15—a great advantage. The pupils went on to make their mark world-wide. One of Gilbert's contemporaries was Herbert Read (1893–1968), who was admitted on his father's death in 1904. Decorated with the DSO and MC in the First World War, he was appointed Assistant Keeper of the Victoria and Albert Museum in 1922 and in his subsequent career was poet, essayist and critic. He was knighted in 1953 for services to literature, having published over 60 books.

Gilbert left the school on 20 December 1905, just before his 15th birthday, having passed his Cambridge Local Examinations with a distinction in drawing. A large work dated 1907 which Gilbert kept on his office wall is evidence of his talent. It survives, but is too faded for adequate reproduction. On leaving, all the pupils were photographed for the school records, in the uniform which had remained unchanged since 1864, and Gilbert's picture is shown in (2.4).

Figure 2.4 Gilbert Briggs on leaving Crossley & Porter's School aged nearly 15. (Photo courtesy of Crossley Heath School.)

Of his time at the school he is mostly dismissive:[7]

> 'All I can remember about Halifax is that it was famous for carpets and toffee; it had quite a nice concert hall to which we school children were occasionally taken to hear choral and orchestral works; and a Glee Party which gave us an annual concert in the school dining room. Of all these attractions, I think toffee was the most popular.'

Chapter Two: 1894–1905

Figure 2.5 Herman Van Dyke and his wife. (Photo courtesy of Crossley Heath School.)

But elsewhere he writes:[8]

> 'The thing I remember most vividly is the piano I heard at boarding school in Halifax around 1904/5, when I used to lie awake at night listening to the instrument being played by the music master after the boys had gone to bed. The strains of music came in through an open window and I was fascinated as much by the sound of the piano as by the actual music. I think I resolved there and then that I would buy a piano as soon as I could save enough money.'

Thus began his infatuation with the instrument.[9] The master in question was almost certainly Mr Herman Van Dyk, a Dutchman who had settled in Yorkshire in the early 1890s, becoming conductor of the Halifax Orchestral Society in 1901. He was an accomplished pianist and taught piano, violin and singing at the school from 1895 to 1922. A picture of Mr Van Dyke and his wife, also an accomplished pianist, is shown in (2.5). Piano tuition would have been available to Gilbert, in principle, had he wanted it. However, this had to be paid for as an extra, at 30 shillings per half year, and by the time his interest was aroused it was probably too late to do anything about it.

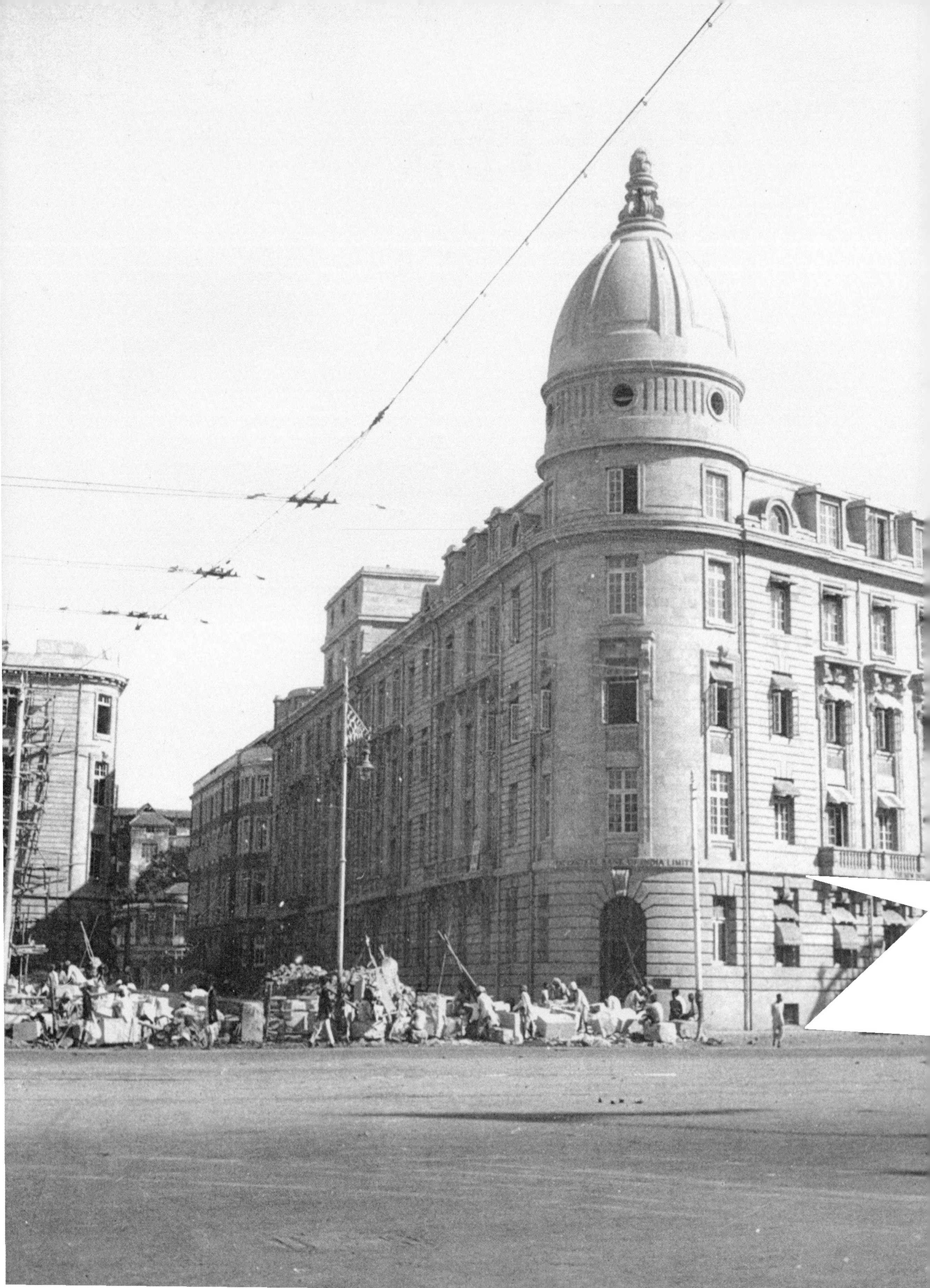

CHAPTER THREE: 1906–1933

Textiles Export Merchant: Bradford, Manchester and the Far East

Gilbert already had a job to go to when he left school; how this was arranged is not known, but almost inevitably it involved textiles. His employer was to be Holdsworth, Lund and Co., a firm of textile export merchants based at 1 Cater Street, Bradford (3.1). So, in January 1906, aged just 15, he started work as an 'office boy' on a salary of six shillings a week. The firm was obviously paternalistic, because this was one shilling more than normal on account of his fatherless status. He received an additional shilling 'tea money' if he worked over until 9 pm.[1,2] Travelling to work was not difficult, as Clayton had a railway station with a regular service to Bradford Exchange Station in Drake Street, only a few minutes walk from Cater Street, on the opposite side of Leeds Road.

The names of Holdsworth and Lund were well established in the local textile industry. In 1906 Holdsworths were doing business as flannel and blanket makers in Cleckheaton, as spinners, yarn and waste merchants, and as cloth merchants in Bradford. Lunds were operating as weavers in Bradford, spinners and textile manufacturers in Cleckheaton and Keighley, with offices in Bradford, and reed and heald makers in Bingley. Some of these firms were long established, but Holdsworth and Co. 'Merchants' of 11 Nelson Street, Bradford started in about 1902. Holdsworth, Lund and Co. started about the same time, most likely a joint-venture formed to capitalise on export potential.

The Office Boy

Gilbert was once again living at 5 Clayton Lane, but the house was less crowded than when he left for boarding school in 1900. His two uncles, Arthur and Phineas Emsley, had married and left, and his younger sister Mabel was at Crossley and Porter's School. His elder sister Claris was working, but living at home, and his grandmother, Emma Emsley, was still alive (she died aged 90 in 1918). His younger brother, Bernard, was only six. Although he had started work, Gilbert's education was far from over. He attended evening classes (night school) to learn about the textile industry he had joined. Some of these would probably have been at Bradford Technical College, created in 1882, which had departments of textiles and chemistry and dying—two of the four initial departments. A new block dedicated to textile studies was built before the First World War, emphasising its importance to the local economy. In 1900 there were well over 600 regular evening class attendees at the College, outnumbering the day time attendees by nearly five to one.

Although the textile output from Bradford was dominated by woollen and worsted goods (nicknames were 'Wool City' and 'Worstedopolis'),[3] Holdsworth and Lund dealt in piece goods in general, including cotton and silk; surprisingly, perhaps, Lister's Mill in Manningham, Bradford was the largest silk factory in the world. There was much to learn about the incredible variety of

Figure 3.1 1 Cater Street, Bradford. © D. Briggs 2008.

cloth types, the raw materials, the manufacturing methods, dying techniques, quality assessment and so on. Since he was hoping to become an export merchant, Gilbert would also have taken evening classes in French. He started at the very bottom—his general factotum duties included making breakfast for the partners/directors—and worked hard.

The trauma of losing his father, rapidly followed by the separation from his family whilst at boarding school, must have developed in the young Gilbert an independent spirit. However, the relatively short school holidays and the occasional visits of his mother to Halifax seem, if anything, to have strengthened his family bonds—he was always very close to his mother and sisters. How his mother managed to generate any income is unclear, but from 1902 she would have been able to work, since Gilbert and Mabel were at Crossley and Porter's, Claris was working part-time and her mother, Emma, was available to look after Bernard. Before her marriage Mary Ann was a 'stuff weaver' and she could have returned to this for a while. Once Gilbert started working he felt obliged, as the only male (Bernard was always the 'baby'), to assume some responsibility for the rest of the family. He would also have been aware during his later years at Crossley and Porter's that his peers in Clayton would, in the main, have been working part-time from about the age of 12. His extended full-time education had given him a number of advantages, not least the chance to escape the textile mills, which employed such a large fraction of the local population, for a potentially rewarding career. He was determined that this opportunity should not be wasted.

Commuting into Bradford from Clayton, working overtime or attending evening classes often made for very long days. Even so, Gilbert found time to learn about music and saved up for the piano he had vowed to buy whilst still at school. It took three years to accumulate the £10 with which he bought his first instrument, a Broadwood upright, and to feel that he could afford lessons, by which time he was 18.[4,5] This episode provides some early indications of Gilbert's approach to life. His first wage amounted to between six and seven shillings a week, so even accounting for some increase over the three years, the cost of his piano represents perhaps 15–20% of his total earnings. Since he was a major contributor to the family budget, this shows that he was very committed to his musical ambition. He was also prepared to wait until he could buy a new quality instrument—John Broadwood & Sons was a very long-established firm, with an impeccable record of supplying to famous composers and royalty. The instrument came from J. Wood and Sons Ltd who, in 1908, had been selling and repairing pianos in Bradford for over 50 years. Gilbert would have spent many hours in the shop over the three years, initially keeping his dream alive and later laying the foundations for an eventual purchase. Wood's came to know him very well in later years, as we shall see.

Working in central Bradford had its advantages. The Mechanics Institute held regular evening lectures on a wide variety of subjects, some musical, and also put on recitals, whilst St George's Hall was the main concert venue. At the latter, in about 1906, Gilbert heard his first famous pianist, Vladimir de Pachmann, and remembered:[4]

'moving over from the right hand side of the hall to the left during the interval so that I could see his hands because it seemed to me to be impossible for human fingers to produce such flowing, rippling effects at the keyboard'.

Like his father, and probably through the continuing influence of his mother, Gilbert was a Methodist and involved in the activities of the Wesleyn Chapel in Clayton, and more generally of a Wesley Guild. This may have been a recently established Clayton branch, but there was

Figure 3.2 Claytonian Entertainers with Gilbert Briggs in the centre, back row. (Photo courtesy of the Briggs family.)

certainly a branch based in the chapel at nearby Horton, the Bradford area in general being very Non-Conformist. He was soon taking part in discussion groups and presenting 'papers', organising weekend rambles over the local moors and taking to the concert platform as a member of the 'Claytonian Entertainers' (3.2). He sang tenor, but also wrote gags to be delivered by the two comics in the group: Alton Ward and Maurice Wilkinson.[5] The Wesley Guilds were as keen on social activities for young people as they were on devotional studies.

The Traveller

The 'office boy' made good progress and by the time he was 20 he was making occasional trips to the continent as a merchant. The letter from John Haigh mentioned in the previous chapter refers to a meeting planned for when Gilbert was returning from Amsterdam on the boat train. This passed through March in Cambridgeshire, where Haigh, now Secretary to the Isle of Ely Education Committee, was living, so Gilbert could easily break his journey there. The meeting never took place, because Haigh was called to an emergency committee meeting and his letter was by way of apology and explanation. He knew that Gilbert had already made several trips to the continent by then, February 1911. It was on one of these in 1910 that Gilbert stopped off in London to go to the Royal Albert Hall. As he entered the hall 'the opening bars of Schubert's Unfinished Symphony were floating around'. The impression this left on him was so profound that he could still recall the moment 50 years later.[4]

The 1911 census confirms Gilbert's occupation as 'commercial clerk and traveller' and shows that 5 Clayton Lane was again bursting at the seams. Incredibly, the listed occupants are Gilbert, his mother, grandmother, both sisters and brother, and his aunt 'Jinny'. Jane Anne and her older sister Emma Briggs were the only surviving siblings of Gilbert's father; all four males had died—two in infancy, Arthur and Phineas in their thirties. Given this degree of overcrowding, Gilbert's travelling probably came as something of a relief. His trips to Europe between 1910 and 1913 introduced him to new societies and languages, so he needed to learn quickly about appropriate etiquette as well as develop his communication skills. As a result, he became a keen observer of people and their reactions to situations—a fascination which stayed with him, together with an acute sense of 'good manners', all his life.

In 1913 he made his first Eastern trip, to India. Before he set off he promised the Wesley Guild that he would give a lecture about the experience on his return. He kept his promise and the original, handwritten manuscript of the lecture, which was illustrated with nearly 60 slides prepared from his photographic record, survives. He departed from Liverpool at about 4 pm on Thursday 2 October 1913 aboard the SS *City of London* and landed at Bombay at 12.30 pm on Wednesday 22 October. He did not return until the end of April 1914. The lecture, entitled 'A Glimpse of the East', was actually given on Monday 2 November 1914, and the Introduction went as follows:

'It feels very strange to me to be giving this lecture here tonight, and I am afraid I am here more by intimidation than intent. Before I went on my trip East the Guild Secretary, with admirable foresight (I might even call it strategy), got me to promise to give an evening after my return, and I can assure you it was much easier to make the promise than it has been to get the thing ready. However, here we are, so let us try, for a brief hour, to forget the war and all our cares and worries, and just take a glimpse of the East.'

Chapter Three: 1906–1933

Of course, the country had entered the First World War in early August, but this is jumping ahead of events. His seven-month trip took in Bombay, Calcutta, Delhi, Karachi and Rangoon with visits to many of the Indian 'sights' on any modern itinerary. He was mesmerised by the Taj Mahal, saying 'one could spend hours gazing at it and then want to look again'. He came home from Calcutta with a brief stop in Colombo (Sri Lanka) and a longer one of two weeks in Egypt. The lecture is a skilful mixture of geography, history, culture, travelogue and funny stories. Only occasionally does he mention anything about the point of the trip, which was to gain orders for cloth pieces for his firm and to come face-to-face with the realities of doing business through local agents in the major population centres. Showing a slide of a 'native street' in Calcutta he comments:

'This is known as Burra Bazaar, or Large Market, and it was to this lovely spot that I had to repair daily and endeavour to persuade the wily native to purchase my wares and not complain about them after he got them!'

Holdsworth and Lund had their Indian office in Bombay, photographed on his first visit in 1913 (3.3), and it was probably nearby that one of his colleagues took a picture of Gilbert. Despite the intense heat, which he found difficult, he is immaculately turned out (3.4).

In his pursuit of textile sales for his firm, Gilbert was not alone in his endeavours. He was in Delhi for about a month around Christmas time and also there, for the same reason, were about 20 other Bradford exporters, so 'they managed to enjoy themselves fairly well'. One enterprise

Figure 3.3 Holdsworth and Lund's Bombay office building, 1913. (Photo courtesy of the Briggs family.)

Figure 3.4 Gilbert Briggs in India, 1913. (Photo courtesy of the Briggs family.)

Figure 3.5 Bradford Travellers, Delhi, 19 December 1913. Gilbert Briggs, in kit, standing first on left. (Photo courtesy of the Briggs family.)

was to form a football team, 'The Bradford Travellers', which took on a Delhi Police team on 19 December and won 2–1 (3.5).

The First World War

He arrived back home at the end of April 1914 with Europe heading towards war. When Kitchener announced the formation of the volunteer army in August, Gilbert responded and expected to be called up. He was stunned to be completely rejected as medically unfit. The reason he cites[2] is that 'his heart was an inch lower than it should be'. The probable reason is that the assessment of his heart through the chest showed some displacement, which was taken to indicate enlargement and therefore an underlying heart condition. For a fit 23 year-old, who had just come through the rigours of over six months on the sub-continent without illness, this seems a strange decision, but under the pressure of screening thousands of volunteers at the recruitment centres, where massive queues formed, the medical assessors had no option but to make rapid yes/no decisions. Whether Gilbert was initially worried as to his real health by this outcome is not known, though he was certainly disappointed over the rejection, but he was surely fortunate. Of the men mobilised about 12% died and about 36% overall were casualties (dead, wounded, missing or prisoner-of-war). Indeed, one in seven of the male population of the UK under 25 died in the War.

So, Gilbert was able to continue gaining experience with Holdsworth, Lund and Co.. for the duration of the War, although activities compared with pre-1914 were heavily curtailed. There was an immediate moratorium on textile exports and the Bradford manufacturers were rapidly turned over to war-related production, which constituted about 50% of all output by 1917. The government took control of the whole industry, from purchasing wool from the Empire to determining the costs at all stages of manufacture, and placed the huge contracts. In driving costs down other producer countries were undercut and the UK ended up supplying uniform cloth, blankets and other materials to the Dominions, United States, French, Belgian, Italian, Portuguese and Russian armies.

Gilbert relates one incident from this period. The company had large stocks of heavy woollen materials suitable for blankets. A government department bought the entire stock and, by mistake, paid one invoice twice. The second cheque, for about £375 (today, the equivalent of nearly £30,000), was returned, but it came back to the Bradford office because 'the Ministry could not possibly have made such a mistake'! In the end, though, they did recoup the money when peace came, through excess profits tax.[2]

A number of letters from friends and work colleagues who were enlisted still survive. Gilbert was obviously concerned for all of them and helped morale by sending cigarettes, pipe tobacco and even food as well as letters. After conscription was introduced in 1916, and no doubt influenced by the fact that his brother Bernard, nine years his junior, was enlisted in the army (ending up in the Royal Flying Corps), Gilbert tried to overturn the original rejection. He clearly thought he could play a useful role in a non-combatant position such as signals or transport. His friends, especially those in France, just hoped that he would avoid the situation in which they found themselves. In the summer of 1918 he was still trying to obtain a medical re-examination when his boss, Mr Lund, was called away to perform some kind of war service. This gave him the opportunity to run a department on his own, which he clearly enjoyed, and before long the War was over.

Musical Lectures and Piano Lessons

Meanwhile, Gilbert was also able to continue his piano playing and increase his knowledge and understanding of music. The Wesley Guild took advantage of this and prevailed upon him to give a lecture in 1918. He chose to give a musically illustrated talk entitled 'An Evening with Beethoven'. The manuscript survives and it is dated 11 November 1918. The Armistice had been signed a few hours earlier, finally bringing the First World War to an end. In his introductory remarks he said:

> 'I make no apology for introducing Beethoven although he was a German. There has been some empty-headed objection during the war to the good music of the bye-gone German composers, but I am pleased that the general good sense of the country has risen above such misdirected patriotism.'

The lecture considered Beethoven's life history, his music and musicianship and included digressions into the development of music, to put Beethoven's compositional style into perspective, and the development of keyboard instruments. To illustrate aspects of Gilbert's exposition his sister Claris, who by all accounts had a beautiful voice and possessed perfect pitch, sang some songs and a Miss Rider played extracts from a piano sonata and one of the variations.

Gilbert had clearly attended similarly structured events in Bradford beforehand, notably at the Mechanics Institute where he regularly attended lectures, although the evening's entertainment was a novelty for the Guild. He concluded:

' … if our Guild becomes an active and bigger organ in the future, I certainly believe there is a useful field to be explored in this direction.'

Unsurprisingly, perhaps, this was a self-fulfilling prophecy and his illustrated lectures on composers became an annual event. He obviously enjoyed doing the background research and organising the 'live' musical illustrations and the events became more sophisticated. However, he also made his own act increasingly difficult to follow and by the fourth event he was confessing:

' …if these evenings are to be continued in future years I shudder to think of the state of nervous collapse to which I shall inevitably be reduced. I suppose this is due to the immensity of the subject in hand, viz. music, for the more one dabbles therein, the more one realises one's ignorance.'

Looking back on these 'Evenings' he told his children that he must have been thought 'an impudent young pup' to even contemplate undertaking them. But they were formative, as we shall see.

The second lecture, in November 1919, was on Mendelssohn and, unfortunately only the first page of the manuscript survives. The third, on 1 November 1920, covered Schubert and in his Introduction Gilbert philosophised about music in general:

'It is thoroughly democratic, second only to Death itself in levelling rich and poor. It is a friend when you are happy and a solace when you feel miserable. Music is one of the finest assets of Home Life. It is absolutely international and rises above limitations of languages. Last but not least the more you cultivate it, the richer it becomes and the more it unfolds its beauties to your wondering senses.'

The form of the lecture closely followed the original Beethoven template. The illustrations were performed by his Clayton friend Clarence Wilkinson, who sang 'The Erl King' and 'The Organ Grinder', his sister Clarice, who sang 'The Young Nun', 'Hark Hark the Lark' and 'Love's Message' and a Mr Tankard, who played one of Schubert's Impromptus, 'The Trout' (a song arranged for piano solo by Heller) and The March Militaire. Gilbert also described his first hearing of 'The Unfinished Symphony' in the Royal Albert Hall, which had so impressed him ten years earlier.

The fourth and final lecture, 'An Evening with Grieg' was given on 13 December 1921. The pianist Edgar Knight (about whom more later) played music from 'The Holberg Suite', 'Butterfly', 'Poetical Tone Pictures' and 'Norwegian Bridal Processions', Claris Briggs sang 'The Birch Tree', 'Subject Two Hazel Eyes' and 'I Love Thee' and Clarence Wilkinson sang 'My Thoughts are Like The Mighty Hills' and 'The Old Mother'. The highpoint was a performance of the Second Violin Sonata with Edgar Knight accompanying a Miss Priestly on violin. Gilbert described this as 'a big work and more ambitious than anything we have hereto experienced here'. Miss Priestly had studied at the Leipzig Conservatorium ('as had Grieg, but not concurrently', as Gilbert dryly pointed out to the audience).

Gilbert probably harboured wishes that he could have been the performing pianist at these events. He had ability and practised religiously every day, but he knew he had started too late in life. As he wrote much later:[6]

'I might put it that I started at the third harmonic [age 18], *and it would have been much better to have begun at the age of 6. The result is that my playing has been handicapped by the lack of this fundamental through more than 40 years. Even at 18 I wasted two more years in futile lessons before I found a teacher who gave me some muscular relief and independence, and saved me from lockjaw of the wrist.'*

This second teacher was Dr J. Charles Henrich, the most respected teacher in Bradford. Gilbert badly wanted lessons from such a good teacher, but could not afford to pay for them. Nevertheless, he asked if he could start having lessons 'on account' and pay back what was owed when he was able. After he had played to him, Henrich asked 'Do you always play that badly?' to which Gilbert replied 'Yes, that's why I want you to teach me!' Despite all this, Henrich was won over, perhaps the first of many during Gilbert's lifetime to be so, by a seemingly irresistible combination of intensity of purpose and sheer charm. This was about 1910–1911, and Henrich had another pupil at the time, aged about 11, who was something of a prodigy—Charles Edgar Knight. Knight had started studying with Henrich when seven, went on to The Royal College of Music in London and was an accompanist at a very early age.[7] He became very well known during the 1930s and 1940s in the North of England as a soloist and accompanist. He also led a trio and made regular radio broadcasts for the BBC (often from the Midland Hotel in Bradford). His wife, Florence, was also an accomplished pianist and they frequently played two-piano concerts together, including at the Mechanics Institute in Bradford at lunchtimes. Gilbert was introduced to Knight through Henrich, although when is not known, and they remained friends for life. When he performed at Gilbert's 'Grieg Evening', Knight was 22 and already playing professionally locally, so it was quite a coup to get him to play at the event.

Peacetime and Partnership

When the 1914–1918 war was over, the textile industry in Bradford quickly got back into its pre-war stride. Exports not only resumed, they expanded rapidly with continental production crippled. 1919 and early 1920 were boom times, but by 1921 the industry was experiencing a severe slump, exacerbated by a coal-miners' strike. At some point during 1921 Gilbert's employer, Holdsworth, Lund and Co., underwent an upheaval. In name, it ceased to exist, and at least part of the business became Thomas B. Lund and Co. (TBL), engaged in manufacturing as well as exporting. Woollens and worsteds were no longer of interest to them and this part of the merchant business was carried on by R.K. Holdsworth, with both companies operating from the original address in Cater Street.

Gilbert was employed by TBL and he already had 15 years' experience in the business under his belt. The proprietor, Thomas Benson Lund, was a Bradfordian by birth, the son of an auctioneer, who turned 50 in 1922. Assuming he was the partner in the original company, he would have been about 30 when Holdsworth, Lund and Co. was established. It is not surprising therefore, to find that

Figure 3.6 Picture taken in a studio near TBL's Manchester office, about 1922. The person standing with Gilbert Briggs is thought to be his Dutch friend and colleague, Kees Groll. (Photo courtesy of the Briggs family.)

Gilbert was made a partner in TBL not long after its formation and certainly before 1923, when he was about 32.

From the outset cotton goods were important to TBL and two cotton goods manufacturers from Nelson (just over the border in Lancashire) were closely involved, as financial backers, in the

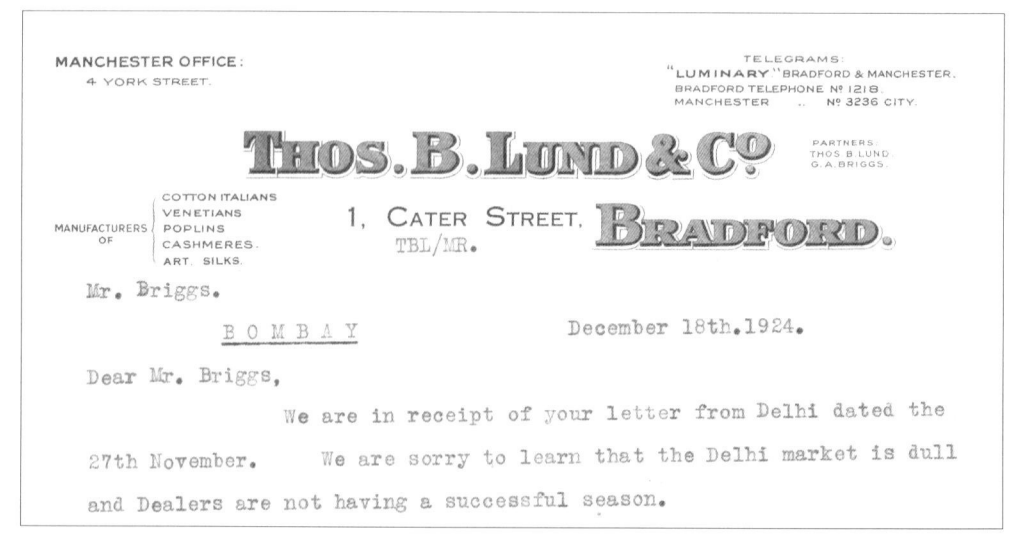

Figure 3.7 TBL Letterhead, 1924, from a letter sent by Lund to Gilbert in response to a negative appraisal of trading prospects in Delhi. (Scan courtesy of the Briggs family.)

formation of the new company: James Nelson Ltd of Valley Mills and H. Ridehalgh & Sons Ltd of Edward Street Mill. TBL had an office in Manchester at 2 Beaver St, moving to 4 York St in 1924. They owned property in Cavell St, Bombay but sold this in late 1924 and moved into offices in Central Bank Buildings in Esplanade Rd. Gilbert was heavily engaged in both the Manchester and Bombay activities (3.6).

A letterhead of 1924 identifies TBL as manufacturers of Cotton Italians, Venetians, Poplins, Cashmeres and Artificial Silks (3.7). Later documents indicate involvement with mohair and alpaca. A common factor in many of these textiles is a glossy or lustrous appearance, an indication of the importance of the Far Eastern market. TBL also had a close relationship with a dying and finishing firm in Yeadon, Naylor Jennings and Co. Ltd, and together they developed a patented lustrous finish, Hachelle, probably for this market.

Romance and Marriage

By 1923 Gilbert had become an eligible bachelor, still in the family home at 5 Clayton Lane, but by now quite well-off and the owner of a car. He was again spending a lot of time abroad, with a winter trip to the Far East in 1922/3. In the summer of 1923 he attended a private tennis party in Ilkley and was rather taken with an attractive younger girl who was also playing. He spotted that her surname (Mart) was on her racquet and managed to find out where she lived. He then wrote to 'Miss Mart' at Yew Bank, Ilkley to ask her for a 'date'. In the social convention of the period, 'Miss Mart' meant the eldest daughter and she was, in fact, Kathleen, whereas the object of Gilbert's attentions was her younger sister Doris Edna (known as Edna). This farcical beginning notwithstanding, the relationship soon flourished. Edna was, according to Gilbert, 'a talented lass', who was working for the Employers Liability Co. in Leeds. She was musical, played the piano, and was an enthusiastic member of the Ilkley Amateur Operatic and Dramatic Society. Her father, John Bradley Mart, was a surveyor who worked for a railway company. The fact that Edna was nearly 11 years younger than Gilbert was not an obstacle, indeed in the years following the 1914–1918 war such relationships were not uncommon.

The winter of 1923/4 again found Gilbert in the Far East and the romance was continued by letter. After the usual couple of months in India he moved on to Malaya and Singapore for another month and then on to Hong Kong, all of which destinations he was visiting for the first time. Edna's letters often played catch-up and reached him out of phase with those of his to her. He could not have been back home very long before he set off on a trip to Canada, returning in early May. They must have become engaged between these two trips, because before he went to Canada Gilbert started renting a place of his own, Rose Cottage, 10 Brook Lane, on the southern edge of Clayton village. The wedding date was fixed for early autumn 1924 to allow Edna to accompany Gilbert on the following winter's trip to India. It very nearly did not happen. In September Gilbert was driving with Edna in his three-wheel Morgan sports car when one of the front wheels came off its axle. In the resulting crash Edna was badly concussed and had still not fully recovered by her wedding day.

The wedding took place at the Parish Church in Ilkley on 1 October 1924 (3.8). Gilbert's best man was J. Holmes and the groomsman Clarence Wilkinson. The bridesmaids were Edna's sisters Kathleen and Nina, and Tony Sarsby was the pageboy. Claris Briggs and Clarence Wilkinson led the singing during the service. The married couple honeymooned in Switzerland, via London and Paris, before setting off to India for the 1924/5 winter sales tour.

Figure 3.8 Wedding picture. (Photo courtesy of the Briggs family.)

Husband and Father

Whether this 'exotic' trip was an ideal start to married life is questionable. Once the three-week voyage to India was over, everything must have been a total culture shock for Edna. However, she was quite undaunted and, in later life, had only fond memories of the adventure. Gilbert had to juggle the demands of the job with looking after his new wife and this must have been a strain. Business was not good. Arriving in Delhi in early November he found he had missed the best trading 'window', markets were lacklustre and woollens, which TBL had abandoned, were coming back into favour at the expense of cottons. In Bombay he had the task of unravelling a complex situation in relation to the property in Cavell Street, owned by TBL but with a mortgage, and sub-let to several tenants. The ground-work for disposing of the property had been laid in the summer by correspondence, for finalisation by Gilbert, in person, in November. However, there were problems and he reported to Lund that 'the whole thing has left me positively ill'. Ending up in Calcutta in the New Year he found his main customer was off to Bombay for a wedding on 8 January and that a main line in artificial silks was very unpopular. The state of business could not justify staying on as long as he had planned and so he booked passages home from Bombay on the SS *City of Canterbury*, leaving 20 January and landing in Liverpool on 11 February 1925.

On their return they lived at Rose Cottage in Clayton whilst they looked for a bigger house. They found what they wanted near Saltaire, conveniently situated between Gilbert's family in Clayton and Edna's in Ilkley and only a few miles from the centre of Bradford. Saltaire was the model community development, initiated by Sir Titus Salt in 1854 to properly house the workers at his massive factory complex constructed to the north of Bradford. The whole project aimed to escape the pollution (air and water) of that city, for which it had become notorious by the mid-19th century.[8]

They moved to 3 Carlton Road, Shipley, a modern semi-detached property just two streets to the west of the original Saltaire village boundary, in early May 1925, with Edna about two months pregnant. This signalled the end of Gilbert's winter trips to the Far East, but not an easier life, because TBL was facing trading difficulties. Sales through the Bombay office had dwindled further and expenses there were not being covered. The partners responded to the changed market by a reversal of policy, sending a specialist salesman in woollens and worsteds to India for the winter season of 1925/6. Just as he arrived there, the Bombay office received the news that Edna had given birth to a son, Peter Rodney, on 22 October (3.9).

Partnership Tensions

The Indian trade did not pick up as much as hoped for and Gilbert was under pressure from his partner, Lund. (At this point TBL had a turnover of about £90,000 and Gilbert's salary was £800 p.a.; today's rough equivalents are £4.1m and £37,000 respectively.) Lund wanted to close the Manchester and Bombay offices and retrench solely to Bradford, but Gilbert strongly disagreed. Whilst he carried the major responsibilities for these offices, his room for manoeuvre was constrained by the need for a joint policy, making his position difficult. Things came to a head in July 1926, when, after 'many sleepless nights', Gilbert wrote to Lund suggesting a dissolution of the partnership as the only solution, with Gilbert forming a new business around the Manchester and Bombay offices. He had apparently promised Lund in the 1921 slump, when TBL was formed, that he would stay for five years and now that period had elapsed: 'the time has come when I have to exert my individuality' and 'I must give my ideas a run or I shall explode'. In practice, this solution was complicated by the financial interests of third parties in TBL, particularly Ridehalgh & Sons. Through their mediation Gilbert was persuaded to stay, but he did so only on the understanding that his policy in relation to these offices was adopted.

Business from Manchester and Bombay was turned around, with Gilbert spending three days a week on Manchester business, but a year later Lund was again demanding closure of the Manchester office to reduce costs. This time his preference was for separation if Gilbert refused, so Gilbert drew up new plans for Lund's and Ridehalgh's consideration in November 1927. The arrangement, partial separation but within TBL, would have been complex and involved Gilbert pulling out of Bradford: 'a considerable sacrifice . . . after nearly 23 years work there'. It was

Figure 3.9 Edna, Gilbert and Peter. (Photo courtesy of the Briggs family.)

not adopted, but the reasons why are uncertain. The partnership carried on, much as before, with Gilbert making trips all over Europe looking for new business. On one of these trips, to Denmark, he became very ill and was found to have a perforated ulcer—a likely indication of the stress he was under. How he got home and exactly what happened is not known.

Lund's attitude, given that his whole life had been immersed in the Bradford textile trade, is understandable in the context of the late 1920's. Business there had been badly hit by the General Strike and Miner's Strike in 1926 and the hoped for boom, when these were over, never materialised. Short-time working in the mills was the norm. Successful, well-run businesses failed regularly and fortunes were lost overnight. There were several reasons for this state of affairs. Changes in lifestyle and public tastes following the 1914–1918 war led to new demands, especially from women, for cheaper, lighter dress fabrics which would only need to last until the next fashion change. Continental suppliers, French and Belgian in particular, were now better able to meet these requirements in terms of textile designs and flexibility of production. There was general overproduction of textiles world-wide because many of the previously importing countries had bought textile machinery, often from the UK, and become producers with lower labour costs.

Exporters were also hit as these same countries imposed steep import tariffs to protect their home markets. Raw material prices were dropping and the value of stocks suffered accordingly. Lund was probably desperate to shrink the business to a small Bradford core in the hope of surviving another slump and reaching retirement with a business of some value (he was 57 in 1928). Gilbert, on the other hand, had become somewhat detached from wool-based textiles and could see a future for a smaller business based on the model of his 'Manchester trade' and India. He had only known one type of work and his family was growing—his second child, Ninetta Mary, was born in the midst of this turmoil on 8 May 1928—so he naturally wanted to continue doing what he knew best.

It may have been shortly after this period that Gilbert started to entertain business associates at home. The burden was entirely Edna's, of course, and it established or reinforced an important facet of their relationship. She was a good cook, but the other aspects of being the hostess were harder to control. The children had to be tucked up in bed when Gilbert turned up with his guests and interruptions from this quarter were unwelcome. The importance he attached to etiquette has already been noted and Edna felt obliged to make sure everything was 'just right' which, she later admitted, was a great strain. However, her role, as she saw it, evolved into that of providing the protective environment that would allow her husband to devote his single-mindedness to whatever he thought was needed to keep them afloat in rather unnerving economic climes.

Financial Crisis

In 1929 the decision was taken to turn TBL into a Limited Company. This was probably a defensive move, as trading continued to be very difficult, to limit the partners' liability in the event of failure, but also an opportunity to raise new capital through the share issue. In order to have equal voting power, Gilbert matched Lund in terms of the new cash for ordinary shares (£1500 each), although initially he intended a much smaller proportion. He could certainly not afford this sum, and Nelson's, who also put in £1500, arranged to loan him the cash. Ridehalgh's were also assigned ordinary shares to the value of £1500 (valuing existing capital in TBL) and one of this firm's directors, a Mr Hindle, put in £500. A further £8000 of Ridehalgh's existing capital was assigned to preference shares and loans to Thomas B. Lund and Co. Ltd.

How optimistic the parties were at this stage we will never know, but these were significant sums (£1500, nearly twice Gilbert's salary in 1929, is very roughly equivalent to £72,000 today). The security for Nelson's loan was effectively a second mortgage on his house. He already had a loan of £2000 plus accumulated unpaid interest of over £900 from the dyers Naylor and Jennings, almost certainly dating back to his becoming a partner in TBL. The situation got worse as the Depression deepened and the partners had to cut their salaries. Gilbert's dropped to £650 in 1930 and £500 in 1931, compared to £800 in 1926. By this time he was extremely concerned about the prospects for TBL and very worried about his own financial position. The crunch came in early 1931, when it was obvious that he could no longer make ends meet. Nelson's were clearly concerned that he might default and he told them that he was trying to sell his house, in order to reduce his outgoings, but the market was depressed. That summer they agreed an exit strategy and Gilbert moved into rented accommodation in Ilkley, at 20 Eaton Road. In a letter to his creditor he wrote that this was an 'old fashioned house, which had been empty for 9 months' but it satisfied his financial criteria (3.10). Nelson's probably acquired the Shipley house, paid off the remaining mortgage and made do with the eventual balance. Gilbert, having paid out on his mortgage for over six years of its 10-year term, lost all the capital he had built up. He was extremely fortunate not to be bankrupted.

Figure 3.10 20 Eaton Road, Ilkley, one of four in the terrace block. © D. Briggs 2008.

Records and Radio[9,10]

Music, the 'solace when you are miserable' as Gilbert had described it in one of his lectures, was never far from his mind during the 1920's. He tried to play the piano for at least 20 minutes every day. On his long trips to the East, when he was often in the same hotel for up to a month at a time, he would arrange for an instrument to be installed in his room. He also took every opportunity to listen to live music, particularly classical concerts and recitals. Technology, however, had been producing new ways to listen to music more-or-less during his lifetime. Gilbert had a lifelong fascination with gadgets and a curiosity about how mechanical things worked, so he would have been aware of the basic history of these developments, which is reiterated here, briefly, to set the scene for his own involvement in the audio business.

Thomas Edison first demonstrated a viable system for the recording and subsequent reproduction of sound in 1877: the phonograph. Sounds (acoustic waves), collected by a horn, caused a diaphragm at its base to vibrate. A stylus attached to the diaphragm indented a thin, tinfoil sheet wrapped round a cylinder, which was manually rotated on screw drive, producing a helical 'track' containing the acoustic information in the impression (the record). Reproduction of the original sound was essentially a reversal of this process; a stylus, connected to a diaphragm, tracked the groove with the cylinder rotating at the same speed. Tainter and Bell's graphophone of 1885 was similar except that the stylus incised the vibrations, using a vertical cut, into a thick wax layer on a cardboard cylinder. It was soon clear that the main market for such machines would be for playing recorded music and the Columbia Phonograph Company was set up to sell cylinders of music in 1890. However, the inventors of neither type were able to successfully mass-produce duplicates of original recordings. Emile Berliner's gramophone of 1888 recorded the sound onto a wax layer on a rotating flat disc, using a lateral cut to place the vibrations into a spiral groove. Mass-duplication of discs was less problematical. Playing times of cylinder or disc recordings, in 1890, was limited to two to three minutes and reproduction quality was very poor.

By 1900 the problem of mass duplicating cylinder recordings had been overcome with Edison and Columbia dominating this market. Eldridge Johnson, whom Berliner originally contracted to build a spring-driven gramophone motor, made so many improvements to both the gramophone and the disc recording/copying process that he ended up owning important patents in his own right. Johnson and Berliner merged their respective operations in 1901 to form the Victor Talking Machines Co. and adopted the famous 'His Masters Voice' trade mark. Reproduction quality improved, playing times increased and sales of records and machines took off. Cylinder sales peaked in 1905, as discs proved more popular, but cylinders were not finally abandoned until 1929.

Berliner had sold a licence to William Barry Owen in 1897 to set up the Gramophone Company in London and it produced its first discs the following year. In a real coup it recorded ten arias by the 28 year-old tenor Caruso in 1902 in Milan and this established the gramophone as a serious new form of entertainment.[11] The company was innovative and in 1912 the first double-sided discs were pressed. After the 1914–1918 war progress was resumed and in 1920 the 'HMV Portable' player was unveiled at the Ideal Home Exhibition, Olympia. This proved to be the biggest selling acoustic gramophone ever and it is probable that Gilbert started 'to dabble with acoustic gramophones' around this time.[4] He would, no doubt, have been frustrated by the frequency response, which was limited to about 200–2000 Hz, and by the range of music available, which was very much restricted by the limitations of horn recording.

Over a similar time scale, a quite different but potentially complementary technology—wireless communication—had reached an interesting stage of development. In 1894 Guglielmo Marconi, building on discoveries relating to radio waves and their transmission made by many researchers, including Heinrich Hertz, during the previous decade, started his experiments with wireless telegraphy. By 1897 he was sending signals over about 10 miles and in December 1901 he achieved transatlantic communication. The first factory for the manufacture of wireless telegraphy equipment was established by Marconi in Chelmsford, England in 1898 and its use for communication using Morse code was quickly taken up around the world, especially by the military. Amplitude modulation (AM) of the radio frequency carrier-wave, allowing audio information to be transmitted, was developed in the early 1900's. Wireless communication progress was accelerated during the First World War, not least by the development of the hard-vacuum triode valve in 1915.

Public broadcasting of speech and music first started in The Netherlands in 1919. The UK followed in February 1920 with very limited transmissions (two half-hour programmes each day) from the Marconi Company's Experimental Section at Chelmsford. Already there was a growing band of amateur enthusiasts, largely amongst technically-minded young men, and many wireless societies. Weekly magazines to serve this interest had appeared, including *Wireless World* as early as 1913. More companies wanted to start broadcasting so that when, in November 1920, the Post Office, which controlled transmission and reception licenses, banned all such transmissions because of interference with military communications, there was an outcry. The Post Office slowly relented and discussions with the 'industry' led to the establishment of the British Broadcasting Company and a network of eight transmitters, which were all operating by October 1922, reaching about half the population.

The earliest type of receiver, used by these enthusiasts to pick up transmissions, was the crystal set. This 'wireless-set' was fairly simple and cheap to make, and needed no power source, so it was ideal for the hobbyist. The simplest device consisted of a very large aerial, earthed through a tuning coil, the detector (a crystal of galena and a fine wire, known as the 'cat's whisker', used to make a point contact on its surface), a capacitor and high impedance headphones. Slightly more complicated circuits gave greater selectivity. These designs were available in the appropriate magazines and the components were readily available; alternatively, kits could be purchased for a few shillings. Gilbert knew all about crystal sets, even though he may not have constructed them himself, and would have spent hours finding a 'sweet spot' on the crystal to give the optimum sensitivity and tuning into the available stations. He would also have wished that there was an alternative to listening to the music on headphones in a very quiet room, which was necessary because of the very weak signal.

When the 1914–1918 war ended, the availability of surplus valves, particularly the generically produced R-type, meant that valve radios started to appear and, of course, these required a loudspeaker. Even in kit form they were expensive. Gilbert recalled[12] how, in these early days of radio, he had carefully studied the contents of *Wireless Constructor*, whose editor from 1923 to 1930 was Percy W. Harris, a pioneer of the publication of complete working drawings for the amateur set builder. In 1927 Cossor launched its hugely popular Melody Maker radio set, available in kit form, and Gilbert built one of these.

1925 was a landmark year for musical reproduction. Following a decade of research at AT & T's Bell Research Laboratories ('Bell Labs') dedicated to improving telephony and sound reproduction, a system for electrical recording of sound was announced. This matched-impedance recorder was devised by Henry C. Harrison and Joseph P. Maxwell in 1924. An acoustic recorder with a long tapered horn was still used but the sound collection was decoupled by using a condenser microphone, the electrical current being fed through a valve-amplifier to a balanced

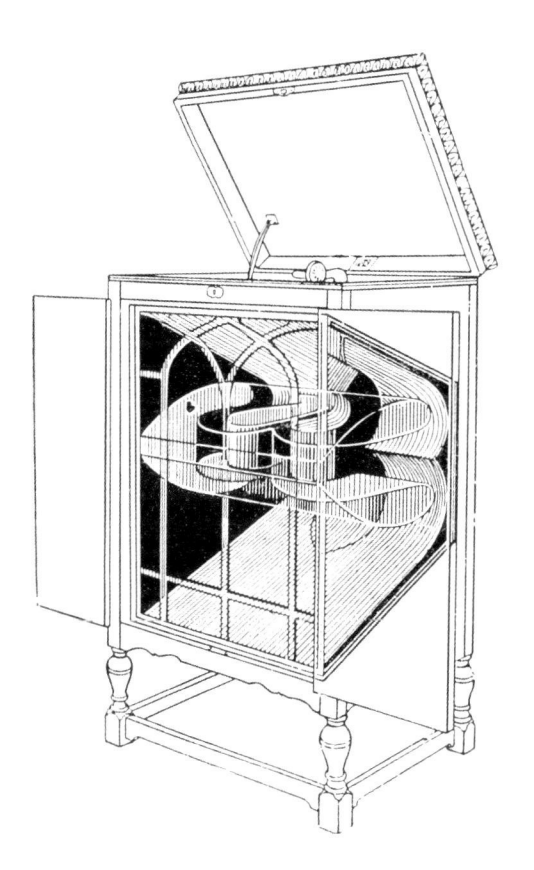

Figure 3.11 Re-entrant horn design of 1927. Reproduced from *A to Z in Audio* (from an original EMI illustration).

armature loudspeaker. The frequency range was 50–6000 Hz. The Gramophone Company rapidly obtained the rights to the technology and issued the first British electrical recordings in 1925, including the first recording of a symphony. This was Tchaikovsky's Fourth, with the Royal Albert Hall Orchestra conducted by Sir Landon Ronald, issued by HMV in December.

Harrison and Maxwell had also applied the principles of matched-impedance to the acoustic gramophone reproduction chain (sound box, tone arm and horn) and the result was the re-entrant exponential horn, a design which allowed an efficient, exponentially-flared horn of many feet equivalent length to be folded inside a cabinet containing the acoustic gramophone. The Gramophone Company's version of this, HMV model 202, was launched in 1927 (3.11).

According to Gilbert[4]:

'… my first impressive reproducer was a sort of enlarged version of the HMV re-entrant horn model which came out in 1927. This was made for me by a friend who knew nothing about exponential theory. (I think he just had a good look at the HMV and enlarged it at home from memory.)'

This seriously downplays his friend's accomplishment, if true, because of the complexity of the design and the craftsmanship required to fabricate the horn itself.

The Bell Labs research programme also produced an electromagnetic pickup in 1925. Together with a valve-amplifier and loudspeaker driver, this replaced the conventional acoustic sound box in gramophones and allowed electrical reproduction from records. Gilbert must have quickly acquired an electric gramophone, because he says of his re-entrant horn:[4]

'I fitted a telephone earpiece type of driver and everybody marvelled at its base response. This was used on radio and records until the Blue Spot balanced armature and then a Magnavox moving coil loudspeaker knocked it for six.'

Into Loudspeakers and out of Textiles

Early loudspeakers used a telephone earpiece as the driver connected to a horn. Balanced armature speakers, popular around 1928–1930, were a significant improvement on earpiece-driven horns, but they still suffered from lack of base response, amongst other deficiencies. Gilbert saw his first moving coil loudspeaker, made by RK, at a radio show in Manchester (where, as we know, he was often working) in 1930. This really impressed him and he concluded that the moving coil approach must the way to realise the desired reproduction of low notes. It wasn't long, therefore, before he purchased the Magnavox speaker to be followed about a year later by a moving coil loudspeaker from Bakers Selhurst:[4]

'…fitted with an ingenious centring device and a cone that looked very much like wallpaper. That was a very good loudspeaker for its day and generation, and I think it was this unit and its cone that inspired—or should I say impelled—me to have a go at speaker building myself … '

It is worth recalling, at this point, that Gilbert was pursuing this rather expensive pastime throughout the period that his finances were deteriorating. It seems to be the case that his drive to access music, either from records or the radio, to supplement what was available live was of over-riding importance in his life. It is almost impossible today to appreciate such a situation.

So, by this time, Gilbert was fascinated by loudspeaker performance and wanted to learn more about how moving coil speakers worked and what determined their characteristics. His experimentation led directly to the setting up of the Wharfedale Wireless Works in his wife's name, to build loudspeakers ('as a sideline which might keep her in pin-money'),[13] but no doubt with one eye on the rapidly deteriorating textile business. This is described fully in the next chapter, but for about six months between late summer 1932 and March 1933 he was effectively doing two jobs, often spending his lunchtimes at his one-man speaker 'factory' not far from Cater Street. If the other directors of TBL were looking for an excuse to reduce their numbers during the difficult times of the Depression, Gilbert provided it. In March 1933 he was sacked, with a pay-off cheque for £200, the equivalent of three months salary. Just before this happened, his daughter Edna Valerie had been born on 14 February.

Gilbert's reaction to his dismissal, despite his bleak financial situation, was:[14]

'I can truthfully say that nobody ever received a more valuable kick in the pants.'

which reflects his disillusion with the job he had been doing for over 27 years. He also commented:[14]

Chapter Three: 1906–1933

'I must admit that my departure from the 'rag trade' caused no stir or commotion: I was scarcely missed. On the other hand, I regret it did not revive the fortunes of the firm I left, which was closed down year or two later.'

In fact, TBL clung on for a little longer than this, until 1936. Lund, the founder of the firm and Gilbert's erstwhile partner, did not live to witness its final demise, as he died in May 1934, aged only 61.

Gilbert himself was dangerously ill towards the end of his time with TBL. A sharp reminder of his own mortality had come with the sudden death of his younger brother, Bernard, in March 1930. He was a few weeks short of his 31st birthday and died of heart complications following pleurisy and pneumonia. On his father's side of the family, all of Gilbert's male relatives who had survived infancy had died in their early 30's from similar causes. Gilbert had survived more than one bout of bronchial pneumonia when younger, but this time he had double lobar pneumonia and could easily have died. Two nurses were in constant attendance and the crisis period, at the height of the fever, was a very tense time for all concerned. He pulled through, but a lasting effect was loss of hair colour which changed from auburn to a pale sandy hue.

14th Oct. 1932.
& 21st "   "

WHARFEDALE M.C. Speakers.—Assemble your own and save 33%; complete kits ready for assembly in an hour; full instructions; first class results guaranteed.
BRONZE Wharfedale Standard Model (2½ watts undistorted); latest 6½lb. P.M. 7in. moulded cone, etc., £1 complete; three-ratio transformer (30 m.amps.), 7/6; carriage 2/6.
SILVER Wharfedale Superpower Model (5-6 watts undistorted); massive fourclaw magnet (57,000 effective lines), linen cone, leather surround, etc., 50/- complete; three-ratio transformer (50 m.amps.), 10/6; carriage 3/6; equals any speaker, regardless of price.
ABOVE Magnets by Swift Levick; components scientifically matched; 3 days approval; guaranteed 12 months.—Wharfedale Wireless Works, 92-96, Leeds Rd., Bradford. [9643

28th Oct. 1932

WHARFEDALE M.C. Speakers.—Assemble your own and save 33%; first class results guaranteed; 3 days' approval; write for catalogue.—Wharfedale Wireless Works, 92-96, Leeds Rd., Bradford. [9782

4th Nov. 1932.

WHARFEDALE M.C. Speakers, conspicuous for their natural reproduction, assemble your own and save 33%, first class results guaranteed; Kits from 20/-; three days' approval; write for details.—Wharfedale Wireless Works, 92-96, Leeds Rd., Bradford. [9842

11th Nov. 1932.

— do —

CHAPTER FOUR: 1932–1937

Part 1. Wharfedale Wireless Works in Bradford

As noted in the previous chapter, Gilbert had personally experienced the improvement in speaker performance as the designs and technology evolved during the late 1920s and by the end of 1931 he had set his sights on gaining an understanding of how moving coil speakers really worked.

The idea of the electro-dynamic, or moving coil, transducer was first patented by Werner von Siemens (Germany) in 1877. A fine coil of wire is supported within the gap between the pole pieces of a cylindrical magnet so that it can move axially. Passage of current through the coil generates a magnetic field which interacts with the static field of the system causing the coil itself to move. Oliver Lodge, in England, patented a major improvement, which anticipated modern designs, in 1898. The small rigid diaphragm, which the attached moving coil caused to vibrate in sympathy with the current variation, required horn-loading to achieve any real volume. John Stroh invented the conical paper diaphragm terminating in a flat section at the rim of the loudspeaker in 1901 and in 1908 the coil-centring spider was introduced by Anton Pollack. Thus the essentials of this loudspeaker type had been established by 1910, with the magnetic field produced by a mains-energised electromagnet. In 1911 Edwin S. Pridham and Peter L. Jensen, working in California, patented a design which they introduced commercially as the 'Magnovox' in 1915 and this was initially successful in public address use.

In 1925 a research paper published by Chester W. Rice and Edward W. Kellogg, who were working at General Electric in the USA, laid out the basic principles of the direct-radiator loudspeaker, in which a small, coil-driven, mass-controlled diaphragm placed in a baffle produced a uniform response over a broad mid-frequency range. In addition, they designed a mains-driven power amplifier which produced about 1 watt of low-distortion output, allowing the volume from the connected loudspeaker to match that of the original recorded sound. This was without horn-loading, which they concluded was critical to the 'naturalness' of the speaker output. This work then set the scene for future incremental developments.

The Experimenter

Early in 1932 Gilbert saw an advertisement in *Wireless World* for surplus moving coil loudspeakers being sold in London. Typically, he immediately followed this up and bought a couple for 7 s 6 d each. He tells us that:[1]

> 'They were of German origin and were fitted with a small energised field and a high resistance voice coil'

> BANKRUPT Stock.—New headphones, 1/11 pair, with adjustable earpieces, 2/9; Dr. Nesper horn loud speakers, complete, original price 35/-, our price 5/11; crystal sets, complete, new, from 2/11; postage on above lines 6d. per article; also moving coil A.C. and D.C. loud speakers, British Magnavox, Amplion, Jedson, Rolo, Siemens, etc., all types, mains units and transformers, all goods guaranteed.—Lisle Surplus Depot, 46, Lisle St., Leicester Sq., London. [7356

Figure 4.1 *Wireless World* advert, 24 February 1932.

Figure 4.2 The 1932 German loudspeaker. Reproduced from *Loudspeakers* 5th Edition.

At this time, British-made loudspeakers almost always had low impedance voice coils, which meant that a transformer was required to match the normally high impedance output of a radio. The advantage of this purchase was, therefore, that Gilbert did not have to buy a matching transformer. The probable advertisement is that reproduced in (4.1). This first appeared in the 24 February issue of *Wireless World* and was repeated in the next four weekly issues. In his Conclusion section of *Cabinet Handbook*, dated March 1962, Gilbert says 'It is exactly 30 years since I started to dabble in loudspeakers' so it seems fairly certain that one of these advertisements caught his attention, particularly as there are no other likely candidates from that period.

Figure 4.3 Gilbert with his 1932 German loudspeaker, taken in his Ilkley office in 1966. Reproduced from *The Dalesman*, (1966, 291) by permission of Country Publications Ltd.

The speakers, which cost Gilbert a fraction of a typical 'list' price, could, therefore, have been made by Siemens. He obviously kept one of them for sentimental reasons because he was able to have it photographed in 1958 (4.2) and again in 1966 (4.3).

Gilbert experimented with his German loudspeakers, winding new voice coils and making new cones from a wide variety of materials and creating many new combinations to test. He says:[2]

'After re-making the speakers two or three dozen times, I announced to my wife with my usual modesty that I knew all there was to know about cones and response, and that I intended to start a sideline in her name, known as Wharfedale Wireless Works, which might keep her in pin money.'

The Sideline in the Cellar

Exactly when and where Wharfedale Wireless Works was started has not been documented. Assuming it took Gilbert a few months of experimentation to reach this point, he probably made the decision to establish the firm in the summer of 1932. This accords with Edna's recollection that:[3]

'he commandeered an attic in the summer of 1932 and played about with paper hats and bits of wire and a few magnets…'

Gilbert goes on:[2]

> *'I went to see Swift Levick's in Sheffield and bought a few magnets. I bought chassis and voice coils from Goodmans, and moulded one-piece cones from Bridger's. I hired a cellar in the warehouse of a friend who put £150 into the venture.'*

Up in the attic at 20 Eaton Road he evolved designs for speakers which he thought could be a commercial success, before he set up business in the cellar premises. Interestingly he remembers:[2]

> *'When my first commercial sample was produced, I decided that my immense knowledge of cone material was of no real value, and I would stick to the professional article in the future. The one-piece cone was three times as 'loud' as my hand made efforts and sensitivity was the main requirement in the moving coil unit at that time: we could not afford to bother about resonances.'*

We know that the premises were within easy walking distance of TBL in Cater Street because of his lunchtime visits. Gilbert had a friend called Harry Sarsby, whose son Tony had been the page boy at his wedding and whose father, Charles, had started a 'Stuff and Fents Merchants' business in Bradford around 1890. In 1932 C. Sarsby Ltd was located at 92 Leeds Road, which is only a short walk from Cater Street, and Harry was running the firm. At this time most of the buildings along Leeds Road out of Bradford were large warehouses, but today only one remains standing on the appropriate side of the road (4.4). It has four floors and a basement and, given the similarity with

Figure 4.4 186–192 Leeds Road, Bradford. © D. Briggs 2008.

Chapter Four: 1932–1937

the more numerous buildings which survive on the other side of Leeds Road, in the area known as 'Little Germany', it is fairly typical. So, the original Wharfedale 'factory' was probably in the basement of 92 Leeds Road.*

In the beginning, the company employees amounted to a full-time assembler working in the cellar (unnamed by Gilbert for reasons which will become clear) and Edna, working at home. As Gilbert says:[4]

'The delicate work of cone and coil assembly needs the feminine touch, so every evening I took home two hat boxes containing cones and coils which Mrs Briggs assembled. I took them back to the Bradford cellar next morning, ready for our day's production. I can say this: when Mrs Briggs put a coil on the neck of a cone it stayed put; we never had a loose joint. She also became a first-class solderer of speech coil wire to eyelets in cones.'

Edna described her contribution more lyrically:[3]

'Each day I laid out my newspapers on a table and got to work on the cones and coils with my soldering iron and flux and tins of adhesive. Like the dwarfs in the fairy story who worked each night making wooden shoes for the cobbler and his wife to sell, at the end of each day I had a row of cones and coils, with neatly soldered leads, ready to be packed in hat boxes and taken to the "factory" next morning.'

What she doesn't say is that she was pregnant when she started doing this in 1932 and carried on with almost no break when Valerie was born in February 1933; this was made possible because her new daughter:[3]

'… was a very good baby. She slept night and day, missed her feeds and had to be wakened by an alarm clock.'

Peter was at school but Ninetta had not yet started. She was between four and five years old during this period and one of her earliest memories is of her mother at the kitchen table wielding a soldering iron.

Whether Gilbert saw this as any more than a sideline, producing a small number of 'quality' speakers for sale locally, is a moot point. Whilst he was a director of TBL and trying to keep that company alive, he had little prospect of doing more, but he had enough experience of previous slumps to know that the one they were in was very serious and possibly terminal (later it was dubbed the Great Depression). Sidelines were part of his thinking, though. For example, he had dreamt up a scheme to send cotton shirting to Southern Ireland to be made up into shirts, which were then sold to Marks and Spencer. The selling point was that manufacture in the clean air of Ireland avoided the 'smuts', the curse of 'washing day', produced by the Lancashire factory chimneys. On the other hand, perhaps indicating greater ambition, he recognised the importance of a 'brand name'. Recalling his 'rag trade' days Gilbert said:[5]

---

*Note added in proof. Adverts which Gilbert placed in the small ads columns of *Wireless World* in late 1932 have come to light in a document which he prepared. This is reproduced, courtesy of the Escott family, on the page facing the start of this chapter. The address for the firm is given as 92–96 Leeds Road.

> '… being a merchant I was constantly undercut in price by other merchants. I decided therefore that I would manufacture something and give it a name, so that if anybody wanted it they would have to buy it from me, either directly or indirectly.'

He wanted something with a 'ring' to it and Wharfedale Wireless Works certainly had the ring of alliteration. Ilkley, where he lived and where the original loudspeaker designs were developed, is in the valley of the river Wharfe in Yorkshire. He used to joke that Wharfedale is the most beautiful of the Yorkshire Dales and this association reflected the beauty of his loudspeakers. The story Gilbert tells in his 'Looking Back' reminiscences, that he dreamt up the name of the company before announcing it to Edna, is rather whimsical. In fact, they spent ages together debating this important question and, in general, Edna was the sounding board for any significant decision in the early days.

The Breakthrough

Just at the time that Gilbert was evolving his speakers in the Eaton Road attic the 1932 Olympics were taking place in Los Angeles (30 July–14 August); it is surely no coincidence that he decided to name his various designs after the medals : bronze, silver and gold. That autumn he had an early breakthrough. He saw an announcement that the Bradford Radio Society was to hold a loudspeaker competition. He entered two of his 'Wharfedale' models, the 'Bronze' (an 8" unit) and 'Silver' (10"). Sixteen entries were judged, in groups of four at a time, mounted behind a screen and hidden from the audience. The 'Bronze' won first prize and the 'Silver' second prize. As Edna put it:[3]

> '…he had hit two coconuts with one shot'

but Gilbert recognised the real element of luck in the achievement which was that:[2]

> '… they were the only two speakers which were—by accident—correctly matched to the amplifier.'

The consequences of this triumph were quite dramatic. A few days later he called in at his one-man factory at lunchtime, as usual, to be informed that a Bradford wholesaler had telephoned and wished to see Wharfedale's representative with a view to placing an order.

The wholesaler was Thomas Dyson Ltd, a well-established firm with a history which must have been replicated many times around the country. Dyson had established a cycles business before 1900 and was soon operating as retail cycles dealer and wholesale cycles factor from two locations in Bradford. By 1905 he had moved into motor car repairs and by 1910 he had added a phonograph and electrical stores, with four premises in total. By 1920 the business had been consolidated into two wholesale outlets, one for cycle and motor factors and the other for gramophone and electrical products. This was still the situation in 1932 with the electrical stores at 79 Manchester Road, not very far from the Leeds Road cellar.

Gilbert walked over there and saw a Mr Priestley, who asked for price, terms and a delivery date for one gross (144) of the new speaker which had been demonstrated at the recent Radio

Chapter Four: 1932–1937

Figure 4.5 Wharfedale Bronze speaker of 1932, reproduced from a contemporary brochure.

Society meeting. He did not need to see or hear the speaker because he already had orders for it; all he wanted was a regular supply. This must have come as quite a shock since the factory output at the time was about 10 speakers a week! Gilbert promised to do his best to deliver a dozen each week.

The 8" 'Bronze Wharfedale', as it was originally designated, is shown in (4.5) taken from an early brochure. The Goodman's chassis was of heavy gauge steel, to support the huge, chrome steel, permanent magnet from Swift–Levick and Sons, which weighed 5 lb. The magnet provided a flux density of 7000 oersteds on a 1" centre pole. The 'universal' transformer, providing nine different ratios, allowed the low impedance voice coil to be matched to the various high impedance outputs of most radios (depending on the type of output valve involved).

The pricing of the speaker was not easy as Gilbert had absolutely no relevant experience to draw on. He recalls:[4]

'We worked on 10% overheads [ratio of indirect to direct running costs] *as merchants, so I plumped for 20% as a manufacturer, but we actually should have had 30% to cover costs'*

However he worked out the price, and this quotation may be post-rationalisation, it was fixed at 39/6 (39 s 6 d or in decimalised terms £1.975, equivalent to about £110 today). In this context, it is instructive to look at the competition in the loudspeaker market he had entered. A fairly good idea of the situation can be obtained from coverage of the 1932 Radio Olympia exhibition in *Wireless World*. Previewing the exhibition, the magazine noted:[6]

> '*The potential purchaser of a loudspeaker at Olympia this year will not have an easy task. That is not to say that he will have any difficulty in finding the right type at the price he is prepared to pay; his embarrassment will arise rather from the extraordinarily wide range of makes offering apparently equal value as regards performance and price.*
>
> *The past few months have produced an enormous increase in the number of small permanent magnet moving coils and this will undoubtedly be the predominant type at the show.*'

This was a period of transition, with many of the established manufacturers offering the same speaker with either a mains-energised electromagnet or a newly introduced permanent magnet (PM), the latter version being typically 30% more expensive. More than 20 companies exhibited loudspeakers at the 1932 Radio Olympia and/or advertised in *Wireless World* around the same time. These included: Amplion, Baker–Selhurst, Blue Spot, Celestion, Epoch, Ferranti, GEC, Goodman's, Hartley-Turner, Igranic, Kloster–Brandes, Loewe, Magnavox, Ormond, R&A, Rola, Stafford Sinclair, Sonochorde, Standard Telephones & Cables and W.B. Prices for a standard (i.e. non-de luxe) PM speaker with transformer varied from around 27/6 to 47/6 depending, amongst other things, on speaker diameter, so Gilbert's pricing probably had more to do with market observation than any serious costing considerations. Doubtless there were other smaller companies also making speakers, so Wharfedale was a small fish in a rather crowded pond.

During March 1933 Gilbert was ousted from TBL as described in the previous chapter. He had to decide whether to turn his loudspeaker activity from a sideline into a significant business, capable of supporting his family, or look for alternative employment. Given his working background and salary expectations and the dire state of most of the Bradford textile industry, it was really Hobson's choice: Wharfedale Wireless Works had to be given his best endeavours.

A Serious Business

At about the same time Gilbert paid a courtesy call to his major customer, Dyson's, when Mr Priestly complained that the soldering on their units was untidy and not up to 'best practice' in the trade. Back at his factory, Gilbert relayed the complaint to his employee and insisted that the joints to soldering tags be improved. His builder replied that, in his opinion, the joints were all right, which led to his instant dismissal!

So, Gilbert and Edna were on their own and Wharfedale was a truly family business. Speaker production had to go on and Gilbert was forced to 'take his coat off' and build them himself. In April the output was 83, about twice the level of the earliest months. However, the business was not going to make any progress without some working capital. Much later, Gilbert claimed that despite the Depression and because he was well known locally to be an honest man, he had no difficulty in raising money.[7] Remarkably, given his personal financial situation at this time, he raised an overdraft facility with a bank. His rescuer, though, was one of his former co-directors from TBL who offered to put £2000 into the business as a 50–50 joint partner, providing he severed the existing, undisclosed, relationship with Sarsby. The identity of this second partner has not been established.

To buy out his first partner Gilbert paid over his severance cheque for £200, producing a gain of 33% on his investment of only a few months previously—a quicker and better return than Sarsby could possibly have imagined! He also moved out of the cellar to premises on the fourth floor of a similar building slightly closer to the centre of Bradford: 62 Leeds Road. This building, another casualty of the transformation of Leeds Road into a dual-carriageway, was opposite Eastbrook Hall, then famous in the North of England as its 'Methodist Cathedral'.[8]

Being in sole operating charge of his business was no doubt an attractive aspect of his new situation compared with the continuing tension at TBL, but having to do the manual work was definitely not to Gilbert's liking. He rapidly looked for a replacement for his sacked employee and found him in 22 year-old Ezra Rushforth Broadley 'who took to the art of speaker building like a duck to water'.[9] Ezra's father knew Gilbert and he persuaded him to give his son, who had just left the army and was unemployed, a chance. (Herbert Broadley is described as a 'stuff warehouseman' on his son's birth certificate, so in all probability he worked for Sarsby's.) Broadley became a key figure in the history of 'Wharfedale', staying with the firm for the rest of his working life except for a period during the Second World War, when he trained commandos. Gilbert's quotation notwithstanding, his early days at the firm were far from easy. He was frequently, literally 'carpeted' by his boss. At the start of the working day Gilbert would make Broadley stand in front of his desk, on a piece of carpet, whilst he laid into him about all aspects of his work for up to half an hour. Presumably, Gilbert had learned his lesson about product quality and was not going to be embarrassed again. When asked, in later life, why he put up with this treatment, Broadley replied that there was no other work available.

Gilbert also urgently needed to learn a great deal more about the technical aspects of 'audio', especially electronics, and he solved this problem by effectively taking on a consultant, Ernest Maurice Price, a young lecturer at Bradford Technical College. They started with weekly, half-day sessions in which Gilbert was intensively tutored, but Price was soon drawn into the Wharfedale net as consultant and, eventually, scientific collaborator for over 20 years.[10]

Early Products and Advertising

Initially, output was concentrated on the 'Bronze Wharfedale'. The 'Silver' was not persevered with, as Gilbert put it:[4]

'The 10 inch unit was actually fitted with a moulded cambric cone sprayed with metal powder, the voice coil diameter being 1.5 inches. It was deficient in top response, and as nobody then knew anything about crossover networks an attractive bass reproducer faded out—having appeared 10/15 years before its time.'

Development must have continued on the third original design (Gold) because at some point during 1933 a 10" de luxe model, the 'Golden Wharfedale', was introduced. This was designed to handle very large inputs and sold initially for 63 shillings. Gilbert managed to get both speakers reviewed in *Wireless World*, the favourable test reports appearing on 8 September 1933. Further additions to the range soon followed, the 'Blue Wharfedale' at 32/6 and the 'Cadmium Wharfedale' at 26/6. These were similar and probably also 8" units, but designed to handle low inputs, with the Cadmium having a smaller magnet. The four models were included in the brochure shown in (4.6). This is undated, but it was probably issued in late summer 1933 in time for the important selling period before Christmas (the reference to 'the last Annual Loud Speaker Competition' presumably refers to 1932; by mid-1934 the Bronze model had been significantly altered). The speakers were 'colour-coded' to match their model names through appropriate lacquering of the chassis.

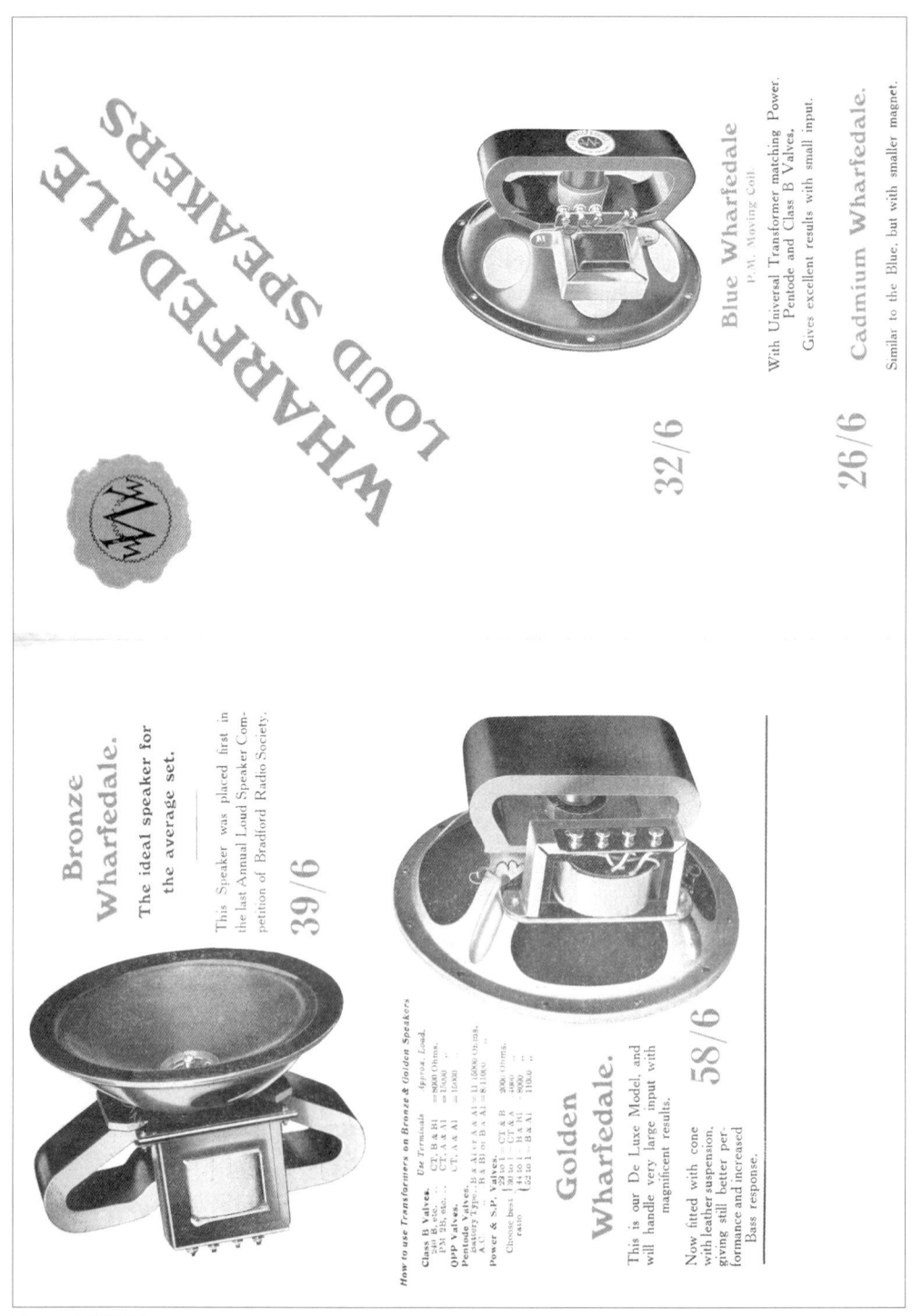

Figure 4.6 1933 brochure.

The *Wireless World* report brought some free national advertising, but the first significant, paid-for national advertisement appeared in that periodical a little earlier on 18 August 1933. As (4.7) shows this was in connection with the Radio Olympia exhibition, where Wharfedale speakers were exhibited on the stand of a J. Dyson and Co.

The advertisement was designed, and placed, by a small advertising agency in Swan Arcade, in the centre of Bradford, owned by Fred Keir Dawson (4.8), who was christened simply Fred, but adopted his mother's maiden name at some point and was known by it for the rest of his life. Dawson had started his own agency in about 1931, aged 28, having worked for other agencies from leaving school. A grammar school boy, he had won a county art scholarship, but was unable to take this up because of family hardship and the need for him to provide an income. However, he had continued his drawing—maybe as part of his professional work—as evidenced by the illustration of the 'Bronze' unit. 25 years later Gilbert used the same drawing in his 'Looking Back' piece.[11] A bigger advertisement was placed in *Wireless World* on 22 September, this time in relation to the Manchester Radio Exhibition (4.9) at which Wharfedale had its own stand. Thereafter, the *Wireless World* adverts, which Dawson placed from 1934, were usually modest 1–2 inch, single column affairs. He and Gilbert became friends, meeting regularly at a conveniently placed café for morning coffee.

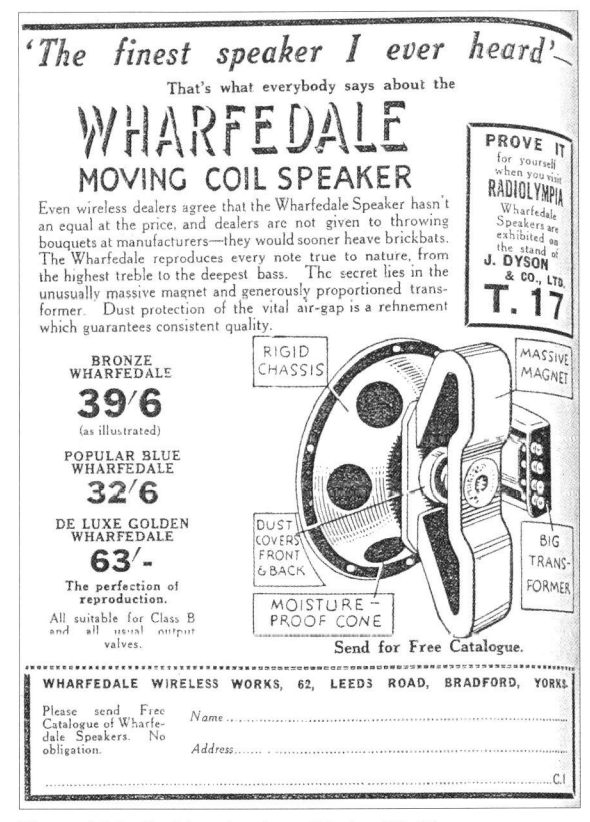

Figure 4.7 Radio Olympia advert, *Wireless World* 18 August 1933.

Figure 4.8 Fred Keir Dawson, reproduced from *Audio Biographies*.

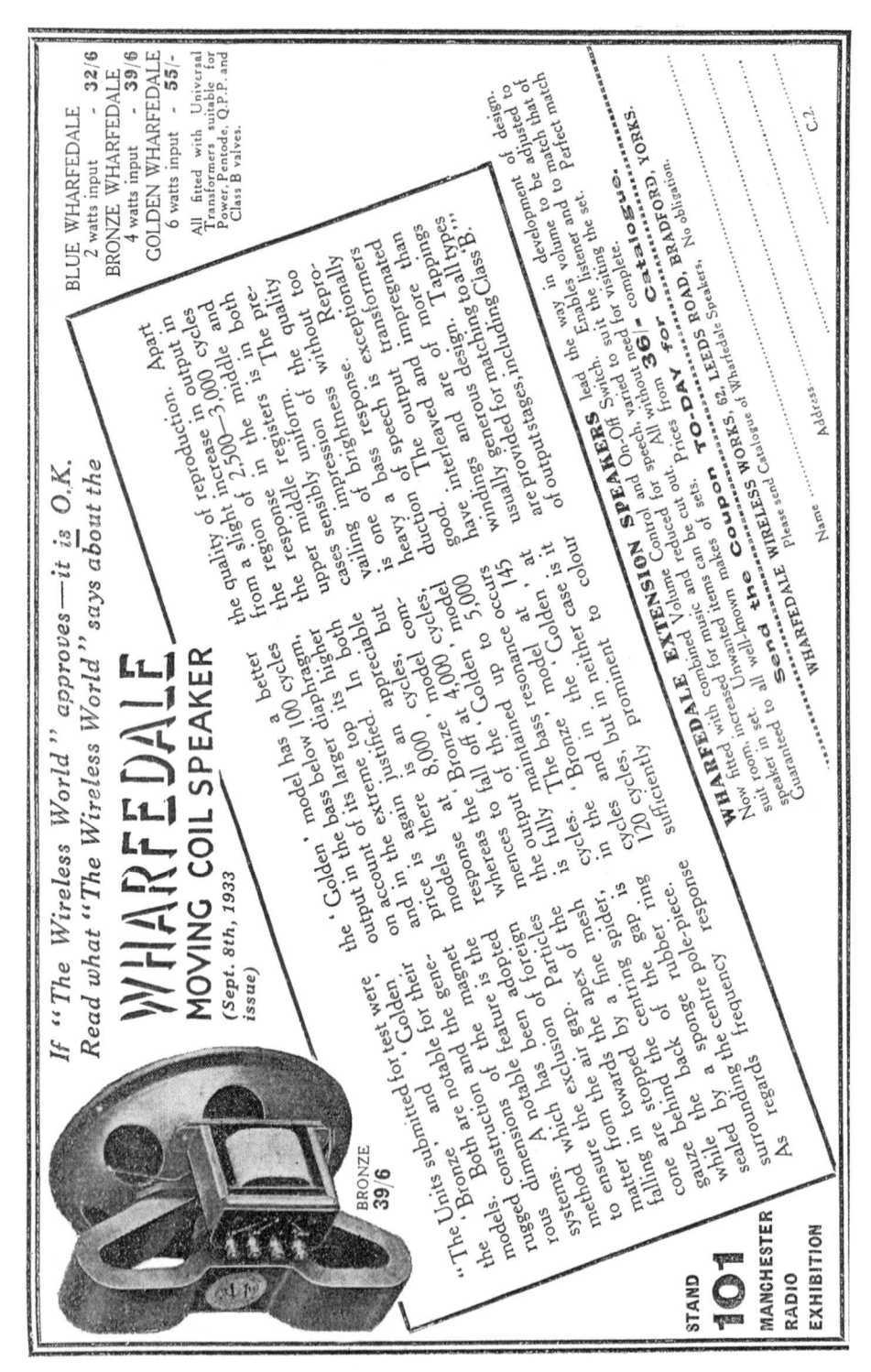

Figure 4.9 Manchester Radio Exhibition advertisement, *Wireless World*, 22 September 1933.

Chapter Four: 1932–1937

As this later advertisement shows, Wharfedale were already offering extension speakers by Autumn 1933. Gilbert kept one of the earliest examples, with its Bronze unit, and was able to photograph it for *Cabinet Handbook* in 1962 (4.10).

Figure 4.10 Extension speaker produced in 1933, reproduced from *Cabinet Handbook*.

Figure 4.11 Nubian cabinet flyer from 1934.

51

During 1934 Gilbert was having to come to terms with a changing industry situation. Home construction of radio or radiogram sets by the hobbyist was giving way to the purchase of complete sets, which meant a decline in the demand for separate speakers. Wharfedale was too small to compete in the market for complete sets so his strategy was to greatly expand the production of extension speakers for radios, which meant designing a range of cabinets. Soon the café became the Wharfedale design studio as Keir Dawson was inevitably drawn into the process of evolving new cabinet designs. Perhaps the first to be separately advertised was the 'Nubian' (4.11). According to Gilbert this was[12]:

> '... a cabinet which F.K.D designed in 1934 and which embodied a rather subtle W for Wharfedale. The cabinet was in ebony with chromium trimmings and never met with the sort of success I thought it deserved.'

No doubt Dawson also influenced other design projects such as advertising leaflets. He is yet another example of someone who fell under the G.A.B. spell early on and remained involved for many years, eventually playing a pivotal role in the story of Gilbert's books (Chapter 12).

Expansion

The earliest speakers were all fitted with transformers and these were bought in. The 1933/4 brochure shows that the three most expensive models could also be purchased without a transformer, as 'Type N.T', for use with radios having low impedance outputs. These were fitted with a four-layer voice coil and a plug and socket board to change the voice coil resistance to allow for better matching of radio output and loudspeaker. These voice coils were probably the first to be made in-house. Production was expanding quite quickly and Gilbert took on additional staff. One of the earliest was Frank Mann, maybe the first coil-winder, and another was Arnold Hatton, who must have started in early 1934. His background is known: he left school at 16 and went to work for Dyson's in 1930, where he may have been (in the electrical stores) when Gilbert started to supply them in 1932. He gained useful experience with Mains Radio Gramophones, also on Manchester Road, who made radios and radiograms before moving to Wharfedale.[13] Broadley, Mann and Hatton were the very long-serving core of the firm.

Another early product was a combined speaker and amplifier unit which could be used to upgrade a battery powered radio set from a class A to a class B (push-pull) amplifier whilst, at the same time, adding a matched, efficient moving coil speaker (one of the 8" units, probably the Bronze). The combination was guaranteed to give a large increase in output volume. The flyer for this unit is shown in (4.12); the reverse side discussed the advantages of the design relative to competitor products, and how to connect the unit to different radio set types. Although the flyer is undated, it probably comes from 1934 or early 1935, because the WWW logo had disappeared from sales literature by mid-1935—although it lived on as a badge on the back of the loudspeaker magnet and on the company letterhead for many years.

It is interesting to read the claim on this flyer: 'the magnificent quality for which Wharfedale speakers are already famous'. This was not a total exaggeration. During 1934 the Bronze model had been improved in several respects. A new Ni–Al alloy permanent magnet, more compact and much lighter than the chrome steel monster, yet with a higher flux density, was used. This increased the response, especially at low frequencies, and improved damping. A new cone with

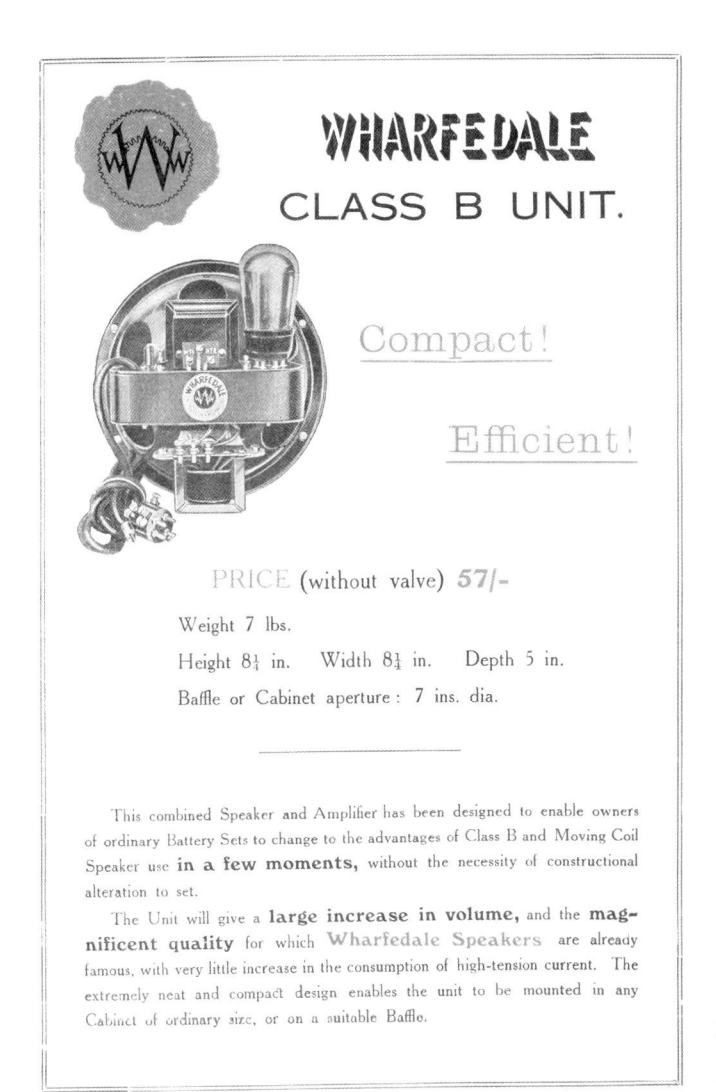

Figure 4.12 Combined speaker and amplifier flyer from 1934/5.

freer suspension reduced the bass resonance to 75 Hz. The price was increased to 42/6. The unit was reviewed in *Wireless World* (14 September 1934) and the report was most favourable (it also mentioned the optional 'Nubian' cabinet). Gilbert capitalised on this and issued a leaflet around the beginning of 1935 entitled 'What others say about the Wharfedale moving coil speaker'. This reproduced the review along with a series of glowing testimonials from purchasers of the 'New Bronze' and 'Golden' speakers, all except one being dated October or November 1934. The geographical spread of the addresses seems to have been carefully chosen to cover much of the country: Todmordon (Yorkshire), Chorleywood (Hertforshire), Banbury (Oxfordshire), Liverpool (Lancashire), Dunfermline (Scotland) and two London districts. Illustrated in (4.13) are the front cover, showing the new-style Wharfedale logo, and three of the letters.

The end of March 1934 marked the end of Wharfedale Wireless Work's first full accounting period, since Gilbert had chosen to align the firm's financial year with the tax year. Some 4600 speakers had been produced and sold in the 18 months or so from starting the business, indicating

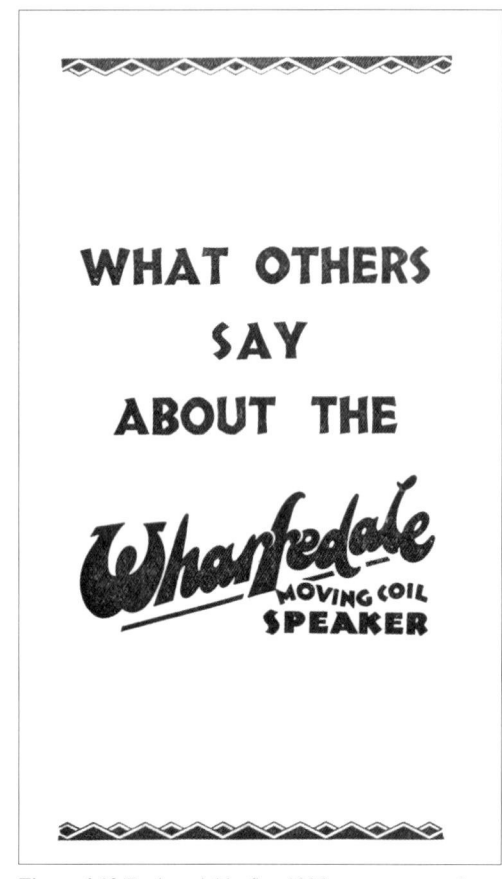

Figure 4.13 Testimonial leaflet, 1935: outer cover and specimen page.

the rapid increase in output. Gilbert, who was acutely conscious of the sales situation, was:[9]

'...very proud of this achievement until I saw our first balance sheet and found that we had lost £1000 in the process, which was half our capital.'

He should not have been too surprised. His speakers were priced at a level which was presumably profitable for much larger companies, whilst he was still building up to optimum efficiency, most of the firm were on a steep learning curve and all his start-up costs were included in the balance sheet. For their quality the Wharfedale speakers offered very good value-for-money; Gilbert later claimed the low prices helped him break into the market when he could afford very little advertising. However, he was seriously concerned. In his 'Looking Back' reminiscences he says:[14]

'Now, to succeed as a manufacturer in a new venture the four essentials are:
Business and organising ability.
Capacity for hard work, integrity and devotion to the job.
Market sense and foresight.
Technical knowledge'

then later:

'When I saw the results of the first year's working I began to suspect that I had lost qualities (1) and (3) completely.'

In August 1934 Wharfedale had its own small stand at the Radio Olympia exhibition. Recalling the impact of the precarious financial situation on the venture, Gilbert says:[15]

'I was going down to London on a train at 2.30 pm. I signed a cheque for £10 and sent my cashier Miss Sykes to the bank and asked her to meet me at the station at 2.20 with the money. Miss Sykes duly arrived at the station looking rather embarrassed because she had to tell me that the bank had refused to cash the cheque and the manager wanted to see me. This made me miss my train and I rushed round to the bank to see if they had run out of funds. The manager then handed me the £10 and gave me strict instructions never again to exceed our overdraft limit.'

During the late 1920's and early 1930's there was a dramatic increase in the number of radio sets. Increased coverage of the country by transmitters, more and better programmes offering cheap entertainment and more homes with mains electricity (still only 25% in 1931) all contributed. Interestingly, *Wireless World* noted on 6 January 1932 that in the previous six months the Bradford area had seen a record 33.3% increase in the number of licences to 45,000, which meant over half the households had a radio set in spite of the Depression. Anticipation of the opening of the Moorside Edge transmitter towards to end of 1931 probably had much to do with this phenomenon. As more homes became equipped with radios there was a demand for extension speakers which would allow the radio to be listened to in other rooms in the house. An alternative to owning a radio receiver was to subscribe to a relay service. The providers operated on a fairly local basis. They fed audio signals of fixed frequencies—in the 1930's limited to a BBC programme and Radio Luxembourg—from their receiver 'hub' down wires to houses, factories, schools etc. The high voltage signal was stepped down in two stages: to around 650 V at 'medium voltage feeders' and then to 'service level' 20–60 V at a local kiosk supplying the surrounding properties. The subscriber paid a rental for a connection, with a station-selecting switch, and a 'set' which was a cabinet, containing a loudspeaker and transformer, with a volume control. Such a set was not that different from an extension speaker, albeit with different transformer requirements (in the relay case for impedance matching at different line voltages). These were the two possible outlets for his speakers that Gilbert decided to concentrate on, hoping to develop a more profitable business.

The Truqual Volume Control

By the end of 1934 Wharfedale employed a 'traveller' or agent covering London and the Midlands—Albert Smith, who eventually became a director. Strangely perhaps, there was also a Dutch agent; before the Second World War The Netherlands was the only country to which exports were sent. This situation was almost certainly the result of one of the many long-standing friendships that Gilbert kept alive through frequent letter exchanges. Just before Gilbert arrived back from his first Far Eastern trip in 1914, a young Dutchman from Gröningen came over to spend time at Gilbert's employers Holdsworth and Lund. The firm dealt with C.E. Groll & Sons there and Kees Groll was to spend time learning about the Bradford trade and improving his English. Since he was also in lodgings in Clayton he got to know Gilbert well and they stayed in touch when Groll returned home. It was doubtless through this contact, directly or indirectly, that the exports started.

Early in 1935 Gilbert met his Dutch agent in London and was shown a five-position 'Yaxley' switch, a US design, which he was ordering from Plessey. Gilbert had been thinking about the problems of volume control on his extension/relay speakers: a control with variable impedance affected the division of output between the main set and extension speaker(s), also the speaker response was affected by the volume setting. He realised that this switch offered a possible solution. In his hotel room that night he designed a constant impedance volume control around the switch and went off to Plessey's in Ilford the following morning to place an order. What followed comes best from Gilbert himself:[16]

'On my return to the works the following day we bought from our local factor a few spare coils of resistance wire—stocked for replacing elements in electric fires—and Mr Broadley knocked 10 nails into a piece of plywood to represent the 10 tags on the control. (He has always been a firm believer in the scientific approach to the problem of testing prototypes.) In this way we were able to measure the exact length of resistance wire needed in series and parallel to maintain the required resistance, and we had our first sample control working within a few hours, the rate of attenuation being decided, not by mathematics, but by listening tests.'

The circuit drawing is shown in (4.14) and the actual product, the 'Truqual' volume control, is shown in its first advertising leaflet in (4.15). A patent was applied for and the control was fitted to all relay speaker cabinets from 1 April 1935. Within a year or so Gilbert was claiming, in sales literature, that many thousands had been sold without a single return for faulty working. The price was 3 s.

Examples of the early single column adverts placed in *Wireless World* by Kier Dawson are hard to find (because of the way the weekly issues of the time have been bound by libraries) but one issued shortly after the 'Truqual' introduction is shown in (4.16)—it occupied two inches of one of the three columns on the page.

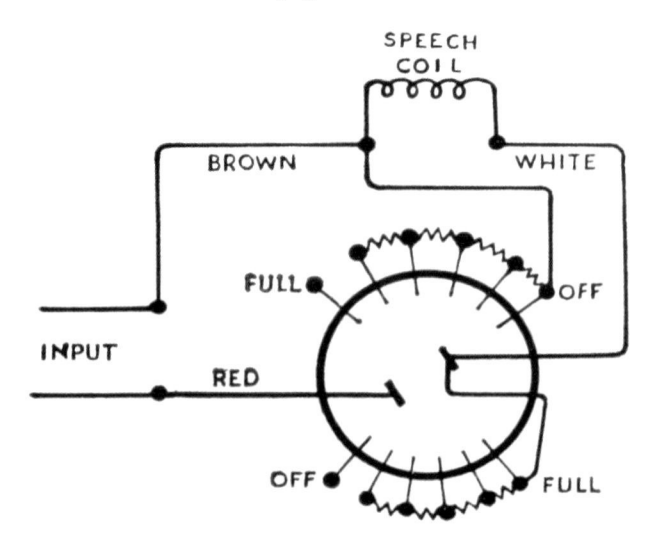

Figure 4.14 Truqual circuit, reproduced from *Loudspeakers* 5th Edition.

Chapter Four: 1932–1937

AT LAST
A VOLUME CONTROL
WHICH DOES NOT DISTORT

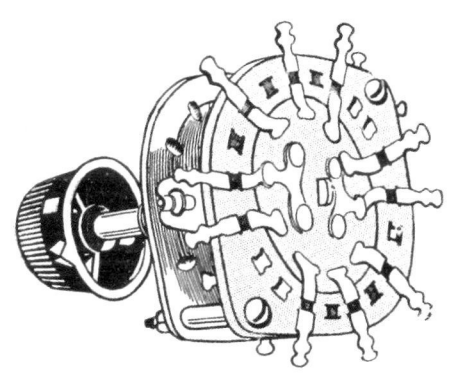

WHARFEDALE "TRUQUAL" COMPENSATED VOLUME CONTROL

Volume can be reduced to a mere whisper yet not a trace of distortion is introduced. The Wharfedale "TRUQUAL" operates on an entirely new system (patent applied for) which does not depend upon the introduction of a high series resistance.

It is the one control which reduces volume without affecting the frequency response of the speaker.

Relay subscribers are provided with only one control—a volume control. The importance of fitting one which really does control without spoiling reproduction is obvious.

Volume reduced in five steps from FULL volume to definite "OFF" position.

- Positioning mechanism works on a separate steel plate, thus removing all strain from contact points, and the construction is robust and practically indestructible.
- Contact points are silver-plated and self-cleaning.
- The Unit is bolted into side of Cabinet and cannot be turned round as a whole by control knob.

PRICE 3/- each. Sample on request.

- This new volume control will be fitted as standard to all Transformer Type

RELAY MODELS
of the

Wharfedale MOVING COIL SPEAKER

as from April 1st.

Prices with "Truqual" Volume Control. Transformer Types.	
Grecian	26/-
E. Bronze	32/6
Nubian	42/-
De Luxe	49/6

WHARFEDALE WIRELESS WORKS
62, LEEDS ROAD, BRADFORD. . . . . Phone 4346.

Figure 4.15 First flyer for the Truqual volume control, 1935.

Figure 4.16 Two inch, single column advertisement from *Wireless World,* spring 1935.

Figure 4.17 Quarter-page advertisement from *Wireless World*, 28 December 1934.

Gilbert did occasionally splash out on quarter-page adverts and a particularly interesting example is reproduced in (4.17). It demonstrates that by the end of 1934, at the latest, Gilbert was being booked to give lecture-demonstrations on a national basis.

The launch of the Truqual volume control coincided with the end of the 1934/5 accounting year; things had improved significantly with a small loss being recorded. A variety of output and impedance-matching transformers was being manufactured for both internal consumption and separate sale, whilst Bronze and Golden speakers were being sold as separate items and in relay and extension speaker cabinets. During 1935 the Bronze speaker was further improved and this new version was again reviewed by *Wireless World* (2 August 1935). The changes included a redesigned magnet, an ingenious dust cover to completely seal the magnet gap and, most importantly, a cone moulded to give an exponential flare. The review confirmed Gilbert's claim that the design had eliminated high frequency focusing, giving a wider radiation of sound, and the remarkable bass output for its size. 1935 also saw the introduction of a moving coil hand microphone, 4.5" diameter, with an AlNiCo magnet and botany wool cone suspension. It was promoted for public address and dance band work or for home use, using a radio set as the amplifier. The price was 27/6 and the recommended Wharfedale 'standard' output (step-up) transformer an additional 7/6.

Enter 'High Fidelity'

Shortly after the 'Bronze' speaker review, in the 23 August 1935 issue, *Wireless World* first used the term 'High Fidelity'. This was in an unattributed article entitled 'High Fidelity in the Home: The Electrical and Acoustical Requirements for Quality Reproduction'. The article set out the requirements for practical perfection in the receiving and associated reproducing equipment taken as a whole. Having discussed receivers and relevant recent improvements, the article went on to discuss loudspeakers as 'the weakest link in the chain':

'…the distortion introduced by the best loudspeaker, particularly when room acoustics are taken into consideration, is so great that the receiver itself can be appreciably below the ideal standard of practical perfection without a noticeable deterioration in the standard of reproduction'

'High Fidelity' had actually been coined several years earlier by H.A. Hartley (founder of Hartley-Turner Radio Ltd) when he was working at Graham Amplion Ltd in 1927. He subsequently came to regret his 'brainwave' saying:[17]

'… the sins that have been committed in the name of high fidelity since that time is nobody's business. I believe that "hi-fi" is pure, unadulterated Americanese.'

Hartley had written a wide-ranging, two-part article on 'Broadcast Reproduction' for *Wireless World* in May 1932. Part 1 was entitled: 'Can We Expect Perfect Reproduction?' and Part 2: 'The Response Limits of the Human Ear'. This was probably the first thorough analysis of the requirements for 'high fidelity' in sound reproduction, although Hartley (4.18) did not introduce the term in these articles. It seems likely, from Gilbert's later approach to writing about the subject, that they were rather influential. Thus, the burden on the loudspeakers makers in the pursuit of high fidelity in the home was clear. On the other hand, there was ample scope for innovation and gaining competitive edge.

Figure 4.18 H.A. Hartley, reproduced from *Audio Biographies*.

Into the Fourth year

Wharfedale Wireless Works posted another small loss for 1935/6 and 1936 saw a further push into the relay market, which was expanding rapidly, with the number of subscribers increasing from 82,690 to 250,978 between 1932 and 1936.[18] Another 8" speaker, a cheaper version of the Bronze chassis was introduced, the 'MR', because there was a perceived market for replacing moving-iron speakers (MR = moving-iron replacement!). Four relay cabinets were offered: the MR (an oak cabinet with the MR chassis); the E 'Bronzian' (walnut with Bronze chassis); the 'Nubian' (ebony and chrome with Bronze chassis) and the 'De-Luxe' (inlaid walnut with Golden chassis). Apart from this last cabinet all the others were only available to relay users and not sold through retailers. Prices for chassis and cabinets were quoted for quantities of 12, 50 and 100. A similar range was available in the extension speaker line with the 'Standard' as the cheaper 8" chassis and the 'Bijou' cabinet being the MR equivalent.

There were two new developments that year. The first was an addition to the loudspeaker range—the 'Twin-cone Auditorium'. This was a unit with twin diaphragms, one of about 3" diameter nested inside a 10", a back centring device and an AlNiCo magnet. Compared with the 'Golden' the flat response was extended from about 6000 Hz to 10,000 Hz and the speaker could handle higher inputs, up to 10 watts. The price was 75 s without a transformer compared with 42/6 for the 'Golden' chassis. The design was based on a P.G.A.H. Voigt patent (413,758) for which Wharfedale paid a license fee. Gilbert may not have liked that, but even so, Paul Voigt was one of his heroes. Whilst at Edison Bell he had worked on moving coil speakers at the same time as Rice and Kellogg and filed for a British patent (238,310) only to find later that they had beaten him with their US application by three weeks. Gilbert considered Voigt's highly sensitive corner horn of about 1933 to be a work of genius. This was an exponential horn with reflectors for enhancing the treble and a resonance chamber for the bass—all from a single 6" unit—designed to overcome the problems of the low power amplifiers of the time. However, it was Voigt's range of 32 patents covering all aspects of audio, including magnet design, which impressed him most.

The second development was the 'Voluphone', a hand-held, moving coil headphone, fitted with an integral 'Truqual' volume control. This is believed to be the first ever moving coil headphone, but it was actually marketed as an aid for the deaf. As Gilbert describes:[19]

Chapter Four: 1932–1937

'The idea was that deaf people could listen to radio without having the volume level from the main loudspeaker too high for the comfort of other people in the room, or next door . . .'

The Voluphone design owed much to that of the existing hand microphone, as can be seen from (4.19), and indeed the flyer issued at its launch identified another use as a microphone by connecting through a radio set, which was used as the amplifier/speaker. The fate of this product is described in the following chapter.

During 1936 *Wireless World* ran a loudspeaker comparison, including the recording of response curves. The latest 'Golden' model, destined to be the 1936/7 offering, was entered and the magazine editor, Frederick L. Devereux, conducted the tests with open-air equipment he had built himself. He later recalled being astonished at the 'Golden' curve which was almost level between 70 Hz and 6000 Hz.[20] The published curve and the accompanying recommendation (14 August 1936) was, of course, seized on for promotion purposes in the product listing for the 1936/7 'season' which was issued shortly afterwards. This was the first real brochure, with 16 carefully designed and illustrated pages, and in it Gilbert broke new ground. One page was headed 'Suggestions on Choice of Extension Speaker by G.A. Briggs' in which he introduced the reader to some of the considerations behind cabinet design and speaker matching, whilst acknowledging the difficulty in choosing which to buy when strong claims were made even for the performance of the

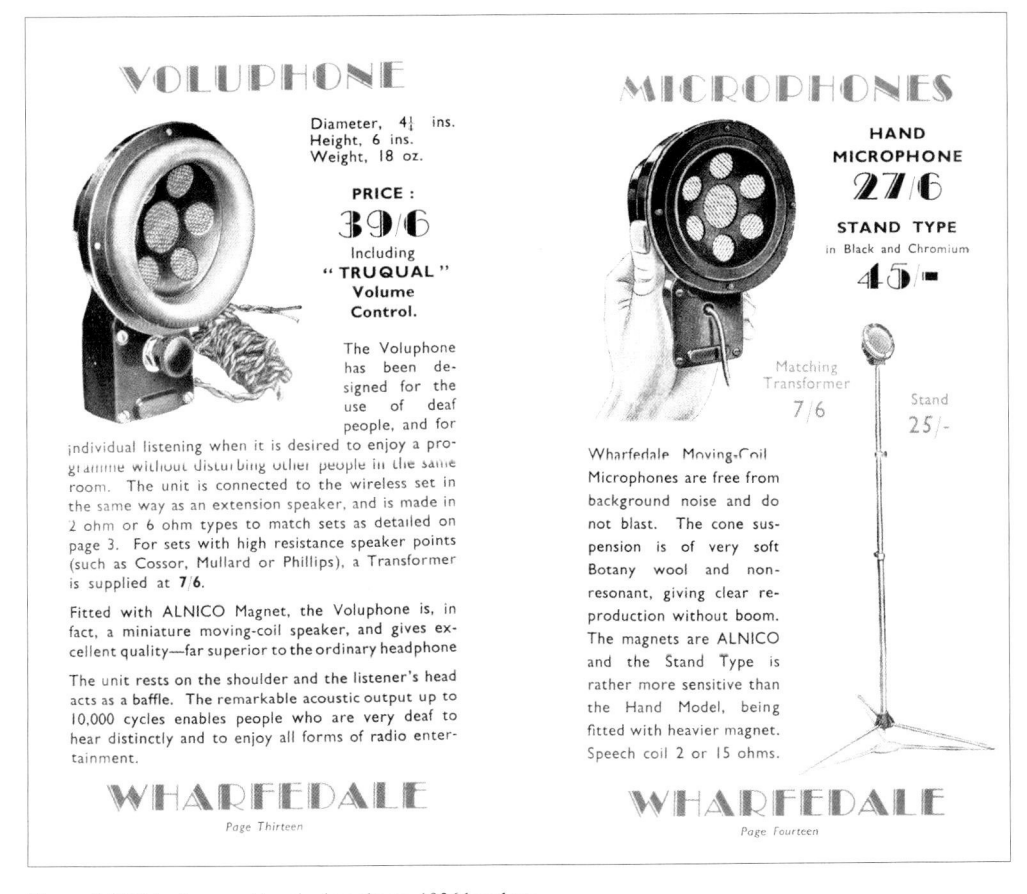

Figure 4.19 Voluphone and hand microphone, 1936 brochure.

cheaper models. His solution was to describe his own home, giving the dimensions of the rooms, and which extension speaker he used in each room and why. The musicality underpinning his personal listening was not ignored. On the 'Golden' used in his sitting-room:

'This gives me a very natural quality with plenty of "top" and I consider it unsurpassed in the reproduction of the piano. A string quartet is also faithfully brought out and this, in my opinion, is the supreme test of any loudspeaker.'

The brochure also announced some dramatic price reductions for the cheaper speakers compared with a year earlier. The 'Standard' chassis was reduced by 9/- to 18/6, whilst the 'Bronze' chassis fell by 7/6 to 27/6 (both without transformer). Gilbert noted:

'The fact that prices are very competitive is due to careful planning and increased production and not, in any way, to a sacrifice in the quality of material or workmanship.'

However, prices for the extension cabinets were little changed so these reductions were probably designed to maintain or increase market share in speaker chassis, which were under pressure compared with the expanding extension speaker (cabinet) market. The radio set makers had gradually been introducing models with low resistance output points, so an extension speaker to such a set did not really need to be fitted with a transformer. Whilst the transformer-less option had always been available, Wharfedale started to make a virtue of the cost saving involved in buying a speaker without a transformer and offered all such extension speaker models with the option of the original 2 ohm voice coils (Type 32) or 6 ohm (Type 98). The brochure listed the names of the set-makers for which each type was appropriate to give the customer confidence that the speaker would be properly matched to the set.

Making Cabinets and a Profit

In late 1936 a drawback to the then product strategy was making itself felt: Wharfedale had to outsource the production of cabinets. The whole industry was very seasonal, with low sales during the spring and summer months and a surge in the run-up to Christmas, hence annual product brochures for the new 'season' were issued after the summer lull. The pressure on the cabinet suppliers to meet the autumn demand inevitably led to a drop-off in quality, which did not help Gilbert in his drive to deliver consistently high quality products, discernible by the customer to both ear and eye. The only way to gain control of the situation was for Wharfedale to make its own cabinets. The firm was still very small, with around half-a-dozen employees including Ezra Broadley, Frank Mann, Arnold Hatton and a cashier who was probably a Miss Sykes. To start cabinet making was a big step, but when the third floor of 62 Leeds Road became available below the occupied premises, it provided the opportunity to expand. A circular saw, a fret saw and a drilling machine were bought second-hand for a total outlay of £28 and installed in the new space. Gilbert hired a non-union cabinet maker and operations got underway, the first cabinets being made in February 1937.

An immediate problem was that standard sized sheets of plywood were too big to go in the building lift, but this aspect of the building's unsuitability proved to be a minor issue compared with what happened next. Gilbert phoned his insurance agent to report the change in production circumstances and request some extra cover. Within half an hour back came the order to stop all woodworking and switch off the machinery immediately. The insurance company did not like the fire risk and would require the whole building to be covered, involving a huge rise in premiums. When an inspector turned up to see what was actually happening, the modest activity he witnessed resulted in a temporary compromise. The machinery would only be allowed to operate until mid-day, and this on the understanding that the company found more suitable premises as soon as possible. As the search got underway the results for the fourth year of trading became known. A small profit of £57 was recorded and Gilbert believed that:[21]

' ….slight as it was, I knew we had turned the corner'

Part 2. Family Life and Music

Gilbert was always interested in people and their welfare and his small band of employees was like a second family. However, he also had a real family which demanded much of his attention. When Wharfedale Wireless Works became his sole occupation in March 1933, Valerie had just been born, Ninetta was nearly five years old and Peter seven (4.20). As mentioned previously, Gilbert also felt a strong sense of responsibility for his mother, then 70, and his two sisters, who still all lived together at 5 Clayton Lane. He visited them regularly and often took them out for the day in his car, in summer to the seaside. One welcome change for Edna was the end of home entertaining, but now she had to contend with the experimentation and other Wharfedale-related work which her husband carried out at home. Since this was crucial to family prosperity, Edna determined that Gilbert would have the peaceful environment that she decided was required. Somehow, a great sense of respect for what he was doing was drilled into the children to achieve the objective, rather than discipline based on fear, although Edna could be very strict. There was no doubt that she was the controlling influence. Gilbert enjoyed playing with the children, particularly activities which had an educational element, such as games based on drawing, music or puzzles, and he was good at making up stories, but as soon as there was any 'horseplay' he would say 'time to go back to your mother!'

Gilbert had a very strong moral compass and instilled his code for what was right and what was wrong into the children. Quite how this had developed is uncertain, but he had been greatly influenced by the 'young men's classes' of his Wesley Guild. At this point he only occasionally went to church (he explained to the children merely that 'his chapel-going days were over') although he was still religious; Ninetta recalls that the only time he really lost his temper with her was when she made a flippant remark about religion. Edna had been brought up in the Church of England and she took the children to Ilkely parish church, which had been her church before the marriage.

There was a great deal of music in the home. Gilbert still had piano lessons, now from Edgar Knight who had become his tutor. These were taken in batches of several months at a time and continued for many years. His approach to playing and Edna's were very different, but they managed to play duets successfully (and pieces for two pianos rather less so). Ninetta recalls how her mother had been roped into playing the piano at a keep-fit class. There were demands for her to play a popular tune of the time with a complicated rhythm which, of course, had to be kept up

Figure 4.20 Gilbert with the children in about 1934. (Photo courtesy of the Briggs family.)

for a considerable time. Edna couldn't manage this and brought the music home in the hope that Gilbert could help her. His intellectual approach to the problem was typical and he wrestled with it in total concentration oblivious to the hysterics of the onlookers. Finally, he remarked 'You know, this really is VERY difficult!' Music was always coming from the music room where Gilbert did all his listening tests or from the radio; as we already know there were extension speakers in every room. A favourite relaxation of his was to listen to the BBC orchestral concert broadcast on Sunday afternoon. Ninetta remembers being summoned into the sanctum to be asked who composed the piece which was being played. In the end she knew because it happened so often! There was also record-playing, mainly limited to solo voice recordings which matched one of Gilbert's musical tastes to the limitations of this medium at the time.

The period of straightened financial circumstances does not seem to have lasted very long, so Gilbert must soon have been able to extract a decent salary from the company, although he seems to have had the feeling that there was a certain standard of living, or comfort, which would be maintained come what may. He had been used to travelling the world on expenses and to devoted female support when at home, first from his mother (later with help from his sisters) and then from Edna, and because he got married relatively late he had enjoyed a prolonged period of being comfortably off. When the move to Ilkley took place in 1931 Peter was of school-age. The state infant's school to which he would have gone was a forbidding building and some distance away, so he went instead to a nearby private school. When Ninetta reached five in 1933, just at the critical point, the school situation was the same, but her fees could not be afforded. The simple solution was not to send her to school. A year later she too went into private education. Gilbert's way of making his children aware of the value of money was to be stingy with pocket-money! About the same time, during 1934, Edna was given help in the home in the form of a maid—the first in a sequence of girls aged 16 or so who came from Durham mining villages and stayed for 6–12 months. They lived in, with a room on the second floor (the site of Gilbert's early loudspeaker experiments) and had to get up at 6.30 am to lay and light fires around the house before taking Gilbert and Edna a cup of tea in bed. It was a lonely situation for the girls and their welfare was added to Edna's list.

Gilbert had owned a car since the early 1920's and he took great pride in the performance of any new model. A weekend treat for the whole family, in fine weather, would be to drive out into one of the 'dales' and seek out a scenic spot for a picnic. Whilst Gilbert and Edna would enthuse about the scenery and Gilbert would take photographs (he had been keen on photography from before his first trip to the Far East), Peter and Ninetta would only be interested in damming the nearest stream. But the outing would not be complete without the car being pitted against one of the steepest inclines in the vicinity—and there were plenty to choose from! The vehicle was always kept some distance from Eaton Road in a lock-up belonging to the local garage. On weekdays the 'constitutional' morning walk to the garage was preceded by the ritual of the children being lined up in the hallway to kiss their father as he set off for work.

The Search for the Perfect Piano

It must have been around 1935 that Gilbert indulged in his first grand piano and began an extraordinary series of trials of different instruments:[22]

'I began by trying to find the perfect piano, but I continued operations to recover the money lost in the early stages, and to accumulate experience which in a vague way I thought might one day form the basis of a book. I never found the perfect piano, but I had some mighty good ones, with which it was hard to part.

As time went on, the frequency with which I changed instruments did not always lead to improved harmony in the home, and certainly caused a good deal of fluttering of curtains in the windows of the house opposite.'

When Gilbert wrote this in 1951 he had gone through a total of 40 pianos: 18 uprights and 22 grands. He had traded in his first Broadwood for a second, better one, before 1914 and his third instrument was a Marshall and Rose upright, bought new in 1921. This remained with him throughout and still lives on. The rest he bought second-hand, mostly in part-exchange deals with Wood's of Bradford, although one or two may also have come from Harrods in London. In *Pianos, Pianists and Sonics*, along with the above quotation, he lists all the pianos in alphabetical order. A penchant for statistics and lists in the recording of such details, evident in all his books, doubtless started in his merchant's days with the myriad of products and the ever-changing pricing, but this is a revealing example. Initially, Gilbert was supposedly being more careful with money; the details of his piano transactions were all recorded but the exercise, on paper or otherwise, was eventually rationalised as a long-term speculation to recover perceived early 'losses'. What really mattered to him, of course, were the characteristics of the piano of the moment and he was capable of recounting these years after the piano itself had passed through his hands.

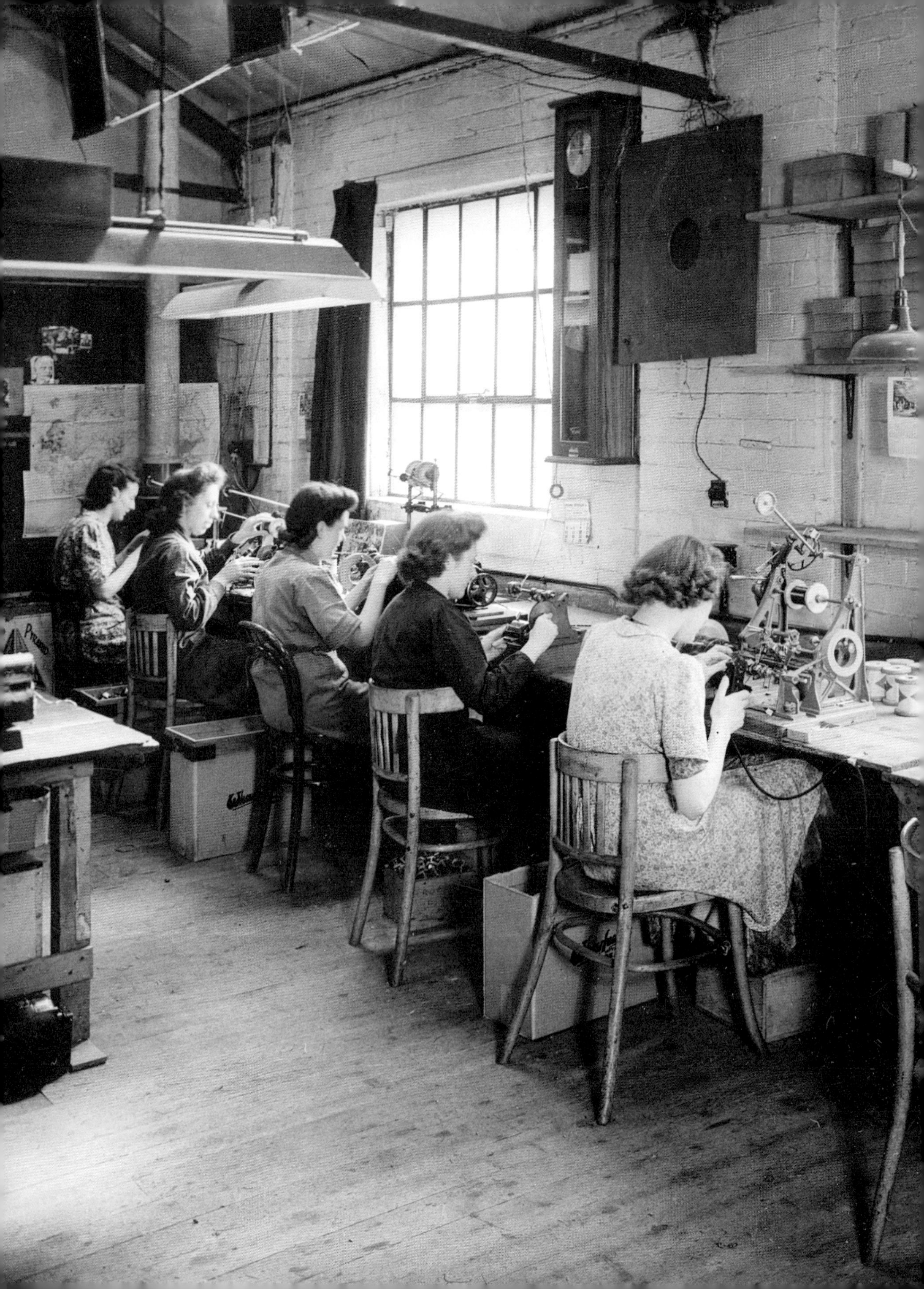

CHAPTER FIVE: 1937–1946

Part 1. Wharfedale Wireless Works in Brighouse

The search for new premises soon led to a possibility in Brighouse, about six miles from the centre of Bradford. This was an unnumbered single-storey building in Hutchinson Lane, a short and narrow street connecting Commercial Street and Lower Oxford Street. A street plan from the period (5.1), on which Hutchinson Lane is not itself named, shows eight connected properties fronting onto the Lane on the left hand side. Wharfedale Wireless Works is believed to have been the last-but-one of these, walking up the slope from Commercial Street, with Ambassador Radio located next door in the last, two-storey building. Today only the lowest part of Hutchinson Lane survives and these sites, along with Upper and Lower Oxford Streets, are now under a bus station. This proximity to another business operating in the same field turned out to be advantageous for several reasons, as will be seen later, but it is quite possible that Gilbert found out about the premises because he already knew the owner of Ambassador Radio, R. Noel Fitton.

The premises had four times the floor area occupied at 62 Leeds Road and the distinct advantage of being at ground level. A disadvantage was the single entrance/exit off Hutchinson Lane (5.1 shows how hemmed-in the premises were) but this might not have been fully appreciated at the time. Once acquired the internal layout was modified for efficient production with all the loudspeaker and transformer manufacture in one large space and cabinet making in another. There was a general office and a smaller office for 'the boss'. Gilbert's management style was literally 'open door' unless his door was closed, which meant that he was attending to letters. Responding to correspondence was afforded a very high priority and his aim, always, was to do this within 24 hours.

The move probably took place during the summer 'lull'; certainly, by the time the 1937/8 brochure was issued trading was from the new address. With this distraction it is not surprising that there was only one new product introduction, the first extension speaker in a corner cabinet. As usual, though, the designs of many of the existing cabinets were altered, and that season's selection is shown in (5.2).

Response Curves

The brochure also illustrated the response curve for the 'Twin Auditorium' speaker for the first time, the source being acknowledged as 'Tannoy'. At that time Wharfedale did not have the means of producing this data and had relied on curves produced during *Wireless World* test reviews for promotional purposes. Gilbert was well aware of the limitations of these curves in terms of assessing speaker performance, listening tests always being his main arbiter, but for experimental purposes they were very useful. He discovered, in late 1936 or early 1937, that Tannoy had developed

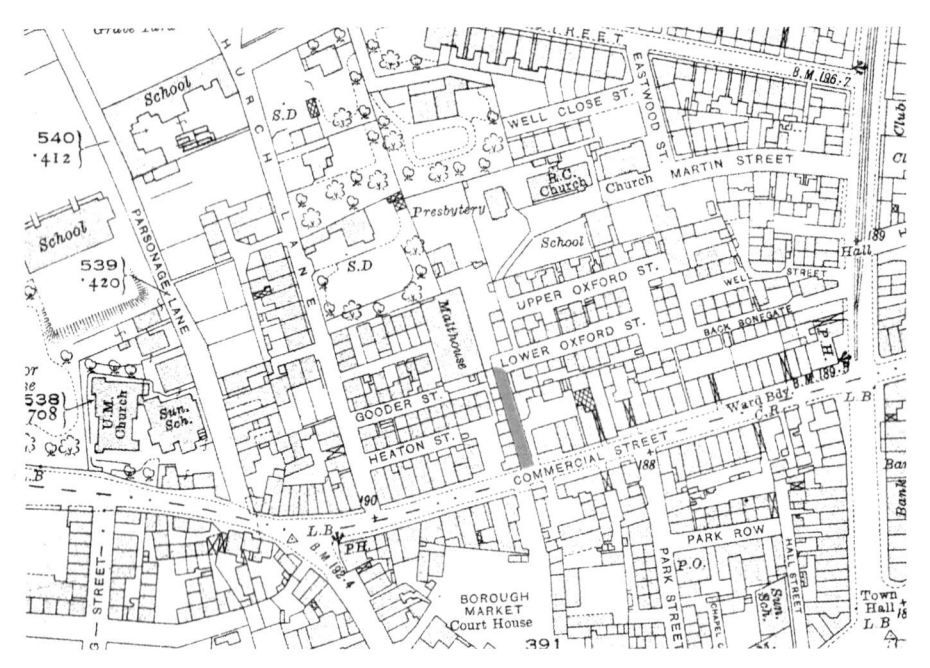

Figure 5.1. Brighouse street plan from the 1930s. Hutchinson Lane coloured in grey.

Figure 5.2. 1937/8 cabinets.

equipment in-house for testing their own products (at that time principally for public address systems) and somehow negotiated to have measurements made on his designs. In 1974 T.B. 'Stan' Livingstone of Tannoy recalled how he first met Gilbert.[1] It was not long after he had started working for Tannoy in West Norwood, London and Gilbert, accompanied by Albert Smith, arrived in his car with a boot full of speaker chassis. It had to be a fine, still day because the measurements were made in the open air above a flat roof. It was Stan's job to screw each chassis onto a baffle, which was then hoisted aloft, whilst Ronald Rackman, the Chief Engineer, did the actual recording of the response curve. When the session was over, Gilbert exchanged pleasantries with Guy Fountain, Tannoy's founder, before leaving. As the car drove away, Fountain said to Livingstone 'Very nice chap Briggs, but I really don't think he's on to a good thing here . . . '

Saved by the GEM

He was almost right! The price reductions made during 1936 and the unavoidable disruption to production resulting from the relocation to Brighouse during 1937 hit profitability; the fifth set of accounts recorded a loss of just over £100. The £2000 capital which Gilbert's partner had injected in 1933 had shrunk to £750 and he decided, in early 1938, to cut his losses. The partnership was dissolved, leaving Edna as the sole owner of Wharfedale Wireless Works. Fortunately, the ex-partner was extremely considerate and left the residual capital in the company on loan account. He must have thought the prognosis was poor—no doubt reinforced by Hitler's annexation of Austria in March and the growing feeling that another war was inevitable.

In fact, problems arose almost immediately which Gilbert summarised as follows:[2]

'The curse of the radio industry before the war was that demand during the summer months always fell off almost to zero. It was difficult not to lose during April–July any profit made during the remaining eight months of the year, unless you had sufficient capital to continue production during the summer and stockpile in readiness for the autumn boom—hoping you had been clever enough or lucky enough to plump for the right model. I recall the position as it was in April 1938. Day after day I received a letter from my London traveller: "Called on so and so. Did not get an order. They had bought a new Celestion extension speaker for 19/6. We cannot compete here." After a week of these dismal reports I wired our traveller to buy one of these monstrosities and send it post haste for our inspection. When it arrived it proved to be an attractive little cabinet speaker with an amazing performance. I took it to Bradford at lunchtime and showed it to our main wholesaler and asked him how he liked it at 21/-. He said "Fine. Send me a gross as soon as possible." I then told him it was a Celestion and they were selling loads in London at 19/6. The best we could do was a rather better job at 21/-. (I hope Celestion will forgive me!) We made a sample within 24 hours, labelled it the GEM, and for the first time worked at full pressure throughout the summer on a line which was easy to sell. In fact the GEM saved the factory....'

Celestion was a major player in the industry. The company originated in 1924 from a collaboration between Cyril French, who owned a plating company in Hampton Wick, and Eric Macintosh, who had invented one of the earliest cone loudspeakers. His original 1925 patent for a free-edge design and their joint 1926 patent for a clamped-edge design were the basis of the first products, with moving-iron balanced armature drivers. 'Celestion' was the name given to the loudspeakers, but soon this became the name of the company, which grew rapidly on the back of the demand for speakers for the early valve radios and subsequently for extension speakers. By 1930 Celestion was operating from a large factory in Kingston-upon-Thames and also making radios, gramophones and radiograms. Moving coil speakers were introduced in 1932 and the company also produced multi-ratio transformers and radio accessories as expansion continued. The worldwide slump which saw Gilbert enter the industry brought the boom times for Celestion to an end, resulting in major restructuring in 1935, but in 1938 the firm was still much, much larger than Wharfedale.

Gilbert's statement that they produced a prototype GEM within 24 hours of gaining their first order for an 'improved' equivalent to the Celestion model is intriguing. The GEM (see 5.4) was fitted with a 6" chassis and up to this point the smallest Wharfedale chassis sold, as such, had been an 8", although they were still selling the Voluphone with its 4" speaker. Celestion had a lot of experience with smaller chassis because of their manufacture of radios and radiograms; they also supplied them to other set-makers. So how Wharfedale were able to conjure up the GEM so quickly is a mystery, unless they had been experimenting with smaller chassis for various reasons and had a design which was suitable for almost immediate production.

Commenting on what happened, Gilbert says:[3]

'We have not made a practise of flattering competitors by imitation, but at the same time we were never strong enough to ignore them completely. Being a small firm, we did however always make a practise of trying to produce a better article at a rather higher price, which kept our competitors happy and enabled us eventually to make a profit without skimming the milk during production.'

The first example of this 'flattery' was probably the combined speaker/amplifier (Class B unit) sold for a short period around 1934/5 (described in Chapter 4), which bore an uncanny resemblance to a Rola product advertised at the same time. However, the Wharfedale sales leaflet emphasised 'two refinements not generally found in competitive makes'—no doubt courtesy of Ernest Price.

The BBC and the General Post Office (GPO)

1938 did, however, see a development which was based on the excellence of their own designs. Perhaps prompted by the *Wireless World* 'Golden' review of 1936, the BBC Research Department undertook their own evaluation of the speaker. It must have been about the time of the relocation to Brighouse that Gilbert received a letter from Mr D.E.L. Shorter saying that they were interested in the speaker for possible adaptation for use in their outside broadcast equipment. This prize, however, was not easily won. There were some specifications to be met: in Gilbert's words:[4]

'They tied us down to a specified resonance frequency plus or minus 10% (the weather often cheated us); they made us put special corrugations in the cone; and they made us take the eyelets out of the cone and put them on the limb of the centring device—a very awkward job requiring much skill to avoid fracture of wire.'

It took six months of 'struggle and strife' to meet these requirements, but the effort was worth it. For one thing, Gilbert needed to liaise throughout with Shorter, who, though only 26 at the time, came to be very highly regarded, and the BBC Research Department team at Balham in London. The interaction was very valuable: not only did he learn a lot technically, but also it connected him with the influential BBC research group at a relatively early stage in his second career. Furthermore, it resulted in an initial contract for 100 speakers in 1938 to be followed by others at regular intervals for at least the next 20 years! The BBC also gave permission for the phrase 'As used by the BBC' to be used in advertising of the 'Golden', which Wharfedale did relentlessly from then on.

Notwithstanding this development work on the 'Golden' chassis, a new model, the 'Portland', was introduced to replace the Twin-Cone Auditorium. Similar to the Golden it was a 10" unit with a concentrically corrugated cone, but with a much more powerful AlNiCo magnet at 14,000 oersteds on a 1" centre pole (Golden 10,000 oersteds) and capable of handling up to 12 watts (Golden 8 watts). *Wireless World* gave it an enthusiastic review in July 1938 and again published the response curve. Sometime after the 1937/8 catalogue was issued Wharfedale made an attempt to sell an extension speaker which could be classified as a high quality piece of furniture. This was the floor-standing 'Console', a large unit of 29.5 × 18 × 10" in highly polished, figured walnut, fitted with a 'Golden' chassis and selling for 140/- including transformer. It was followed by the slightly smaller 'Console-Junior', fitted with the Bronze chassis. Neither model survived long enough to feature in the following year's catalogue.

1938 also saw an interaction with the Post Office Engineering Department. A plan had been mooted to nationalise radio relay services and the GPO was to run a pilot scheme in Southampton. Two of their engineers, Messrs West and McMillan, had designed a superior loudspeaker cabinet which would be part of the equipment. Along with other suppliers, Wharfedale were invited to submit a unit built to the GPO specification for evaluation. This they did and the result is shown in (5.3). The new 10" Portland chassis was fitted with a free-edge cone supported by four small cloth segments and the cabinet was lined with felt, lead foil and corrugated cardboard, resulting in an inaudible bass resonance of 35 Hz irrespective of whether the cabinet was fully enclosed or not. The specification called for a cone design which would deliver a rising response of around 3 dB per octave in the middle and upper registers to counter-balance absorption of furnished rooms. The unit was difficult to construct, great skill being required to fix the cloth segments without pulling the cone-coil assembly out of alignment. Ezra Broadley was the maestro when it came to this type of work. Eventually the relay project was abandoned, but Gilbert made an arrangement to market the speaker unit under licence, as described later.

Whilst all these new developments were being worked on, attention had to be given to an existing product which had turned out to be less than the brainchild Gilbert thought he had conceived. The Voluphone moving coil headphone, introduced in late 1936, was intended to act as the equivalent of an extension speaker for the deaf. As Gilbert later recalled:[5]

'Unfortunately, we priced it at the ridiculously low figure of 39/6, as we did not then know that the cost of selling a unit to deaf people can easily amount to more than actually making it.'

Figure 5.3.
GPO cabinet, inner linings parted on the right for observation. Reproduced from *Loudspeakers* 5th Edn.

An attempt to improve the margin was made by redesigning the unit somewhat, so that the volume control was no longer integral and pricing the new version at 45/-. An optional remote volume control box was then sold for an additional 7/6, raising the total price by a third. However, the real problem was that they were simply not geared up to market the product effectively—aids for the deaf being a completely different business. This was finally accepted a year later and the product was dropped during 1939.

Improving Fortunes

The 1938/9 catalogue was significantly different from the previous year's with several new items: the 'Gem' extension speaker in dark walnut and a similar 'Moderne' unit in light maple; the new 6" chassis developed for these; the Portland chassis; a new large 'W66' cabinet designed for large rooms, hotels, schools etc. which could be fitted with Bronze, Golden or Portland chassis and the revamped Voluphone. By 1939 annual production of speakers had risen to about 9000, some three times the number of 1934. Transformers and volume controls also made a solid contribution to turnover. Replacement-type transformers alone sold at the rate of about 10,000 per year. Quality was a key selling point and returns from all causes averaged less than 0.5%.

Cabinet making was now a significant activity, of course, and Gilbert had struck up a mutually beneficial relationship with neighbouring Ambassador Radio in this regard. Effectively the two firms pooled resources, making each other's cabinets according to the demands on men and equipment and the ups and downs of their order books. Although this was a completely informal arrangement based on verbal agreements between Gilbert and Noel Fitton, it lasted the whole time that Wharfedale was in Hutchinson Lane. Gilbert also became very friendly with Ambassador's Chief Engineer, Frank H. Beaumont. Since Ambassador made radios and radiograms, he was involved in all the aspects of the audio business, including designing loudspeaker enclosures, and a great source of technical and market knowledge for Gilbert to have literally 'next door' during the formative years of Wharfedale. Ambassador Radio was a significant concern, with its own retail outlets in several West Yorkshire centres, which advertised and exhibited nationally.

The number of Wharfedale employees was slowly advancing towards the 20 mark and one of these was Gilbert's elder sister Claris. Strangely, this fact is never mentioned by Gilbert himself, although she played an important role for at least ten years as office manager with responsibility for orders, shipping and book-keeping. When she joined the firm is not known, but around the time of the move to Brighouse is the most likely. Unlike Gilbert and her younger sister Mabel, Claris did not benefit from the trust which paid for their education at Crossley and Porter's. By the time this possibility was being considered Claris was over 11, an age when children in mill towns and villages were typically starting to work part-time. By 1911 she was a tailoress but later, however, she went into office work like her sister Mabel who had left Crossley and Porter's, with a Pitman shorthand qualification, to become a shorthand typist. The sisters never married and continued to live with their mother at 5 Clayton Lane.

The 1939/40 catalogue, issued just before the outbreak of the Second World War, is reproduced in full in (5.4). Two new chassis are listed but not illustrated: a 5" unit and an additional 8" unit, the 'Coronet', which sat between the Standard and Bronze models in price and performance. The major new introduction was the extension speaker resulting from the GPO design criteria of 1938. It was named the 'Langham' and was offered at almost twice the price (150/- without transformer) of the next most expensive cabinet in the range, the De-Luxe fitted with the Golden chassis. The Langham was deliberately pitched at the purchaser wanting 'the best possible reproduction regardless of price' and it seems to have replaced the W66 cabinet. Wharfedale claimed that their extension speaker range, with the additional flexibility in terms of cabinet shades and finishes offered, was the most comprehensive on the market. A new Speaker Switch was also offered for fitting to a receiver cabinet, allowing selection of the set speaker, the extension speaker or both speakers—the price was 3/6. Both the Voluphone and the related hand microphone had been discontinued. The catalogue illustrates two response curves which, for the first time, did not acknowledge a source, suggesting that by this stage Wharfedale had developed the means to record them in-house. 1938 and 1939 were good years for the company and the accounts for the 1939/40 financial year made good reading. The £750 capital deficit of two years previously had been replaced by a £424 capital surplus and Gilbert 'began to sleep better at night'. It had taken seven years to put the business on a reasonable footing.

The Impact of the War

By March 1940, with the Second World War entering its seventh month, a different set of uncertainties had to be faced. Almost immediately after the outbreak of war the cost of materials started to increase, plywood in particular. By early 1940 the prices in the catalogue issued the previous autumn could no longer be sustained and a new catalogue was issued with prices from 1 February. The price rises were typically 10% rising to nearly 15% for the smaller cabinets. In July another price-list was issued with prices for the more expensive chassis items further increased. A new 10" unit, the 10" Bronze, was introduced and priced between the 8" Bronze and the 10" Golden units. The Langham cabinet was no longer included. Sales had been disappointing—'only a few dozen'—and it may be that if Ezra Broadley had been called up by this time then the key constructor

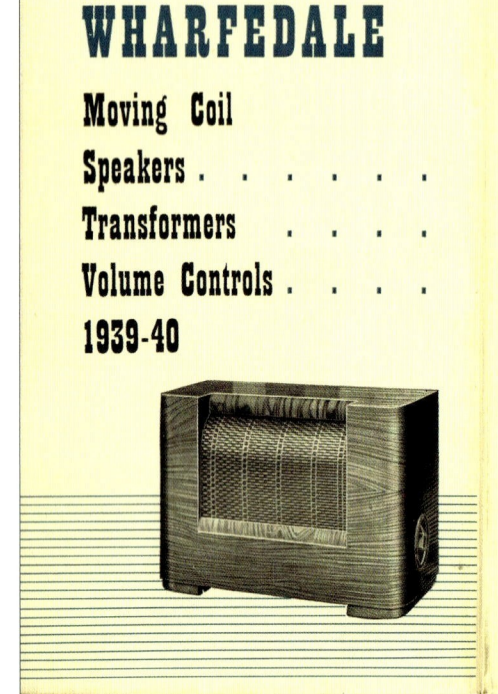

Figure 5.4a 1939/40 catalogue.

Chapter Five: 1937–1946

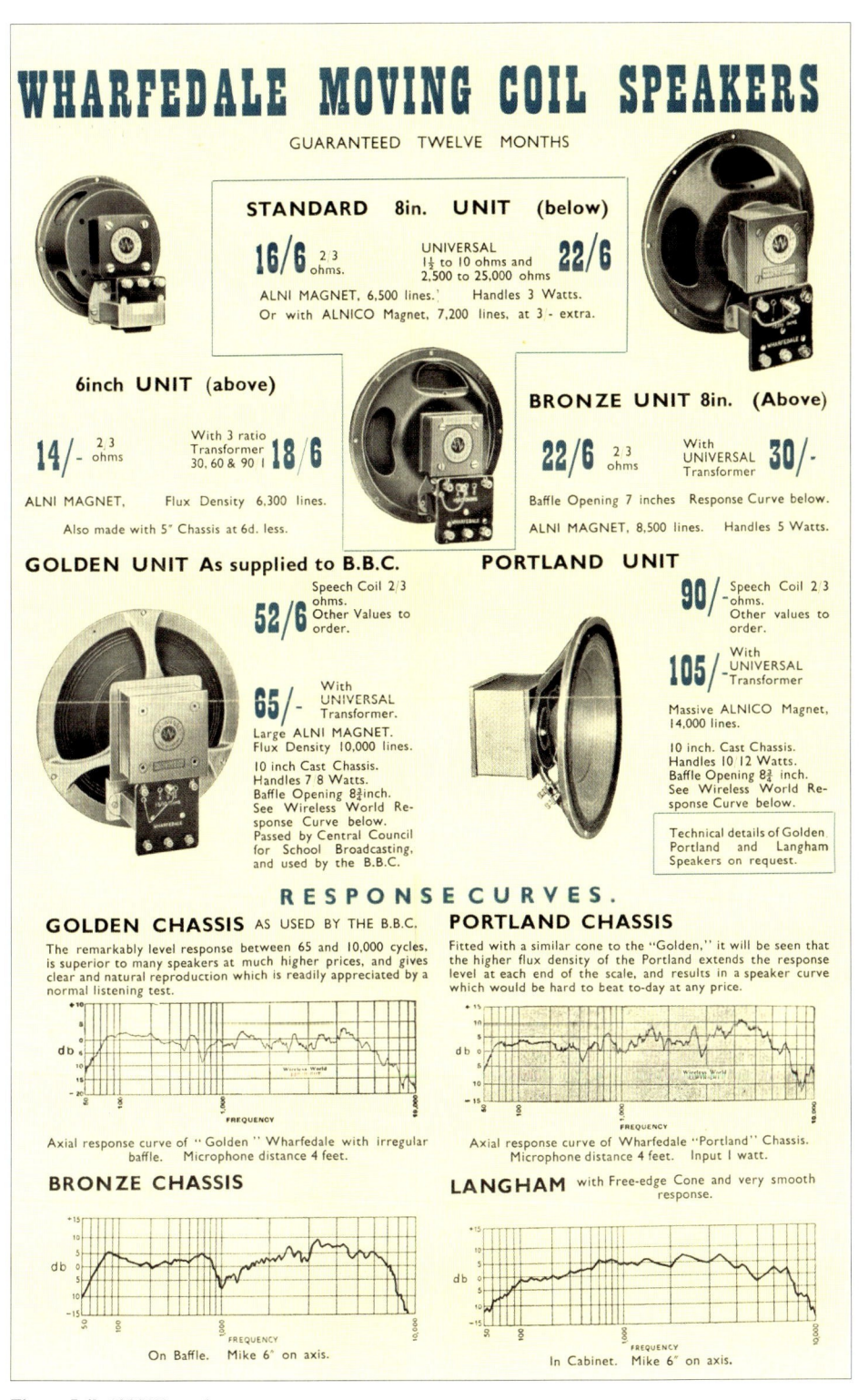

Figure 5.4b 1939/40 catalogue.

Figure 5.4c 1939/40 catalogue.

Chapter Five: 1937–1946

WHARFEDALE EXTENSION SPEAKERS — To match any set

MODERNE

Size of Cabinet: 9½ × 8 × 4¼ inches.
PRICE:
22/6 (1½ to 5 ohms)
Also Type 98: (6 to 10 ohms)
27/- with 3-ratio Transformer

Light Maple Finish

Equivalent to the Gem in size and performance, the Moderne is beautifully polished in Light Maple Colour and Satin-wood finish, edged with Black. Ideal for rooms with light furniture or for Kitchens with white or pale-coloured fittings. Volume Control, 3/6 extra.

MERITOR

Size of Cabinet: 12½ × 10¼ × 5¼ in.
PRICE:
26/- Type 32, 64, 98
32/- UNIVERSAL

Walnut. (Oak or Mahogany 2/6 extra.)

Fitted with six inch Chassis, the Meritor has a very good performance with surprising punch and depth for its size. In common with all Wharfedale Models, the Cabinet is solidly built and hand-polished. Handles 3 watts. Guaranteed 12 months. Supplied with "Truqual" Volume Control at 3/6 extra.

CORONET

With "TRUQUAL" Volume Control.

Size of Cabinet: 14 × 11 × 5½ inch.
PRICE:
42/- Type: 32, 64 or 98
49/6 UNIVERSAL 1½ to 10 ohms and 2,000 to 18,000 ohms

Walnut. (Mahogany 2/6 extra.)

Similar to the Bijou in design, the Coronet is rather larger and is fitted with a special unit with a heavier Magnet with a Flux Density of 7200 lines. Handles 3/4 Watts.

BRONZIAN

With 'TRUQUAL' Volume Control
15 × 12 × 6½"
PRICE:
50/- Types 32, 64, 98
57/6 UNIVERSAL

Walnut. (Oak or Mahogany 2/6 Extra).

This modern and attractive Cabinet is fitted with the famous BRONZE Wharfedale Unit. The performance is first-class, and the sensivity is high. Handles 4/5 Watts.

DE LUXE

With "TRUQUAL" Volume Control.

Size of Cabinet: 16" × 12" × 8"

A very handsome Cabinet with metal grill.

Available with Bronze unit or with Golden unit as supplied to the B.B.C.

Walnut. (Mahogany 2/6 extra.)

	Less Transformer	Universal	Max. Input
With BRONZE UNIT	60/-	67/6	5 Watts
With GOLDEN UNIT	80/-	95/-	8 Watts

LANGHAM

With "TRUQUAL" Volume Control.

14" × 16" × 8"
PRICE:
150/- Types 32, 64 or 98,
165/- UNIVERSAL

Walnut (Mahogany 2/6 extra).

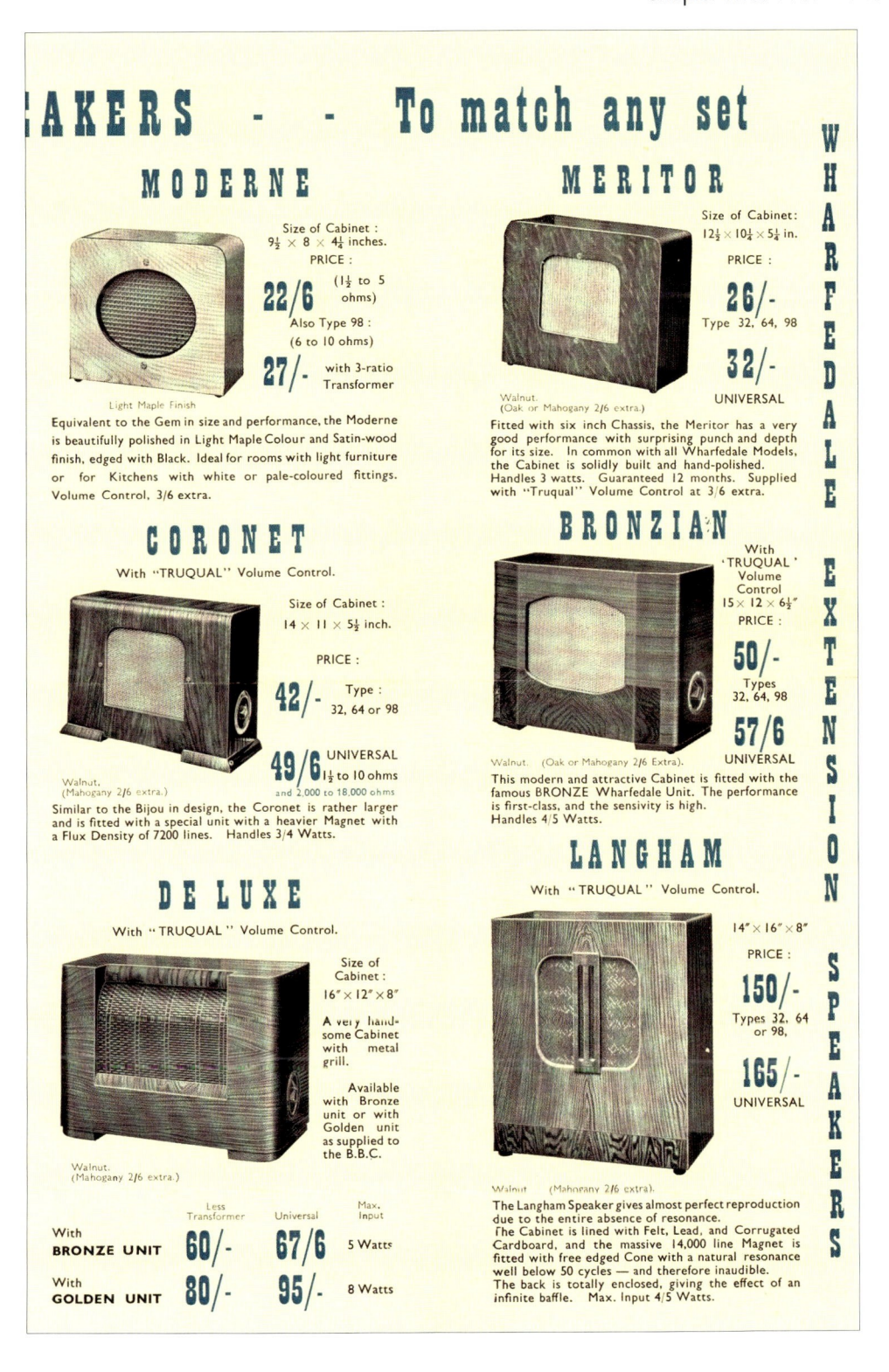

The Langham Speaker gives almost perfect reproduction due to the entire absence of resonance. The Cabinet is lined with Felt, Lead, and Corrugated Cardboard, and the massive 14,000 line Magnet is fitted with free edged Cone with a natural resonance well below 50 cycles — and therefore inaudible. The back is totally enclosed, giving the effect of an infinite baffle. Max. Input 4/5 Watts.

Figure 5.4d 1939/40 catalogue.

was not available. Gilbert reckoned that the unit was 'about 10 years before its time'.

The early months of the War saw a significant change in the transformer part of the business. Wharfedale had started producing output transformers for their own chassis and extension speaker units before 1935 and by 1936 they were being offered for sale as separate items. There were three models all with centre-tapped primaries: Standard with ratios 36, 60 and 72:1 (7/6); Universal with ratios 30, 35, 45, 60, 70 and 90:1 (9/6) and De-Luxe, a larger version of the Universal for higher inputs (15/-). An additional output transformer was introduced in 1937, specially aimed at the replacement market; the Service model with ratios 30, 60 and 90:1 at 6/-. In July 1940 a major expansion of the range was unveiled, as illustrated in (5.5). The Service model was uprated to four ratios and renamed Type P whilst a new model was introduced as MR (Multi-Ratio) Service, with four primary and three secondary ratios (12 in all). In addition two small output transformers (OP3 and CT3) and five inter-valve transformers (Permalloy LF and QPP, Class B Driver, QPP Service and Service LF) were announced. Prices for these new units were between 4/3 and 7/6.

The number of coil-winders employed may well have increased as transformer production increased, but there was certainly one new employee taken on during the War. In 1943 Dorothy Stevens, aged 18, was recruited straight from completing a course at Stead's Commercial College in Brighouse to become the 'office junior', working for Claris Briggs. A photograph taken some time later (5.6) shows the three members of the Hutchinson Lane office: Claris (sitting, centre), Dorothy (at the filing cabinet) and Mrs Housten, the typist (sitting, left).

About this time an attempt was also made to break into the 'factory' market with two products which were not included in the normal, domestic, catalogue. The first was an indoor speaker unit consisting of a reversible cabinet, which could be ceiling suspended, fitted with a Bronze chassis (either the normal 8" model or a new 10" version). The second was the 'Flare', a large flared horn of rectangular cross-section fitted to a small cabinet containing an 8" Bronze chassis, for either factory or open-air use where wide-angle and directional propagation was required. The remote volume control introduced in 1938 was optionally available for use with these units.

From mid-1940 the problems with materials shortages became more severe. Lack of suitable plywood led to the introduction of small extension speakers covered in 'rexine' leather cloth, replacing veneered models, by the beginning of 1941 and the reduction in the number of cabinet speaker models from eight to four by that autumn, when the last wartime catalogue was issued. This did introduce a new chassis, however, called the Midget—a 3.5" unit which used the newly available Alcomax magnet material. A revised price-list dated September 1942 shows that a new speaker chassis had been developed for public address use, the W.12—a 12" unit with a 1¾" centre pole. Gilbert gives a strong hint that this was modelled on the Rola G12 in a story about a repair request during the War:[6]

> 'A Mr Corrigan called with a couple of 12 in speakers which had blown out by overloading, and begged us to fit new cones and coils. They were Rola G12's—a world-renowned 12 in unit—but we had long ago given up the repair of other makes of speaker as a mug's game. However. . . . . we agreed to help him. (It so happened that our 12 in cone assembly fitted the G12—I think we had copied it because of the big demand!)'

This opens up the fascinating possibility that the first naming of a Wharfedale unit in this way (viz W.12) was also a direct analogy.

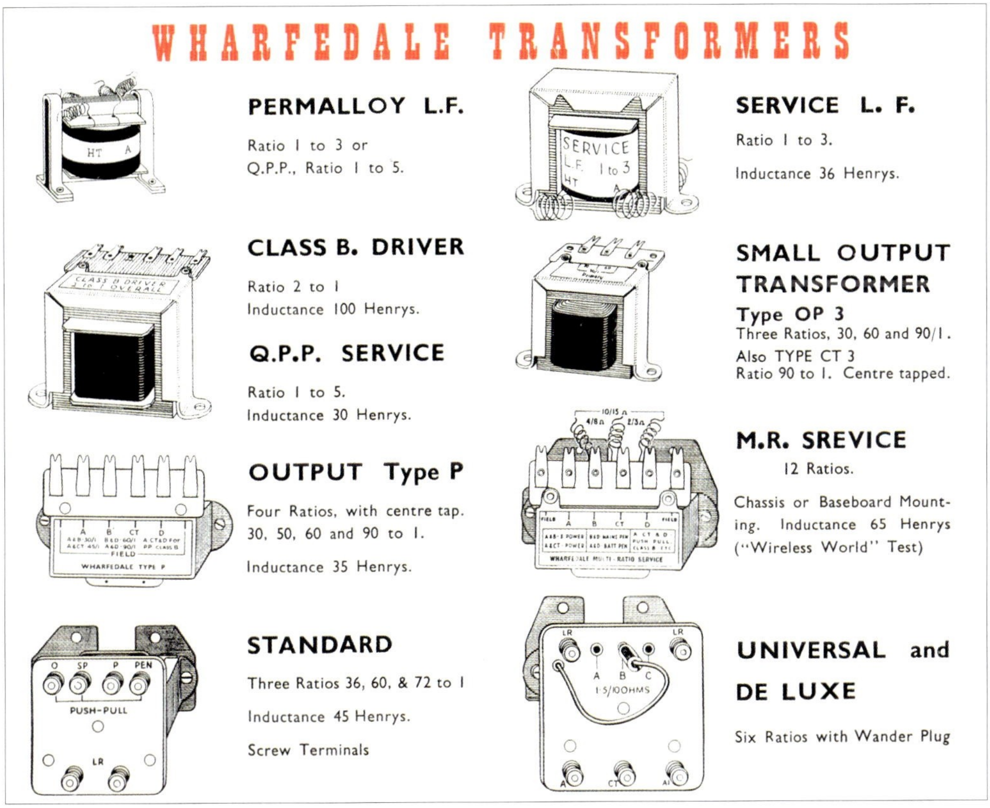

Chapter Five: 1937–1946

WHARFEDALE TRANSFORMERS

PERMALLOY L.F.
Ratio 1 to 3 or
Q.P.P., Ratio 1 to 5.

SERVICE L.F.
Ratio 1 to 3.
Inductance 36 Henrys.

CLASS B. DRIVER
Ratio 2 to 1
Inductance 100 Henrys.

SMALL OUTPUT TRANSFORMER
Type OP 3
Three Ratios, 30, 60 and 90/1.
Also TYPE CT 3
Ratio 90 to 1. Centre tapped.

Q.P.P. SERVICE
Ratio 1 to 5.
Inductance 30 Henrys.

M.R. SREVICE
12 Ratios.
Chassis or Baseboard Mounting. Inductance 65 Henrys ("Wireless World" Test)

OUTPUT Type P
Four Ratios, with centre tap.
30, 50, 60 and 90 to 1.
Inductance 35 Henrys.

STANDARD
Three Ratios 36, 60, & 72 to 1
Inductance 45 Henrys.
Screw Terminals

UNIVERSAL and DE LUXE
Six Ratios with Wander Plug

Figure 5.5 Transformers made in mid-1940.

Figure 5.6 Brighouse office about 1944. (Photo courtesy of Dorothy Stevens.)

79

Contracts and Sidelines

From about 1941 production for the domestic market was determined solely by the availability of materials and anything that could be made was easily sold. Priority orders for speakers were regularly received from the BBC, the GPO, other public utilities and government agencies, but the Brighouse factory was kept busy with large Admiralty contracts for output transformers to be used in equipment manufactured by Marconi's Wireless Telegraph Co. Ltd. Doubtless the range and quality of Wharfedale's transformers was the key to winning these contracts. An eight ratio general purpose model (ranging from 12–72 : 1) was developed (GP 8) and over 50,000 of these alone were supplied between 1942 and 1945.

The reduction in cabinet output as the War progressed must have caused Gilbert to be concerned that he could not keep all his cabinet makers employed; laying off any of his employees, through no fault of their own, would have been anathema. The solution he came up with, around 1942, was to turn them to making tables, which were useful furniture items in their own right, but marketed primarily as stands for a radio or an extension speaker. There were several designs, some in a range of sizes, all 'substantially built with moulded legs, walnut veneered and hand-polished to a piano finish'. The most expensive, the W50S, was actually a table-speaker (the Bronze speaker being mounted in a cabinet beneath the table-top) at 90/-, with transformer. A related, yet characteristically Gilbertian sideline, was to market a 'Polishing Outfit', which consisted of a tin each of wood filler, smoothing compound and polish, bottles of walnut and black stain (each with brush), one duster and two cotton cloths (5.7). How successful a gambit this was will probably never be known!

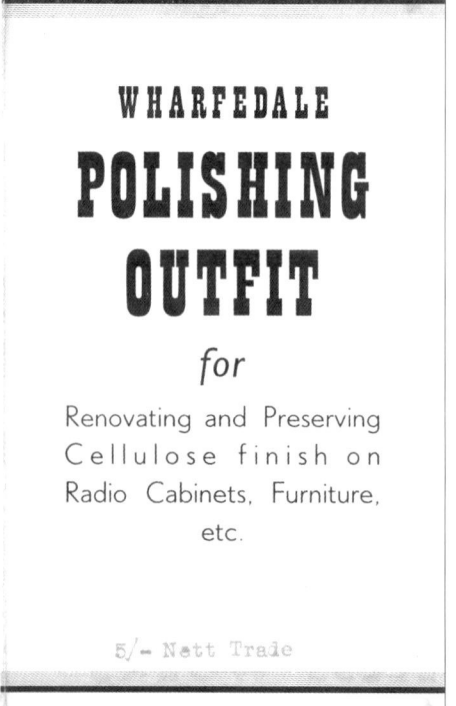

Figure 5.7 Polishing Outfit leaflet, about 1942.

Peacetime

Advertising in *Wireless World* recommenced towards the end of the War, with a quarter-page placement in every monthly issue during 1945. The only new non-transformer product introduced during the War was the Midget chassis and this featured in some of the advertisements. A curiosity of these is the inclusion of the line: 'Sole Proprietor D.E. Briggs' (Edna) as shown in (5.8). Why this was done is a mystery.

Of the wartime effort Gilbert recalled:[7]

> *'As our entire staff at the time was only about twenty—including three in the office and two elderly French polishers—I thought they did quite well.'*

Perhaps Gilbert was acknowledging their contribution when he designed the first peacetime brochure, issued in January 1946 and printed on very poor quality paper. This included six pictures of factory operations (and one of himself) taken during the War (5.9). The six pictures of factory operations date from 1942.[7] Some originals, or good copies, have survived and three of them are reproduced below (5.10–5.12).

Fig 5.9 must include nearly all the employees. The picture of Gilbert himself (5.9a) was taken in August 1944, probably at the same time as (5.6). It is interesting to compare this cropped image with the original from which it was derived (5.13). Gilbert often had to spend nights at the factory on 'fire-watch' duty, to minimise damage caused by possible incendiary bomb strikes, and the shaving kit above the hand-basin confirms this. This reminder of wartime has been eliminated and retouching of the negative has hidden the fact that the office walls were simply painted brickwork.

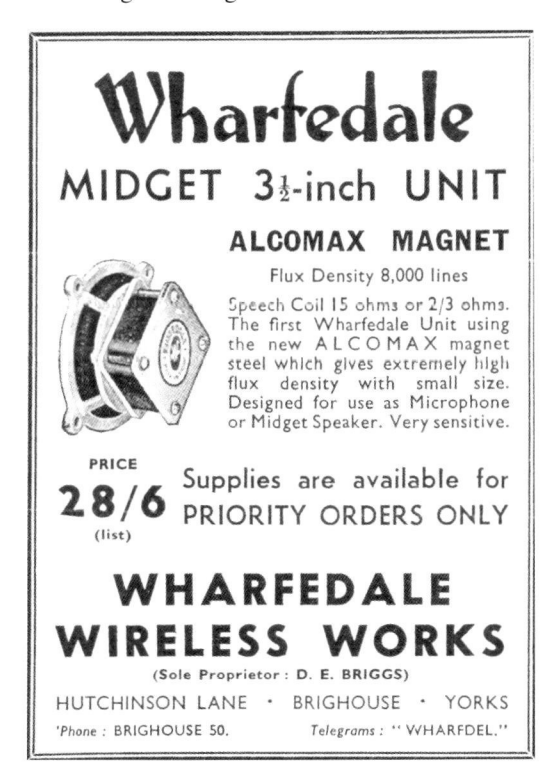

Figure 5.8 *Wireless World* Advert, June 1945.

Figure 5.9a 1946 brochure.

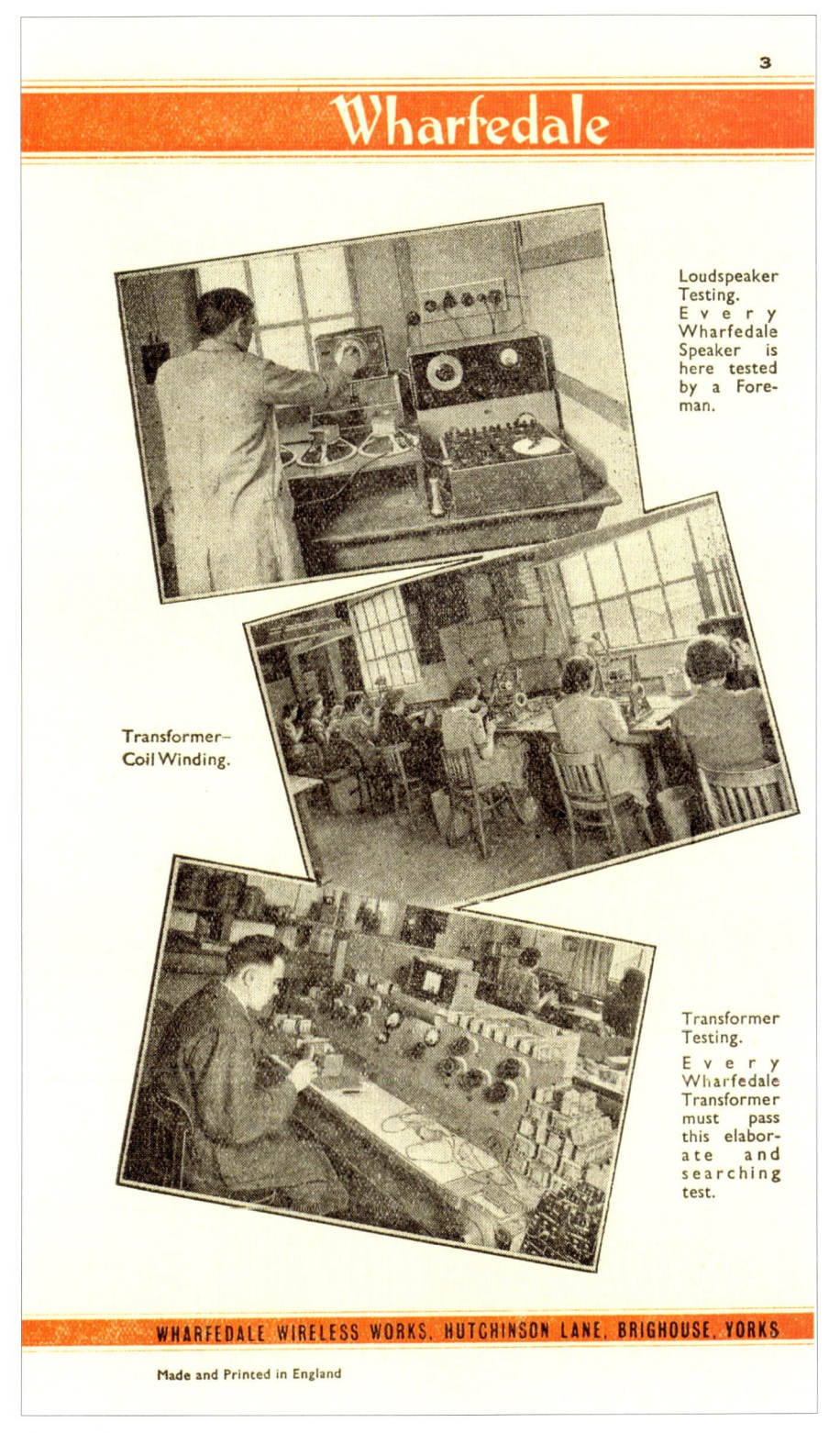

Figure 5.9b 1946 brochure.

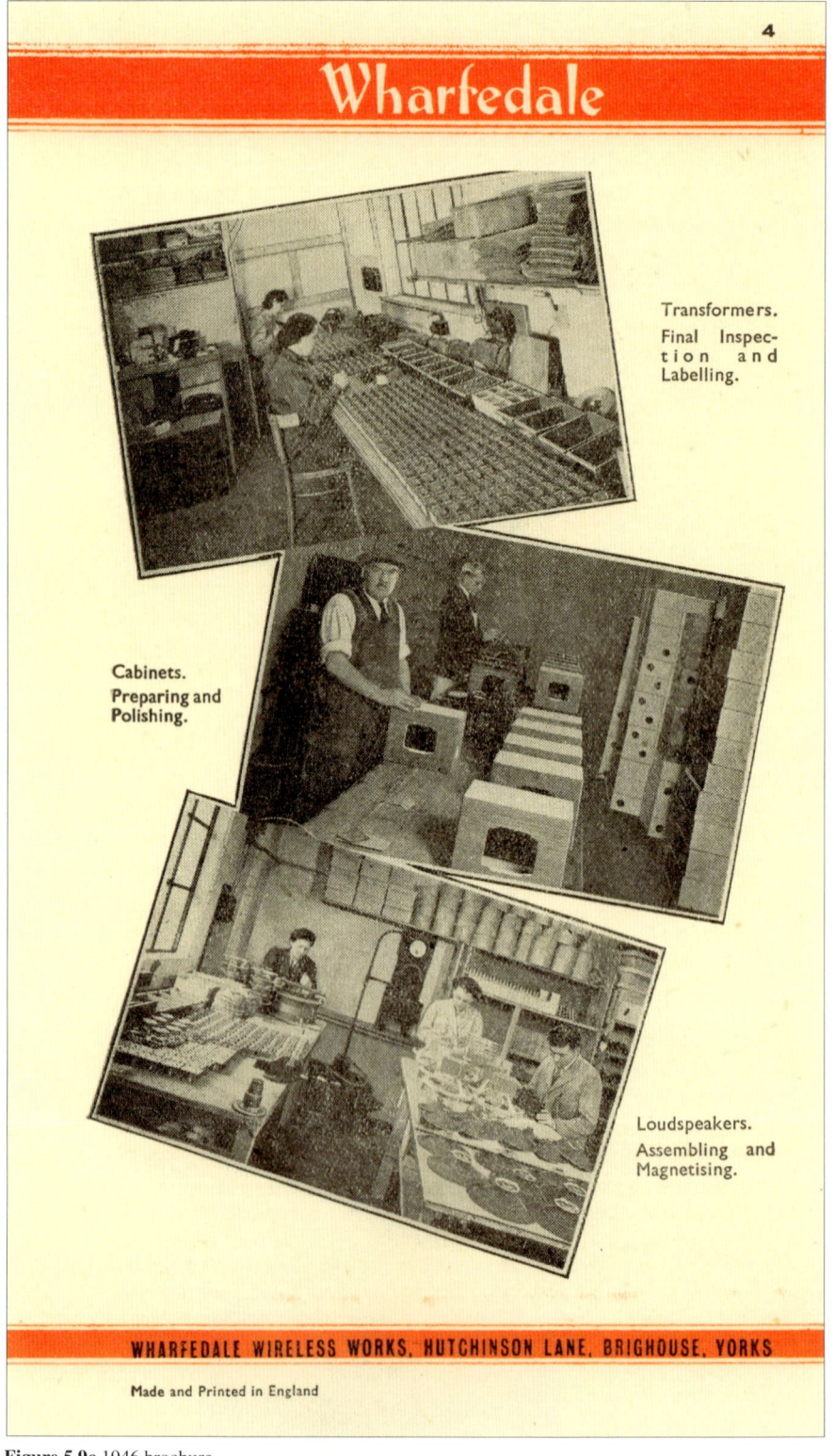

Figure 5.9c 1946 brochure.

Figure 5.10 Coil-winders in the Brighouse factory, 1942. (Photo courtesy of the Escott family.)

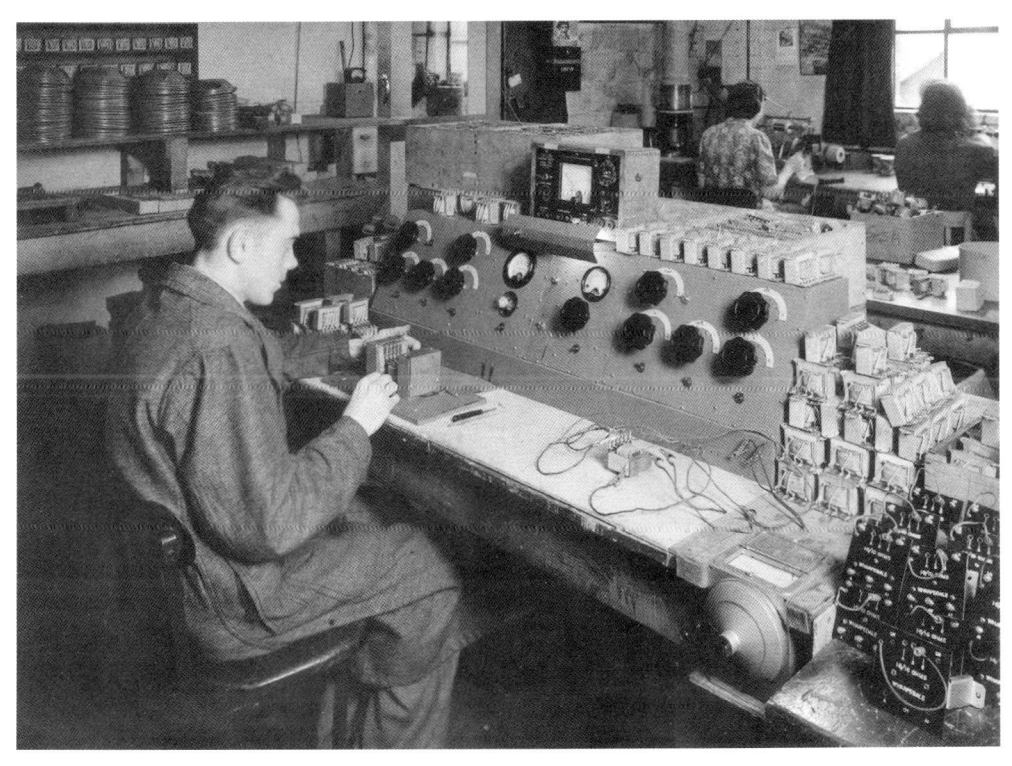

Figure 5.11 Transformer testing, Brighouse, 1942. (Photo courtesy of the Escott family.)

A Pair of Wharfedales: The Story of Gilbert Briggs and his Loudspeakers

Figure 5.12 Assembling (right) and magnetising (left) loudspeakers, Brighouse, 1942. (Photo courtesy of the Escott family.)

Figure 5.13 Gilbert in his office, 1944. (Photo courtesy of the Briggs family.)

Chapter Five: 1937–1946

Looking to the Future

The products on offer in 1946 are collected in the price-list from the same catalogue (5.14). The list appears to be almost the same as that of 1941/2, but as far as the speakers are concerned this is deceptive. Research and development had continued in the interim and all the units had been redesigned. Most models now had open die-cast chassis replacing pressed steel, which reduced audible resonances and gave greater coil-centring precision, whilst magnets now employed the recently developed Alcomax and Ticonal materials, which gave significantly higher flux densities whilst reducing size and weight. New bakelite coil-centring devices had also been designed and manufactured. Overall these changes led to significant improvements in performance but prices were, on average, little changed from late 1942. Gilbert's pricing policy had evolved as his products became

Wharfedale

Price List
JANUARY 1st, 1946

CHASSIS

	Flux Density	Less Transformer	With Transformer
Midget, 3½"...	9,000	25/–	—
Five-inch W.5.	9,000	22/6	27/6 O.P.3
Six-inch W.6.	8,000	25/–	31/– P.
Standard, 8"	8,000	27/6	37/6 Universal
Bronze, 8"...	10,000	32/6	42/6 Universal
Bronze, 10"	10,000	37/6	47/6 Universal
Golden, 10"	12,500	75/–	90/– De Luxe
Portland, 10"	14,000	120/–	135/– De Luxe
W.12, 12" ...	13,000	130/–	147/6 W.12

CABINET MODELS

	Unit	Less Transformer	With Transformer
Tyny, with V.C.	5" W.5.	42/6	47/6 O.P.3
Gem, with V.C.	6" W.6.	47/6	57/6 Universal
Meritor, with V.C.	8" Standard	57/6	67/6 Universal
Bijou, with V.C.	8" Standard	60/–	70/– Universal
Bijou/Bronze, with V.C.	8" Bronze	65/–	75/– Universal
Bronzian, with V.C.	8" Bronze	85/–	95/– Universal

TRANSFORMERS

	Ratio	Price
Permalloy "C," L.F.	1 to 3	8/6
Permalloy "C," Q.P.P.	1 to 5	9/–
Class B Driver	2 to 1	8/6
O.P.3	3 ratios	5/6
P. Type	4 ratios with C.T.	6/6
G.P.8	8 ratios with C.T.	9/6
Universal	6 ratios with C.T.	12/6
De Luxe	6 ratios with C.T.	17/6
W.12	3 ratios with C.T.	17/6

TRUQUAL VOLUME CONTROLS

Heavy Duty	Less Escutcheon	With Escutcheon
Type 32 for 1½ to 5 ohms	7/6	8/6
Type 98 for 6 to 15 ohms	7/6	8/6
	Price	
Speaker Switch	5/6	

WHARFEDALE WIRELESS WORKS, HUTCHINSON LANE, BRIGHOUSE, YORKS

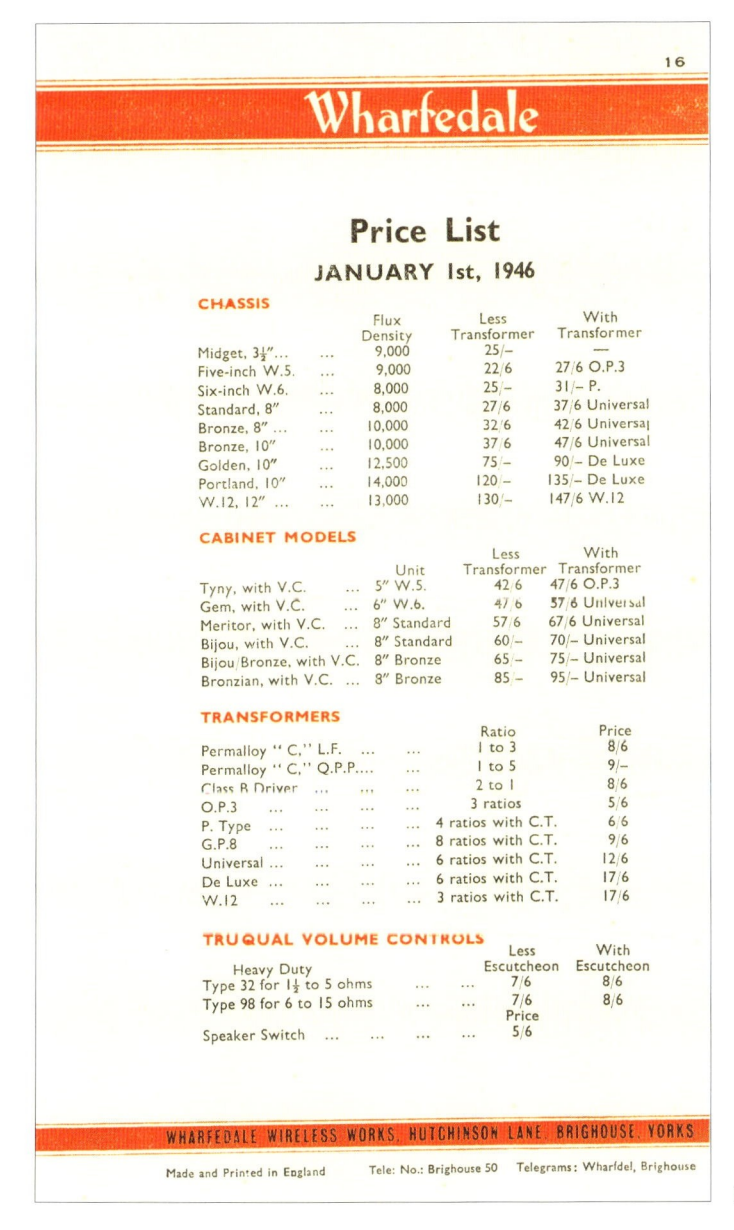

Made and Printed in England Tele. No.: Brighouse 50 Telegrams: Wharfdel, Brighouse

Figure 5.14 Price-list, 1946.

more familiar and the reputation of Wharfedale grew. Once a product had been introduced he aimed to use gains from improved manufacturing efficiency or reduced material costs to improve quality/performance rather than reduce the price. On the other hand, he only felt justified in increasing his prices when material costs forced him to do so. He was quite transparent in this regard, using catalogues and advertisements to expound his philosophy and include specific examples whenever possible. He occasionally made mistakes in an original pricing and thought nothing of explaining this to his potential customers in a hastily revised price-list!

The all-new products in 1946 were the 5" chassis (W5) and the 'Tyny' extension speaker (5.15), which incorporated this in a small pressed-steel cabinet finished in cream enamel ('for kitchen, bedroom or office'). The 'Gem' extension speaker, withdrawn for a period during the war because of material shortages, had been reintroduced and, despite these problems, over 12,000 had been sold since the dramatic launch in 1938. The larger, mid-priced 'Bijou' model first introduced in 1935 was, however, the consistent best-seller.

The manufacturing difficulties of wartime did not end when peace came in May 1945; indeed they went on, because of continuing shortages, or controls over the supply, of key materials for several years. However, there was now the opportunity to think constructively about the future of Wharfedale. Without even contemplating increased production, the situation at Hutchinson Lane was unsatisfactory. By 1942, as shown in the photographs above, the factory was already cramped and as time went on the problem of the single entrance from Hutchinson Lane became increasingly apparent. A factory in which the inflow of materials/components collided with the outflow of finished products was always going to be inefficient; combining this with shortage of storage space seriously compounded the problem. So, in late 1945, Gilbert once again started searching for a more suitable home for Wharfedale. He soon found what he was looking for in vacant premises next to the Jowett cars and vans factory in Idle on the northern edge of Bradford.

$7\frac{1}{2}" \times 7" \times 3\frac{1}{2}"$

Figure 5.15 Tyny pressed-steel extension speaker.

Part 2. Home Life during the War

Whilst the war years were difficult for Wharfedale, there were no real scares and the firm came through unscathed. Unfortunately, this was not the case for the Briggs family.

Almost at the outset, in the winter of 1939/40, Gilbert went down with pneumonia which, for him, was always likely if he had a bronchial infection or influenza. Fortunately, this time, his GP was able to procure M&B 693 (sulphapyridine), the first antibacterial drug found to be effective for this type of infection. First used in a seemingly hopeless case, on a Norfolk farmer, in 1938 it was a true breakthrough for the manufacturers May and Baker in Dagenham. Gilbert would have been pleased to know that his restoration was shared by Winston Churchill, later in the War, who described the drug's effect on the bacteria thus:[8] 'This admirable M and B, from which I did not suffer any inconvenience, was used at the earliest moment, and, after a week's fever, the intruders were repulsed.'

By this time Peter and Ninetta had progressed to the boys' and girls' grammar schools in Skipton. Peter had developed a great interest in radio and electronics and he gradually extended his 'workshop' from his attic bedroom to most of the available attic space, thus carrying on the experimental tradition there, since this was where Gilbert had first 'played' with his second-hand German speakers. Used to loudspeaker experiments being carried out in the drawing room (the Bradford and Brighouse factories had no space for a 'listening room'), with radios connected to extension speakers and wires all over the place, and magazines such as *Wireless World*—essential reading for his father—always around, Peter could not but be strongly influenced one way or the other. Gilbert also appreciated having 'young ears' available to compensate for his own declining sensitivity to high frequencies, so Peter and later Ninetta became accustomed to oscillator tests and being asked to opine on relative speaker performance.

Arrivals and Departures

Ninetta's early interests turned out to have far-reaching consequences for the whole family, but particularly her mother. Coming home from school on the bus she became fascinated by the animals in the fields and eventually decided one day to get off at a farm close to home and talk to the farmer. She became a regular visitor, arriving home with a dirty uniform from farmyard activities such as milking cows. Far from being angry, Edna was encouraging because it reawakened her own love of animals which had been buried for years. Ninetta sought out strays and tried to introduce these into the household. The first, a cat, was not difficult, but Gilbert would not countenance the second, a red-setter; perhaps he sensed this was the thin end of the wedge! The third attempt was a collie and Edna somehow found a way to persuade him to allow this a home. Edna had responded to the 'dig for victory' campaign and taken on an allotment (followed later by a second) a short bicycle ride from Eaton Road. Being of small stature she rapidly gained the admiration of the mostly male fellow plot-holders for her energy and determination. At various times during the War the produce supplied not only her own family but also evacuees at 20 Eaton Road. During the Blitz Gilbert and Edna voluntarily took in the family of his London traveller, Albert Smith, his wife and three young children, and later on two teenage boys evacuated from Leeds were looked after. Additional beneficiaries were her aging parents (her father was 75 in 1940, her mother nine years younger) who lived very close by in Yew Tree Terrace. Furthermore, Gilbert had been put on a low-fat, vegetable-rich diet as a result of continuing gastric problems following his perforated ulcer and the allotment output ensured this could be maintained. The allotment work kept Edna very fit and it got her out of the house. Altogether it was very satisfying and she developed a taste for working outdoors. Soon she was keeping chickens—a neighbour who owned a small area of woodland just beyond the bottom of their garden allowed them to be kept there—and before long she acquired a pair of goats. These were ostensibly to provide milk, although only Edna actually drank it, and they were kept in a hut at the bottom of the garden. For Edna and Ninetta this was all great fun; Gilbert looked on with bemused tolerance!

The next animal arrived courtesy of an ironic incident. In May 1940 Gilbert was travelling between home and work when a cow shot out of a field in front of the car. In the ensuing collision he sustained the typical head injuries from striking the windscreen as seen in (5.16). This photograph was taken as evidence for a compensation claim which he won, worth £20 (equivalent to about £840 today). By now Ninetta was clamouring for a pony and Gilbert stumped up all but the last £1 of his 'winnings' to buy one for her. No greater love....

Before the War family summer holidays had usually involved a fortnight in a B&B in the Morecambe Bay area, about the easiest drive to the seaside from Ilkley. Gilbert was unable to leave the factory for the whole time, of course, so the rest of the family would go by train and

Figure 5.16 Accident photograph, 1940. (Photo courtesy of the Briggs family.)

he would drive over for a few days at a time. Another destination was Keswick in the Lake District. The War brought such holidays to an end, but Edna came up with a novel alternative. Somehow she discovered an unoccupied cottage on a moorland estate not far from Blubberhouses, on the opposite side of the Wharfe valley from Ilkley. She negotiated to rent this for the school holidays and she and the children, plus animals, moved in for the duration. Conditions were extremely basic: no internal sanitation, no power, cooking over a fire—in short a slight improvement on camping. Ninetta, free to ride her pony for miles in any direction, thought it idyllic. Peter and young Valerie were less enamoured. Gilbert, fastidious from years of being at least reasonably comfortable in hotels when away from home, could not cope with this holiday concept and drove the few miles to see them, just for the day, every so often! This adventure was repeated every summer until the War ended.

In February 1940 Gilbert's mother died aged 77. He was with her at the end in 5 Clayton Lane where she, Claris and Mabel (except for her schooling in Halifax) had lived continuously since the death of Phineas 40 years earlier. Gilbert may have been spending nights there periodically because of the difficult travelling between Ilkley and Brighouse in the blackout when the weather was bad, or because of fire warden shifts. These problems would be the same, or worse, for the foreseeable future and Claris and Mabel needed to 'move on' from the rented house without their mother so Gilbert helped them to buy a modern house, built about 1926: 1 Holly Bank Grove in nearby Great Horton. For significant periods of the War Gilbert stayed there during the week, returning home for the weekend.

Ilkley Distractions: Scenic and Musical

One of the attractions of Ilkley for Gilbert had been the local moors and walking, similar to the moors between Clayton and Halifax which he and his friends had tramped before the 1914–1918 war, but rather more dramatic with their extensive, weathered rock structures. On some Sunday mornings, as an alternative to church, he would take the children for a hike. A favourite was to climb up Ilkley Moor to White Wells, the site of the original spring which launched Ilkley as a spa town around 1840, where the café served hot chocolate (5.17). Sunday lunch was something of a ritual; always a formal occasion with Gilbert carving the roast at the head of the table. Manners had to be impeccable and, as the older children entered their teens, the conversation would sometimes be conducted in French.

The dreariness of the war years was enlivened in Ilkley by the birth of two musical societies whose descendants are still flourishing today. The first was established around 1940/1 by an officer of 125 OCTU (Officer Cadet Training Unit) which was based in the King's Hall theatre. This was a gramophone club, which played mainly classical 78's, but which was open to the public. When the organiser moved on and the club was about to fold a local committee was formed to continue the venture and this became the Ilkley Gramophone Society. It held weekly events, including concerts of classical recordings, musically illustrated talks and visits to and by similar nearby societies, from September to May starting in 1943. This has evolved into the Wharfedale Recorded Music Society, operating to much the same format due, in no small measure, to the fact that the current Chairman, David Pyett, has been involved from the outset and has held this position, on and off, since 1953. He recalls that at a very early stage Gilbert supplied a 10" Golden Wharfedale speaker and advised on how best to baffle and position it in the hall that was being used. The second society was the Ilkley Players' Concert Club, formed to put on recitals with top performers, its first event being a

Figure 5.17 Scenes from the moors around Ilkely. A postcard sent by Gilbert to daughter Valerie not long before he died, recalling the Sunday walks.

piano recital by Cyril Smith on 8 March 1942. The prime mover of this society, as artistic director, was a local GP, Dr Arthur W. Gott. He was himself a fine pianist, with a vast knowledge of music and musicians, and no doubt this was partly why so many famous pianists performed for the Club during the War. Of course, this was a feast for Gilbert, who went to as many of the recitals as he could, usually with Ninetta. An indication of the quality on offer comes from the fact that between September 1943 and December 1944 piano recitals were given by Solomon (twice), Louis Kentner, Dame Myra Hess, Moura Lympany, Moiseivitsch, Clifford Curzon and Irene Scharrer. By September 1945 there had been 31 concerts in all. In 1946 the Club was reconstituted as a subscription society, The Ilkley Concert Club, with patrons whose donations created a reserve fund in order to allow high quality programmes to be maintained. The first recital under this regime, which continues unchanged today, was given by the pianist Denis Matthews on 28 August. The concert was such a success that the artist was immediately rebooked for the following season. He also made a great impression on Gilbert, as described in Chapter 11.

In order to be able to continue his own piano playing whilst away from home and stuck in Brighouse on firewatch duty, Gilbert installed a grand piano in his office for a considerable part of the War. A picture taken in 1942 (5.18) shows the piano and his other sources of music in that part of his office not captured in (5.13).

Figure 5.18 Steinway grand piano installed in Gilbert's Brighouse office during the War. (Photo courtesy of the Escott family.)

Tragedy

In late 1944, when he was just 19, Peter was called up. Somehow, when taking School Certificate examinations aged 16, he had failed to reach the required grade in one key subject and could not matriculate, preventing University application. Instead, he was studying electrical engineering and radio at home through a correspondence course. He became a private with the Royal Army Service Corps and joined a training battalion in Scotland. In March 1945 he came home for some welcome leave and embarked on a project to provide music in Edna's goat shed at the bottom of the garden. This involved the connection of wires from a first floor bedroom window and in the process of doing this, by himself, on the evening of the 22nd, Peter fell out of the window onto stone flags 20 feet below. He was

Figure 5.19 'Woodville', Easby Drive, Ilkley © D. Briggs 2008.

killed instantly. The ensuing trauma was made worse by the fact that Gilbert was not at home at the time—it was a Thursday and he was staying with Claris and Mabel. The loss was devastating for the whole family. Gilbert and Edna kept their grief very private, determined that life must go on as normal as humanly possible for the sake of the two girls. Dorothy Stevens recalls that everyone at Wharfedale knew and felt for the family, but it was never mentioned by Gilbert at work.

When the War ended, two months later, the lady who owned 20 Eaton Road (Gilbert's landlord since moving there in 1931) decided to sell the house and asked Gilbert if he would like to buy it—a timely catalyst to make a fresh start after the tragedy. The family could well afford to buy a larger property and more extensive grounds would be useful. By now Edna was breeding Saanan goats and they were constantly on the move around Ilkley, being distributed around large gardens and orchards for forage and brought back for milking every day in Edna's Jowett Bedford van. Also, Ninetta's pony had to be kept wherever a suitable field could be found. 'Woodville' on Easby Drive was on the market, having been occupied by military personnel during the War. Its decorative state had suffered somewhat, so it was probably a bargain, but it had all the required attributes, including a tack room and loosebox! (5.19)

The move was quickly accomplished, but Ninetta had just that summer to relish having her pony on the premises. That autumn she went off to Liverpool University to study veterinary science. Edna, however, had more room for her goats (5.20).

Figure 5.20 Edna and her Saanan goats at 'Woodville'. (Photo courtesy of the Briggs family.)

Varnadale Wireless Works

CHAPTER SIX: 1946–1952

Part 1: Wharfedale Wireless Works in Idle

The new factory which Gilbert found was the Blake Hill (or Blakehill) Works on Bradford Road, Idle. It had been the manufacturing base of Bristol Tractors (coincidentally also started in 1932) since the company take-over by Jowett Cars Ltd in 1935. Jowett had occupied an adjacent site on Bradford Road, the Springfield Works, since about 1920. After the War Bristol Tractors expanded significantly and completed their relocation to a new factory in Earby, near Colne, Lancashire in January 1946. Gilbert wasted no time, even retaining the existing telephone number—Idle 461—and immediately started to reconfigure the existing buildings to meet his requirements. There was a complex of traditional stone buildings and possibly several sheds. No pictures of the whole site at this time have come to light but (6.1) shows an early picture taken of one shed, probably the cabinet works, which must have faced onto Bradford Road. The available space, according to Gilbert, was about twice that of the Hutchinson Lane factory and his plan was to create working conditions which were lighter and more spacious than in Brighouse; expansion of capacity was not high on his agenda. The move took place in May 1946.

   Gilbert's senior employees Ezra Broadley, Arnold Hatton, Frank Mann and Claris Briggs all moved, as did Dorothy Stevens, who had taken over responsibility for handling orders and shipping (referred to as 'sales') from Claris. So, although some turnover of employees was inevitable, disruption to production was not a major problem. Someone who benefited from the situation was Winifred Hatton, Arnold's wife, who had been working part-time assembling volume controls, at home, since the end of the War. She obtained a permanent job and went on to play an important role in the USA ten years later, as we shall see. A few months after the relocation, however, Dorothy Stevens told Gilbert that she was thinking of having a break from work to concentrate on taking some final examinations which would qualify her to become a teacher of dance (a parallel career which eventually led her to becoming a well-known and respected figure in the world of classical ballet teaching). Gilbert's response was to ask her to stay until a replacement could be trained, with the hope that she would return to Wharfedale when her goal had been accomplished. Exactly how Gilbert framed this job advertisement is unknown, but the successful applicant was one William Swires Escott. Bill had left school when he was about 17 and following 'wireless' training had joined the merchant navy as a radio operator. This pitched him into the last two years of the War, spent mainly on voyages to and from the USA, and he had stayed on until he was 21. At the interview Gilbert seemed happy with his School Certificate qualifications but Escott expressed uncertainty because he knew little about loudspeakers (Wharfedale Wireless Works, he now understood, did not make radios!). Gilbert dismissed this with words to the effect that he would teach

Figure 6.1 Part of the Bradford Road site, Idle, at an early stage of occupation. Probably the cabinet shop. (Photo courtesy of IAG.)

him all he needed to know, and gave him the job. Dorothy showed him the ropes for a few weeks and then took her leave of absence. Escott never looked back. His potential was quickly recognised and after Dorothy Stevens resumed her job in 1947, he was taught office administration; Gilbert was no doubt preparing for the time when Claris would retire and her skills would be lost. As Wharfedale expanded so did Escott's responsibilities, firstly in export sales and then more generally, until he was essentially Gilbert's second-in-command.

Although 'Wharfedale' was the accepted brand name for the company's products, surprisingly it was not until the move to Idle that this was actually registered as the trade mark. The first manifestation of the registration was on a flyer for factory speakers issued in June 1946 (6.2).

Once production was fully re-established at Idle, Gilbert discovered that the improved factory layout had delivered a 50% increase in output, compared with Brighouse, without any increase in his workforce.

At the end of the previous chapter the improvements in speaker performance in 1946, relative to the outbreak of the War, were briefly noted. It is worth examining these in a more detail in order to take stock of the situation, as Gilbert might have done, as production settled down in the new factory. In the 1946 brochure issued before the move Gilbert included 'Some Loudspeaker Axioms and Hints' of which the first three were:

- Flux density is the most valuable quality in a permanent magnet speaker
- High flux and low resonance are worthy partners
- A die-cast chassis eliminates resonance usually audible with a steel pressing

Chapter Six: 1946–1952

Wharfedale
(REGD. TRADE MARK)

FACTORY SPEAKERS
available

with or without Volume Control with or without Carrying Handle

FACTORY / BRONZE
10″ Unit 10,000 lines Flux Density
5 watts maximum input

FACTORY / GOLDEN
10″ Unit 12,500 lines Flux Density
8 watts maximum input

Available in 15 ohms without Transformer or with 3 ohm Speech Coil and P-Type transformer (30, 45, 60 and 90 to 1)
Cabinet in Wood and Metal sprayed Grey, fitted with metal Grille back and front

15½″ x 12¾″ x 6½″
FACTORY / BRONZE & FACTORY / GOLDEN

FACTORY / W 12
12″ Unit 13,000 lines Flux Density
15 Watts maximum input

Available in 15 ohms without Transformer or with W.12 Transformer (45, 22 and 15 to 1— all centre tapped)
Plywood Cabinet, sprayed Grey with Fret back and front

21″ x 20″ x 8″
FACTORY / W 12

WHARFEDALE WIRELESS WORKS
BRADFORD RD., IDLE, BRADFORD
Telephone: IDLE 461
Telegrams: Wharfdel, Idle, Bradford

JUNE 1946   **TRADE PRICES**

	15 ohms	Transformer	Truqual V.C.	Handle
Factory/Bronze	55/-	5/- extra	6/- extra	2/6 extra
Factory/Golden	75/-	5/- ,,	6/- ,,	2/6 ,,
Factory/W12	135/-	12/- ,,	6/- ,,	2/6 ,,

Figure 6.2 First example of registered trade mark, 1946.

Figure 6.3 George Horsburgh (left) and Frederick Tetly. Reproduced from *Audio Biographies*.

Magnet Developments

Magnet performance was therefore one of Gilbert's obsessions. When he first started to build his own speakers in 1932 there were two Sheffield firms that were prominent suppliers of relevant permanent magnets: Darwins Ltd (Fitzwilliam Works, Tinsley) and Swift–Levick & Sons Ltd (Clarence Steel Works). Gilbert chose the latter for the massive chromium-steel magnet used on the first 'Bronze' model. Whatever the reasons behind this choice, it certainly turned out to be a good one in the long term. When Wharfedale Wireless Works suddenly became a serious business in 1933 and Gilbert had greater need of magnet expertise, he was fortunate to find that there were two new people in key positions at Swift–Levick. George D.L. Horsburgh (6.3), who was five years older than Gilbert and a director, had recently moved from London to direct production, particularly of the Magnet Department. He had already made great contributions to progress in permanent loudspeaker magnet design, patenting the cross magnet with mild steel pole pieces (in 1928) which spelt the demise of electromagnets in loudspeaker construction. Although an engineer by training, his route to this current position was almost as unlikely as Gilbert's and they struck up an immediate friendship. One of Horsburgh's first acts in Sheffield was to appoint a metallurgist as an assistant for research into new magnet materials—Frederick W. Tetley (6.3). Their collaboration continued for most of Gilbert's association with the firm and resulted in many industry 'firsts'. Tetley also became a director of Swift–Levick and an equally long-term friend.

Through this association, and more generally the Permanent Magnet Association, based in Sheffield, Gilbert was able to stay abreast of developments and introduce new magnet materials rather quickly. The speakers produced immediately after the 'Bronze' employed somewhat smaller, higher flux density, cobalt-steel magnets (Fig 4.6), whilst much smaller nickel–aluminium steel (AlNi) ring magnets were being used by 1935 (as in Fig 5.4b). The next advance was to

Figure 6.4 Speakers with die-cast chassis, 1947 catalogue.

nickel–cobalt–aluminium–iron alloys (AlNiCo), first used on the 'Twin-cone Auditorium' in 1936, then in 1941 the variant Alcomax was introduced with the 'Midget' unit. With this advance the option became available, for smaller units, to employ the magnetic material as the centre pole in a mild steel pot as opposed to the traditional ring magnet with a steel pole, resulting in the elimination of stray magnetic fields from around the assembly. To produce the same performance as the original (1932) chromium-steel magnet now required an Alcomax magnet with only about 6% of its weight.[1] During the War further, though less dramatic, improvements in performance became available through Alcomax II and TiCoNAl magnets which Gilbert first used on the 1946 'Golden' and W12 units respectively. For any given loudspeaker design, particularly in respect of the coil diameter and the depth of the gap, increasing the total magnetic flux within the gap resulted in greater sensitivity (increased audio output per watt of input) and better damping (increased attack and improved transient response)—hence Gilbert's first axiom and his constant drive to use new magnets offering higher flux densities.

Chassis and Cones

Achieving low resonance involved aspects of both chassis and cone design. Early Wharfedale loudspeakers used mass-produced pressed-steel chassis from Goodmans. An interesting footnote and another illustration of Gilbert's charming interaction with people is given by this recollection:[2]

> *'It was fortunate that I had gone to Goodmans for my chassis, because Mr Newland* [the MD] *warned me that we must use heavy gauge steel which would not bend with the weight of our huge magnet, otherwise the speakers would go out of centre and we should have them coming back faulty. As I was—in my ignorance—thinking of changing to a cheaper and lighter chassis, which would have ruined us completely, I am in the unique position of having to thank a competitor for my economic survival. The boot is usually on the other foot.'*

As illustrated in Fig 5.4b similar chassis were still being used in 1939 except for the 'Golden' and 'Portland' units which had die-cast chassis. These had first been introduced for the 'Golden' and 'Twin-Cone Auditorium' in late 1936. Die-cast chassis had several advantages: greater acoustic transparency and absence of audible resonance; more rigidity and improved dimensional stability; no interaction with stray magnetic fields from ring magnets (the alloys being non-ferrous). However, they were two to three times more expensive. Nevertheless, after the War, pressed-steel chassis were phased out and by 1947 only the W6 chassis was still of this type. Examples of the new chassis are illustrated in (6.4)

As for the cones, Wharfedale was reliant on the specialised manufacturers whose output was dominated by the demands of the mass market but whose designers had to be able to respond to a very wide variety of requirements. At this point, however, all cones were made of paper and produced by pressing pulp into a mould. The variable design parameters for any given size were shape, weight and texture, which led to a wide choice of products and performance characteristics. In any particular situation the choice would be a compromise between a low fundamental resonance frequency, smooth response over the widest frequency range, overall efficiency etc. and in general the incorporation of corrugations or flexible couplings into the body of the cone was central to this optimisation, but the interface between the cone and the chassis, the surround (or suspension), was

6A

Wharfedale

REGD.

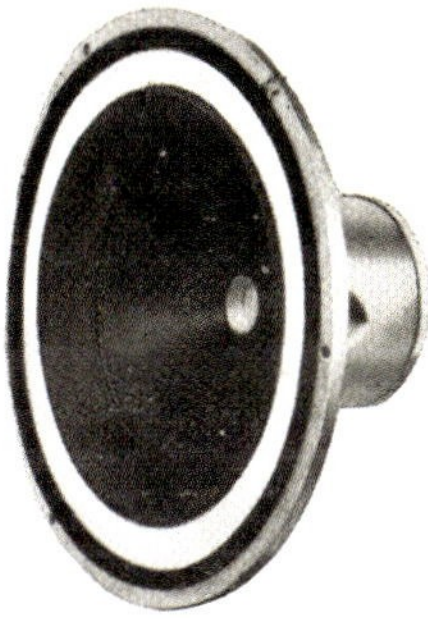

W.10/C.S.

Flux Density 14,000 lines
1″ centre pole
Total Flux Density 74,000 lines

SpeechCoil 2/3 ohms or 12/15 ohms
Peak input ~~7/8~~ 5 watts

The W10/C.S. is a new model fitted with the same magnet and chassis as the Portland, but with cloth suspension around the cone in place of the usual corrugations, at an extra price of 10/-.

The bass resonance is below 60 cycles and the top note response is effective up to 18,000 C.P.S. This unit will certainly appeal to those interested in the finest quality of reproduction, and the response curve gives a pretty good picture of the benefits of soft suspension.

This model is only produced in very small quantities and is not available for Export.

RESPONSE CURVE

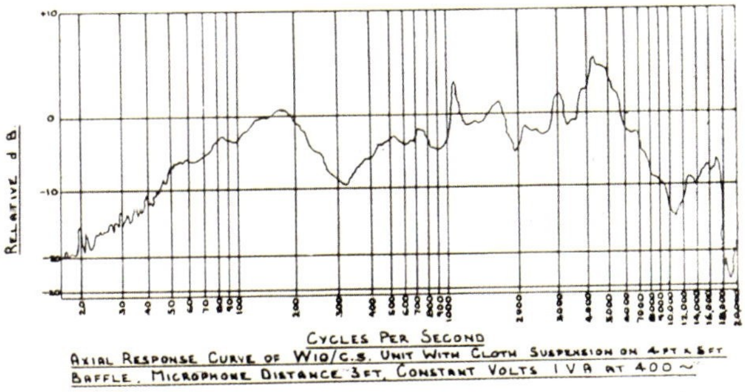

WHARFEDALE WIRELESS WORKS, BRADFORD RD., IDLE, BRADFORD

Telephone No.: Idle 461 | Telegrams: Wharfdel, Idle, Bradford

Figure 6.5 W10/CS 1947 catalogue.

of key importance. As Gilbert observed:[3]

'The corrugated, one-piece cone is obviously the cheapest proposition and is essential for mass production; but it can be criticised on three counts:
It is difficult to obtain sufficient compliance to give low resonance without making the edge too thin.
The corrugations usually have resonances which produce irregularities in response with audible colouration.
Untreated corrugations do not possess the correct mechanical impedance to absorb the high frequency energy which reaches the edge of the cone, and therefore it is reflected back to produce standing wave effects.'

But he went on:

'The paper corrugation trouble can be overcome by fitting a soft surround to the cone.'

Cloth Surrounds

The benefits of soft suspension, in general, had long been recognised and the better, early moving coil loudspeakers had used rubber, leather or cloth for the surround. Indeed, soon after its launch the 'Golden' speaker was improved by the fitting of leather suspension (see Fig 4.6), although for how long is unclear. The ill-fated 'Langham' extension speaker cabinet incorporated a 'Portland' unit with cloth segment suspension. Both these speaker units were built by Ezra Broadley. It required skilled workmanship to ensure that the surround was correctly tensioned both to avoid pulling the coil and cone out of alignment and to achieve the desired low resonance. This 'hand-built' element necessarily lengthened the assembly time and increased the cost.

With the improvements resulting from incorporation of the latest magnets and die-cast chassis now established across the range, revisiting cone surrounds was an obvious development option in 1946. Ezra Broadley, now the works foreman, had the track record and the idea required serious discussion. Eventually, in September, it was decided to test the water with a modified 'Portland' 10" unit having a plain cone with a cloth surround and to sell this at a 10 s premium (140/- compared to 130/-) to cover the extra labour costs. However, there was one specific aspect of the general problem with supplies which threatened the project, as Gilbert later recalled:[4]

'Clothes rationing was still in force, and I was humiliated to find that my 28 years' experience in the cloth trade did not enable me to buy suitable material for cone surrounds without coupons. We therefore used yellow dusters, and I bought up all the stocks to be found in the shops in Ilkley. We then bought regularly in gross lots from a wholesaler, who must have thought that we ran the best dusted factory in the country.'

The new unit was dubbed the W.10/CS and the catalogue issued in April 1947 had the rather striking illustration shown in (6.5) in which the yellow-duster surround is obviously present. The response curve shows that the desired characteristics of low bass resonance, below 60 Hz, and extended frequency range, up from about 8 kHz for the Portland to 18 kHz, had been achieved. This was the first use of appended abbreviations to designate specific variations in a particular speaker model—here CS for cloth suspension.

Since the initial aim was only to test the market, output in the first year averaged about 10 units a week and Ezra Broadley built and tested them all. Gilbert still kept in touch with old colleagues in the Bradford Dyers Association and he discovered that khaki cloth which had been damaged during dyeing, being essentially worthless, could be obtained from them without coupons. For use as a surround material, however, it was ideal because the damaged areas could be avoided when the segments were stamped out; so this cloth replaced the rather garish yellow dusters at some point. It was not until textile rationing ended in March 1949 that the khaki was itself replaced by a specially commissioned grey flannel.

New Transformers

By April 1947 the transformer range had been revamped and those on offer are illustrated in (6.6). The GP8 had been developed early in the War and the W12 towards the end. A completely new speaker was also developed about this time, the largest to date. The W15, which had a chassis diameter of 14.5" and an Alcomax II magnet with a 2" centre pole, was designed

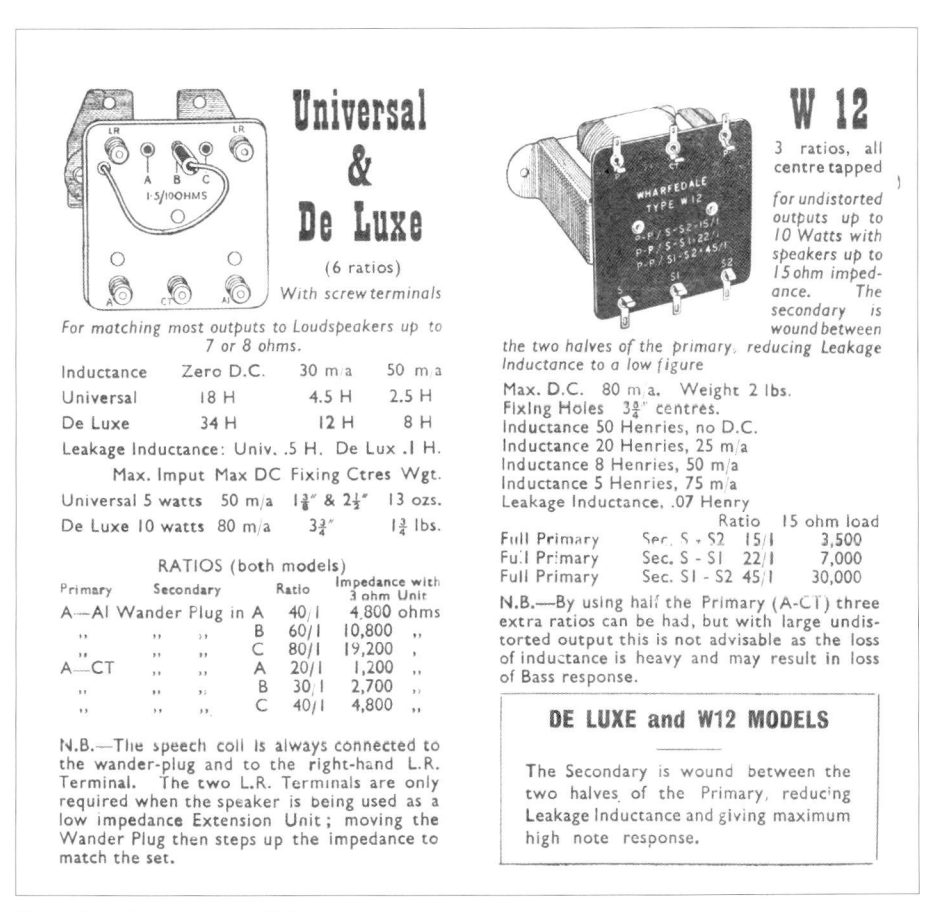

Figure 6.6a Transformers in 1947.

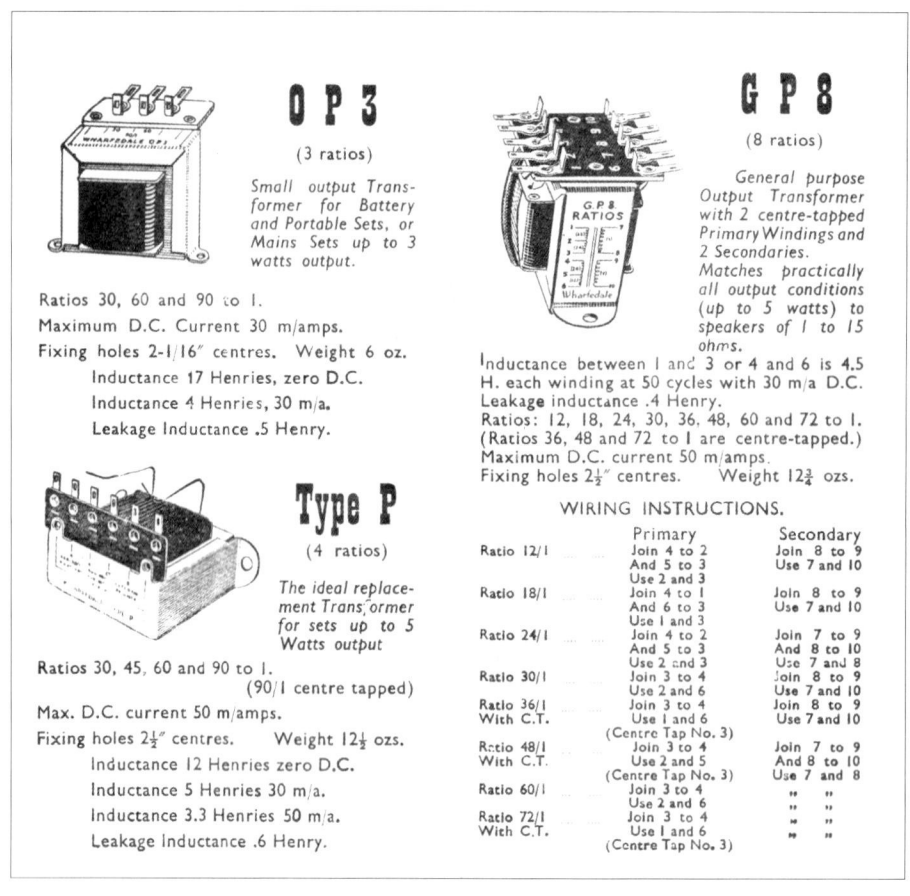

Figure 6.6b Transformers in 1947.

to handle large inputs (up to 20 watts) with a smooth response from the cone resonance of 60 Hz up to 5 kHz.

The First Two-way System

The next development came completely out of the blue, courtesy of Frank Thistlethwaite (6.7) of Excel Services (Radio Engineers) in nearby Shipley. Frank's operation was rather more than this title suggests; he was another local audio pioneer who can be credited with the invention of the first mobile radio communication system in 1932 and its subsequent development for the Police, the setting up in 1938 of one of the earliest private recording studios in the UK, using lacquer-coated glass discs and the production of an early high quality 15 in/sec tape recorder just after the War. Excel Services designed, built and repaired audio and radio equipment, in addition to the recording activity, and was a Wharfedale customer. Gilbert's first high quality amplifier had been designed and built by Thistlethwaite in about 1946.

One day early in 1947 Thistlethwaite turned up at the Idle works and asked Gilbert if he would like to hear two Wharfedale speakers performing really well. Gilbert agreed, of course, and recalls:[5]

Figure 6.7 Frank Thistlethwaite. Reproduced from *Audio Biographies*.

'... so Frank T went out and returned with two 12" speakers in cabinets, and a third object about the size of a small coal scuttle which proved to be a crossover network.'

At this time crossover networks, for dividing the frequency spectrum between two or more loudspeakers, were extensively used in cinema sound systems but were unknown in domestic sound reproduction. After the War, Thistlethwaite, who had been doing research work for the Ministry of Aircraft Production, returned to his audio business and had a project to develop a tape recorder for coupling up to cine equipment to provide the sound track. During this project, which was completed by 1947, Frank was exposed to the cinema sound world and this was surely when the seed for a 'domestic' crossover unit was sown. It was Gilbert's good fortune to be in pole position for the first demonstration and he was certainly impressed.

The Thistlethwaite demonstration unit is shown in (6.8). The central switch is on/off for the unit as a whole, whilst the other two switches are on/off for the bass and treble speakers. The unit's chassis is presumably one that Frank had to hand, probably for an amplifier. Gilbert reckoned a commercial version would carry a price tag of about £15 and such a unit would be too large and expensive to attract customers. What kind of crossover circuit was used is unknown, but the size of the unit and its likely cost suggest it was at least second order. However, Gilbert recognised an opportunity and struck a deal: Wharfedale would take up the idea and pay Thistlethwaite a royalty on sales, but would redesign the unit to make it commercially viable. Gilbert thought the selling

A Pair of Wharfedales: The Story of Gilbert Briggs and his Loudspeakers

Figure 6.8 Thistlethwaite demonstration crossover and commercial model (1947). Reproduced from *Loudspeakers*, 5th Edition.

price needed to be no more than £5. Ernest Price was probably given this challenge and the resulting unit is also shown in (6.8). It was essentially a first order network (referred to as a 'quarter section' type) with a crossover at 1 kHz which actually sold, as a 'loudspeaker separator unit', for under £4 (75 s).

The bigger task was to design a two-speaker unit with the desired superior reproduction properties. Fortunately, the recently developed W10/CS with its sensitivity up to 18 kHz provided the ideal treble speaker to couple with the W12 for the bass. Gilbert drew on all his experience and came up with the corner unit shown in (6.9); his reflections on the design philosophy accompany the pictures in the brochure. During the evolution of the bass enclosure, listening tests suggested that soft suspension for the W12 speaker would be advantageous and so a cloth surround was used on this too—leading to a W12/CS variant being offered for use in acoustic chambers, again at a 10 s premium. The corner cabinet was fitted with a new type of volume control, based on a tapped choke, designed to be compatible with the crossover and avoid frequency distortion. This was state-of-the-art sound reproduction 'regardless of cost' and the price was a hefty £48 10 s (the equivalent of about £1400 today). The choke-type volume control (6.10) was sold as a separate product from the following year, along with a switch box which had been developed to facilitate the comparison of speaker systems, particularly in demonstrations of this model.

Figure 6.9a Twin-speaker corner cabinet, 1947 brochure.

12D

Wharfedale
REGD.

Corner Cabinet
(continuation)

The best position for a loudspeaker is in the corner of the room, for two main reasons : (a) the walls act as reflecting planes and improve low note radiation by 100 per cent., and (b) everybody in the room is in front of the speaker and is therefore well placed for the top note response, and for clear reception of speech.

It is also generally admitted that the most realistic effect is attained, especially with vocal and instrumental solos, when the upper register comes from a source about 3 or 4 feet from the floor. It was therefore decided to adopt a corner cabinet to embody these essential features.

A good deal of attention has been paid to the design of the bass chamber, which is fitted with fibre-board sound reflectors, and a glance at the two Impedance curves on Page 12B will show that all bass resonance above 50 cycles has been eliminated. The main cone resonance is below 50 c.p.s.

Exploded view of Units in Corner Cabinet.

W10/CS.

Choke v.c.

Separator.

W12/CS.

The treble chamber is also fitted with a reflector, and a diffusing cone is placed in front of the top speaker to spread the usual high note beam.

Both Magnets have very high flux density, and even with inputs of the order of half a watt the Corner Cabinet gives the impression of stepping into the Concert Hall when compared with the usual type of loudspeaker, and quick-fire programmes such as Itma are much easier to follow.

Demonstration models are being arranged for in London, Leeds, Bradford and other towns, as it is appreciated that a thorough listening test will be desired before purchase.

EXTENSION SPEAKER

N.B.—It is quite satisfactory to use the 6 ohm model as an Extension Speaker on a set fitted with the usual 3 or 4 ohm extension points.

MATCHING

For 2 to 8 ohms use 6 ohm model.
For 9 to 20 ohms use 15 ohm model.

See page 16 for prices.

WHARFEDALE WIRELESS WORKS, BRADFORD RD., IDLE, BRADFORD

Telephone No.: Idle 461   Telegrams: Wharfdel, Idle, Bradfor

Figure 6.9b Twin-speaker corner cabinet (continuation), 1947 brochure.

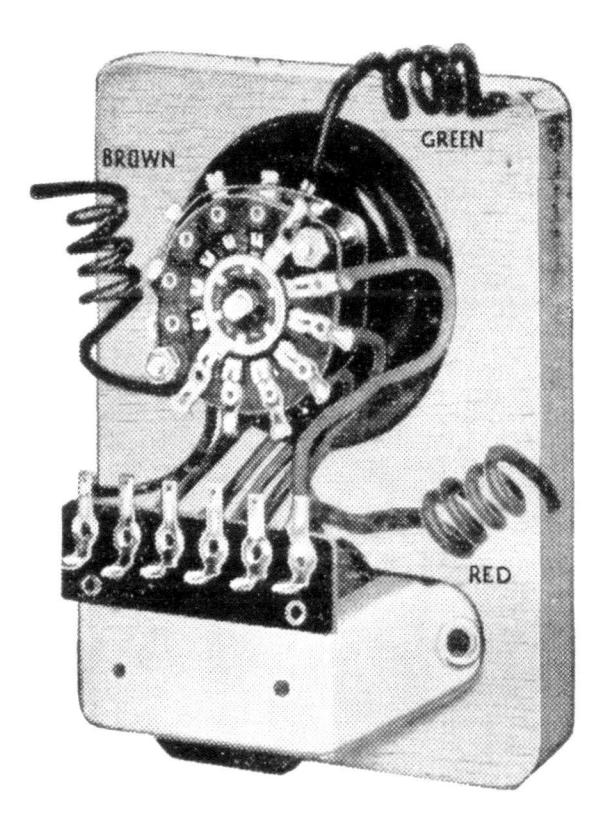

Figure 6.10 Tapped choke volume control, 1947.

The Winter of 1946/7

The fact that these developments and the comprehensive, 22 page catalogue which detailed them in April 1947 happened at all is something of a surprise. The winter of 1946/7 was one of the worst on record. Spells of cold weather in December and January were just a taster before the seven week freeze which lasted from 21 January until 16 March. February was the coldest on record: temperatures dropped as low as −21°C and the persistent snow drifted to bring transport to a standstill. Soon there were fuel shortages, power cuts of five hours each day (coal could not reach the power stations) and gas pressures were reduced to dangerously low levels. Production in many industries was severely disrupted or halted completely. The television service was stopped and radio transmissions reduced. Traffic lights were dimmed or turned off. Ports were frozen up and food rationing became more severe than during the War. Damaging gales and heavy snow in early March exacerbated the problems; in the Pennines the drifts reached five metres. Then the weather changed. Heavy rain and the rapid thaw in the second half of March led to widespread flooding—the Wharfe was one of the Yorkshire rivers to burst its banks.

The repercussions were very serious. The loss of output was a huge blow to the UK economy as it struggled to recover from the War, leading to savage cuts in public spending and prolonging the deprivations of that period. The materials shortages to which Gilbert had grown accustomed continued and output was limited as a result. In a way, though, this may have been beneficial, long term, because it reinforced his philosophy of 'quality not quantity' and the scarcity value of Wharfedale products increased.

G.A. Briggs—Author

In the autumn, Gilbert found himself on a routine trip to London to visit some of Wharfedale's important customers in tandem with Albert Smith. A chance remark during one such visit had life-changing consequences. Whilst they were in Webb's Radio in Soho Street:[6]

'A man came in and asked for a book on Loudspeakers, but none was available. The assistant came over to us and said: "Every day somebody asks for a book about Loudspeakers. Why don't you write one?" We went upstairs to see the manager, Mr Pickard, and asked him how many copies he would buy if we brought out such a book. To my amazement he said he would place an order for 72 copies there and then, provided the price did not exceed 3s 6d. Mr Smith promptly got out his notebook and entered the order.'

To a certain extent this was a case of 'chance favouring the prepared mind'. Gilbert had been hankering after publishing a book on pianos since the late 1930's and had actually investigated the practicalities not so very long before this occurred. What he had discovered was not favourable; even if he took the risk and bore the expense, a professional publisher would only quote a two to three year timescale for printing. The then severe paper rationing was a major factor, and this was going to be the case whatever the subject matter. Gilbert's typically telescoped account implies that there was the inherent assumption in the discussion with Mr Pickard that Wharfedale Wireless Works would be the publisher, but the problem of printing still had to be solved.

At this point Fred Kier Dawson, Gilbert's pre-war advertising agent, re-enters the story. Around 1937 he had abandoned advertising for printing, but continued his informal design work for Wharfedale. During the War he joined the RAF (Fighter Command), but his artistic talents eventually led to his move to a unit which prepared, from aerial photographs and maps, precise scale models of targets. Gilbert must have stayed in touch because he knew that, after the War, Dawson had returned to Clifford Briggs Ltd. (printers—no relation) in Keighley. He wrote to him to ask whether his firm would have the paper to print a book, if he wrote it. By return came the reply "Yes, provided you take fine art paper. We have a store room full of it." Since this high quality, glossy paper was best for high definition reproduction of technical figures anyway, this was not a problem. All that remained, therefore, was the actual writing:[6]

'I soon discovered that I had no talent for writing, and I would struggle for half-an-hour in the formation of a single sentence, ending up with little more than a headache. Mr Smith continued selling the book, and orders for dozens came pouring in almost every day. I began to wish that the book and Mr Smith were in Jericho, when I hit on the bright idea of writing page after page of padding, leaving it for a day or two to simmer, then cutting it down about 50%. A further pruning the following day left a fair extract of anything worth printing.'

Once this regime was established Gilbert made good progress. He had some excellent sources to draw on, notably Harry F. Olson's *Elements of Acoustical Engineering* (Van Nostrand, 1940), *Applied Acoustics* by Harry F. Olson and Frank Massa (The Blakiston Co., 1939), Alexander Wood's *The Physics of Music* (Methuen, 1944), *The Psychology of Music* by Carl E. Seashore (McGraw-Hill, 1938) and *The Radio Engineering Handbook* by Keith Henney (McGraw-Hill, 1941). He asked his friend Frank Beaumont of Ambassador Radio to go over his manuscript and check/correct technical details—which inevitably led him into the less passive 'technical sub-editor' role—and Edna did some general editing. Keir Dawson combined his artistic and printing skills by taking on the task of making the drawings and the printing blocks to reproduce them. In order to illustrate certain points about speaker design and performance Gilbert generated his own data, which required his cabinet makers to produce 'weird and wonderful models for purposes of test'. So, the project took on many aspects of the wider Wharfedale *modus operandi*.

The title of the book evolved into *Loudspeakers. The Why & How of Good Reproduction*. Gilbert set out his approach in the Introduction as follows:[7]

'My qualifications for writing this booklet are not very extensive, as I am neither a scientist nor a mathematician, but I have been making loudspeakers for fifteen years, and all my life I have been fascinated by sound. In my search for the perfect piano, more than three dozen uprights and grands have darkened my doors during the last 30 years. My hobby is music and I make a practise of attending concerts and playing the piano regularly in order to keep my hearing fresh, as I think that the tonal quality of music is quite as important as its melodic and harmonic structure. My approach to the subject is, therefore, as much from the musical angle as the technical.

The book is written in non-technical terms throughout, and many readers may find some chapters extremely elementary. Nevertheless I hope they will be patient for the sake of those who have no knowledge or experience of the subjects concerned, for whom the book is primarily intended.'

The book itself is discussed in more detail in Chapter 12.

Destination USA

In early 1948 another chance remark led to the story taking a quite unexpected turn. Gilbert was visiting Swift–Levick's in Sheffield when the conversation with his old friends Horsburgh and Tetley turned to developments of magnets in the USA. Gilbert commented that he would like to visit that country one day, but was none-the-less very surprised to receive a phone call from Tetley in March offering just such an opportunity. He and Horsburgh had been invited to visit the Indiana Steel Corporation in May and had procured a three-berth state room on the *Queen Elizabeth* leaving Southampton for New York on 7 May—would Gilbert like to join them? The book was by now well advanced and Gilbert thought he could have it proofed in time to try and interest a New York publisher in a US edition, so he did not hesitate to gratefully accept.

Financing the trip was not trivial. Currency restrictions were in force and the move to Bradford Road meant that Wharfedale was once again reliant on a bank overdraft. Given that the transatlantic crossing took five days, Gilbert wanted to spend three to four weeks in the USA to justify the time and cost of getting there. Besides trying to sell his book, he was keen to find out more about the audio business, and particularly the loudspeaker situation there, in the hope that an export market might materialise. He hoped to visit Dr Olson, from whose books he had derived so much benefit, at RCA's research facility in Philadelphia and he intended visiting a cone manufacturer near Chicago. He had a cousin, Lloyd Emsley (the son of one of his mother's five brothers), who had left Clayton in 1914 to emigrate to the USA and was now working for Ford in Detroit, and this was an opportunity to see him again. In his spare time Gilbert expected to sample the New York music scene and he had obtained a letter of introduction to the Vice-President of Steinways, with a view to trying their latest piano models. He needed about £500 and had to persuade his bank manager to extend the overdraft accordingly; he succeeded, but not without difficulty. Dollars were the next problem and he was only able to secure a minimum daily allowance of $30, which worked out at £7 10 s. He booked flights with American Airlines from New York to Chicago then on to Detroit and back to New York. With the essentials in place he wrote to the publishers McGraw-Hill in New York and received an invitation to discuss his book on arrival there.

The final galley proofs were ready just in time, but the page layout remained to be done, so Gilbert took two dummy books with him and 'pasted up' the pages from the proofs during the first two days afloat (6.11). He had not been on a serious voyage since his final trip to India in 1924/5 and compared with the 'Indian steamers' the size and opulence of the *Queen Elizabeth* took his breath away. Before leaving he had promised to keep a diary and send reports back to Valerie during the trip. This survives, in 34 handwritten foolscap pages, and offers a fascinating blow-by-blow account of an innocent in the USA. It turned out that Frank Tetley had a relative high up in the Cunard company and he had wangled their first class state room. His companions were amused to find that Gilbert had brought along a dinner suit and shirt which dated back to his Indian trips. They still (just) fitted him and the soft pleated shirt style had recently come back into fashion! His companions were made to read the book, however, and both genuinely thought it to be excellent. Horsburgh said he had no idea Gilbert knew so much about loudspeakers, to which he responded "I did not until three months ago when I started reading other books!" In fact, as the book had taken final shape in his hands, he had become plagued by self-doubt and their reaction gave him a boost.

Figure 6.11 On board *Queen Elizabeth* en route to New York, May 1948. G.A.B. centre with George Horsburgh (holding papers) on his left. (Photo courtesy of the Briggs family.)

Chapter Six: 1946–1952

The Long Road to British Industries Corporation

The setbacks started as soon as Gilbert disembarked. Once in his hotel he realised that his dollar reserves were going to be seriously tested. Everything was much more expensive than expected and he had probably made a mistake in booking into the Plaza. The first morning he contacted McGraw-Hill and was told that the previous week a policy decision had been made that they already had enough books on loudspeakers, so a meeting was pointless. Suddenly 'the outlook was pretty grim':[8]

'The only thing to do was to go out and try and sell the book. I visited bookshops, stores, publishers and libraries. After five days of total failure I decided that anybody who wrote a semi-technical book in his spare time, published it himself and tried to sell it as well, must be a jackass. Then quite suddenly, I recalled having had, about a year previously, some correspondence with one of the technical editors of the journal 'Tele-Tech' so I decided to pay him a visit and ask his advice.'

The editor's business manager provided the breakthrough by suggesting that what he should have done from the start was to 'tackle the jobbers downtown'. These were the wholesalers dealing in radio and audio components and Gilbert obtained a list of the best firms. He got nowhere with the first (chosen because the manager's name was Briggs!) but the second was interested and suggested he really should be talking to Mr Carduner of British Industries Corporation (BIC) who, as a large importer, would be able to cover the whole country. Gilbert was able to meet Leonard Carduner just before he was due to leave for Chicago. He was well received and Carduner was interested in principle, but wanted to have the book checked out by an expert. It was agreed that this could be done by the time Gilbert got back to New York a week later and a dinner date was booked. However, it was understood that, if the 'expert' review was satisfactory, then an initial order might be for 250 copies if the selling price in the USA was $1.50 or 1000 copies if it could be $1.00. So, this gave Gilbert hope but also something to chew on whilst on his travels.

On his last day in Chicago a copy of the published book, specially airmailed hot off the press, caught up with him. When he met Carduner again in New York, Gilbert could show him the 'real thing', which he had decided to sell to BIC at the cost price of 2 s 2 d for the first printing of 5000, of which Albert Smith had already taken orders for about 2000. This allowed BIC to price it at $1 and still make a decent return—the official exchange rate was $4 to £1 (20 s). The review Carduner had commissioned was favourable, so the deal could be done. BIC was appointed sole agent for the book in the USA; 1000 copies would be shipped immediately with the option on a further 5000 from a reprint if sales went as anticipated. Since the reprint costs would be much lower than for the initial printing, Wharfedale would then also be making money. Carduner also promised to send a sample order for various speakers and a corner unit. On 27 May 1948 Gilbert recorded in his 'diary':

'I can assure you that anybody who succeeds in getting dollars out of these hard-boiled business men is some guy. So far they have soaked me good and hard, and I have been a few hours behind them most of the time, but I feel that I have now caught up after a slow start. Even so I shall be satisfied if I eventually collect as many dollars as I have spent.'

In the years to come he had every reason to be satisfied. Falling in with BIC, through the book, gave Wharfedale an early entry into the USA market for loudspeakers. Carduner was a shrewd operator and one wonders whether he quickly saw the book deal as the means to capturing the agency for Wharfedale speakers, which is what happened once he had heard them. BIC had been started by Bill Carduner, coming from a background in the radio business, in 1935 and in 1937 he was appointed agent for the British company Garrard, which was already famous for its gramophones, turntables and record changers. In 1938 their RC100 record changer, which could play both sides of each record, and also mixtures of 10" and 12" records, was produced for sale in the USA and this led to his brother Leonard joining the expanding operation. Leonard was an important force in the drive to establish a distribution network for high fidelity equipment—a long struggle to persuade radio parts jobbers (the wholesalers Gilbert had to deal with on his trip) to become high fidelity equipment retailers. BIC were also assiduous attendees at the audio fairs across the USA which were becoming familiar by 1950. Gilbert had certainly ended up with the right people although, as Leonard Carduner told him, he would have been directed to them straight away had he gone to the British Consul for advice. That information alone gave Gilbert a head start in tackling other markets in the old Empire later on.

Radiolympia

Wharfedale had a stand at the 1948 National Radio Exhibition at Olympia (Radiolympia) in September and this resulted in the interesting photograph reproduced in (6.12). Queen Mary, widow of King George V, was visiting and her apparent interest in the stand was recorded by an opportunist News Agency photographer. The Queen's daughter Mary (then Princess Royal) was the Countess of Harewood and lived at Harewood House, which is about 10 miles east of Ilkley, in Wharfedale. This, rather than the loudspeakers, was probably what caught her attention! Nevertheless, Gilbert treasured the picture and was fond of using it as the basis for a humorous take on the incident.

The following year a bigger stand was taken and a picture of this also survives (6.13). All the products are on display and the range demonstrates that, despite the continuing difficulties with materials procurement, the pace of development had not slackened. A comprehensive catalogue issued at the same time detailed the new introductions since Gilbert's eventful trip to the USA, allowing all the products to be identified in the picture.

On the left are all the chassis, details of which are given in a table taken from the September 1949 catalogue (6.14). A new introduction was the Super 8, built around the latest Alcomax III magnet and setting new standards for an 8" unit ('Super' was then used to indicate models which used an Alcomax III magnet). A feature of its corrugated cone was the bakelised apex. The cloth surround experiment of late 1946 had proved to be a success; as word spread demand increased and other chassis were offered with this more expensive option, including the new Super 8/CS. Behind the chassis collection are three cabinets designed for factory use.

In the centre, on the front display, are the extension speakers. The smallest, also new, is the 'Bantam' which used wood-filled plastic for the cabinet and contained a special 4.5" chassis (not sold separately). The next largest is the 'Tyny'. Further examples of this and the 'Bantam' are also hung on the back wall of the stand. The triplet of cabinets behind the support pole are, increasing in size, the 'Bijou', 'Bronzian' and 'De-Luxe'. The other two cabinets on this display section are also

Chapter Six: 1946–1952

Figure 6.12 Radiolympia 1948. (Photo courtesy of the Briggs family.)

new and described as 'streamlined baffles'. These had curved backs made of expanded aluminium to reduce resonance at high volume. The smaller is the 'Gem Baffle' (W5 chassis) and the larger, on the extreme left, is the 'Sylvan Baffle' (8" Standard chassis). On the right-hand display are examples of transformers, volume controls and crossover units and a baffle designed specially for school use (with a 10" Golden chassis). Two wall-mounting cabinets for school use are hung on the back wall: the WM8 reflex cabinet (8" Bronze chassis) on the left and the WM10 (10" Bronze chassis) on the right. Five floor-standing cabinets are arranged along the back wall. From right to left these are: the W12/Reflex (a rugged mobile unit designed for school use, particularly for reproduction of gramophone records); the Varitone [this is obscured and it is described in (6.15)]; the Reflex Cabinet (designed to take advantage of the Super 8 chassis); the Corner unit (now fitted with two volume controls—one for each speaker—effectively providing 'tone control', see (6.16)), and a new corner unit design.

The new corner unit was not being sold as such at this point, but the design details were offered:[9]

117

Figure 6.13 Radiolympia stand 1949. (Photo courtesy of the Escott family.)

'... *as a guide to those who may contemplate the construction of a non-resonant corner reflex cavity. Magnificent results are possible with concrete, marble or bricks, using a W15/CS unit for bass, and a W10/CS for treble, with Separator.*'

The cavity volume was nine cubic feet and this housed the 15" bass unit. As seen in (6.13) the treble unit was housed in a separate baffle sitting on top of the corner cavity and this was being sold as the W10/CS Baffle. Gilbert had been extolling the virtues of brick construction, in print, for some time and had one built into a corner of his office (6.17). This gave rise to one of his best-loved jokes:[10]

Data Sheet

Model	Chassis Dia. ins.	Bass Resonance c.p.s.	Flux Density	Total Flux	Peak Input Watts	Weight less Trans. lbs.	Baffle Opening ins.	Depth ins.	Fixing Holes P.C. Dia. ins.
Midget	$3\frac{5}{8}$	220/230	9,000	24,000	$\frac{1}{2}$	$1\frac{1}{2}$	$3\frac{1}{8}$	2	$3\frac{15}{16}$
W5	$5\frac{1}{4}$	160/170	9,000	24,000	$1\frac{1}{2}$	$1\frac{3}{4}$	$4\frac{1}{4}$	$2\frac{1}{4}$	$5\frac{3}{8}$
8" Standard	8	80/90	8,000	32,000	4	3	7	$3\frac{1}{4}$	$7\frac{3}{4}$
8" Bronze	8	75/85	10,000	39,500	5	$3\frac{1}{4}$	7	$3\frac{1}{2}$	$7\frac{3}{4}$
Super 8	8	70/80	13,000	54,000	6	$4\frac{3}{4}$	7	4	$7\frac{3}{4}$
Super 8/C.S.	8	60/65	13,000	54,000	4	$4\frac{3}{4}$	7	4	$7\frac{3}{4}$
10" Bronze	$10\frac{1}{4}$	70/80	10,000	39,500	6	4	$8\frac{3}{4}$	$4\frac{1}{4}$	$9\frac{7}{8}$
Golden	$10\frac{1}{4}$	65/75	13,000	54,000	8	$5\frac{3}{4}$	$8\frac{3}{4}$	$4\frac{3}{4}$	$9\frac{7}{8}$
W10/C.S.	$10\frac{1}{4}$	50/60	14,000	74,000	5	9	$8\frac{3}{4}$	$5\frac{3}{4}$	$9\frac{7}{8}$
W12	$12\frac{1}{2}$	60/70	13,000	145,000	15	12	$10\frac{7}{8}$	$6\frac{1}{4}$	$12\frac{3}{8}$
W12/C.S.	$12\frac{1}{2}$	40/50	13,000	145,000	10	12	$10\frac{7}{8}$	$6\frac{1}{4}$	$12\frac{3}{8}$
Super 12	$12\frac{1}{2}$	55/60	17,000	190,000	18	$18\frac{1}{4}$	$10\frac{7}{8}$	$6\frac{3}{4}$	$12\frac{3}{8}$
Super 12/C.S.	$12\frac{1}{2}$	35/45	17,000	190,000	12	$18\frac{1}{4}$	$10\frac{7}{8}$	$6\frac{3}{4}$	$12\frac{3}{8}$
W15	$14\frac{1}{2}$	50/55	13,500	180,000	20	17	$12\frac{3}{4}$	$7\frac{1}{4}$	$14\frac{1}{2}$
W15/C.S.	$14\frac{1}{2}$	30/35	13,500	180,000	15	17	$12\frac{3}{4}$	$7\frac{1}{4}$	$14\frac{1}{2}$

N.B.—Golden, W.10, W.12 and W.15 units: Transformers supplied separately, not mounted on chassis.

Figure 6.14 Details of chassis from 1949 catalogue. Although the dimensions are in inches, the flux density and total flux are in cgs units (lines/cm^2 or oersteds and maxwells respectively).

One lunchtime, as often happened, a friend dropped in to see him whilst construction was in progress and, intrigued, he returned for a demonstration when it was finished. He was so impressed that he decided he must have one at home as soon as possible and immediately went off to a builders merchants to order the necessary materials. They were duly delivered and dumped in his front garden. His wife, who had been away from home whilst this was going on, returned and asked 'What is all this stuff for?' When told, there was a moment's silence followed by 'If those bricks come into my house, I'm leaving'. Gilbert continued: 'and this is when my friend made his big mistake—he got rid of the bricks!'

The picture of the Radiolympia stand also reveals that Gilbert's second book, *Sound Reproduction*, was on sale. This had just been published, in July 1949; it is described in Chapter 12, which details the saga of the many audio books which Gilbert published through Wharfedale. Altogether, including the two books, there were 41 separate products for sale. Sometime shortly thereafter the bakelised cone apex modification was applied to the two most expensive 10" units—identified as the W10/CSB and Golden CSB—to further increase their frequency range.

Hand-polished veneered Oak or Walnut. Truqual V.C. 3 or 15 ohms Speech coil. Size 30" x 15" x 8". Weight : Oak Model 22 lbs. ; Walnut Model 23 lbs.

Designed in the first place for use in schools, the Varitone sets a new standard in reproduction from an 8" Unit.

When the "PORT" is open, the Cabinet works as a phase invertor and the bass resonance of the Bronze Speaker is lowered from 80 cycles to about 45 c.p.s. This gives really excellent bass reproduction which is only possible by correct design of acoustic chamber and porthole.

On the other hand, it is often found that reproduction of speech is easier to follow when the lower frequencies are subdued, and closing the door of the Varitone improves the clarity of many forms of speech and it is found to be a most useful control in schoolrooms to counter reverberation due to absence of carpets, curtains, etc. The Walnut Cabinet is made with rounded corners in solid wood, and costs **£4** extra.

The Varitone has been approved by School Broadcasting Council.

Figure 6.15 Varitone cabinet, 1949 catalogue.

Chapter Six: 1946–1952

Figure 6.16 Corner cabinet with two volume controls. Photographed alongside the TSR108 of 1980–81 for size comparison. (Photo courtesy of IAG.)

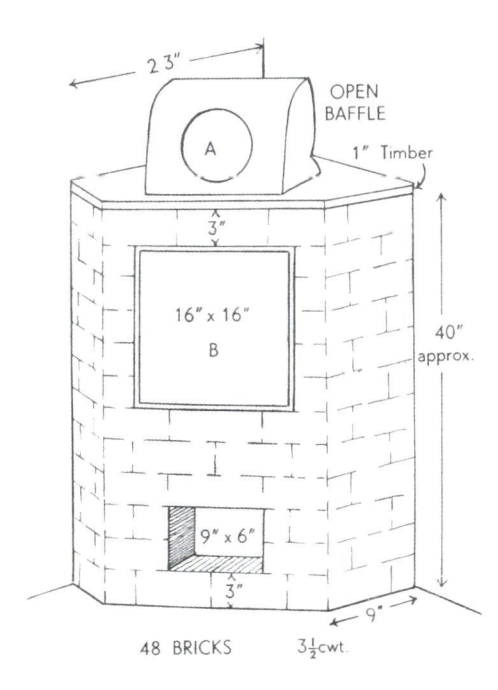

Figure 6.17 Design for a nine cu ft brick corner enclosure. Reproduced from *Loudspeakers*, Second Edition.

Cabinet Systems and Records

The rapid emergence of the floor-standing cabinet systems is intimately bound up with the writing of *Sound Reproduction*. The runaway success of *Loudspeakers* had several unforeseen consequences, one of which was a stream of letters from readers wanting more information and posing interesting questions. This proved to Gilbert that there was scope for a supplement to *Loudspeakers* which addressed the issues raised, many of which concerned loudspeaker enclosures (the chapter on cabinets and baffles, whilst the longest in the book, was still only 10 pages). Many readers of the book were enthusiasts who wanted to construct their own systems—with or without a Wharfedale loudspeaker! As described in Chapter 12, this led Gilbert to embark on a major experimental programme to gain a better understanding of the design and construction factors affecting performance. He did this largely from empirical testing but guided by theory and judged ultimately by listening tests. For these he was joined by Bill Escott and Ezra Broadley, who were both naturally gifted at this but unafraid to disagree with Gilbert's opinion. This was research driven by the requirement for data for publication in the book, but it naturally led to ideas for products, hence the cabinet designs which appeared in 1949.

It was also pointed out, in the correspondence received, that '*The Why and How of Good Reproduction*' in the first book's subtitle was not reflected in any coverage of factors relating to record reproduction. This criticism was somewhat misplaced, since the focus was Loudspeakers, nevertheless it indicated a frustration with available information (or lack of it) and so Gilbert decided to widen his scope accordingly. However, this took him well outside his comfort zone and he needed help from experts in the recording, production and playing of records. During the early years of Wharfedale the reproduction focus had been very much on radio, where even modest transmission quality far exceeded that available from records and where loudspeaker performance was clearly the weakest link. By the end of the War, however, the situation was changing. These were still the days of shellac 78's, as far as the mass market was concerned, with their short playing time and inherent surface noise, but quality, especially frequency range, was improving as a result of wartime projects. Moreover, direct recording onto cellulose lacquer-coated discs was becoming more common and, although the number of times these recordings could be played before they were damaged was much lower than with harder shellac discs, the quality and frequency range could be exceptional and surface noise was almost eliminated. They showed what might be achievable commercially. In 1948, at about the time Gilbert was himself in the USA, Columbia launched the vinyl Long Play (LP) record, a 12" disc with microgrooves played at 33.33 rpm and providing over 20 minutes of music per side. Whole symphonies, requiring an album of five 78's could be accommodated on a single LP. Although the market had to be developed from scratch in the UK, producers like Decca were soon making and exporting LP's to the USA and the technology was revealed when Gilbert made his enquiries. This research, for what had now become the *Sound Reproduction* project, really opened Gilbert's eyes as to the likely future for disc recordings and the importance of working with this industry (see also Chapter 11). The 1949 cabinet promoted as being specially designed for the reproduction of records in schools may be an early indication of this recognition.

The emergence of the floor-standing cabinet systems, involving larger and more complicated enclosures than the extension speakers, placed new demands on the cabinet makers. Optimism about the future was growing, the workforce was slowly expanding, and so the decision was taken to build a new cabinet shop. This was brought fully into operation in mid-1949.

Chapter Six: 1946–1952

Group Photographs

Gilbert did not have a personal secretary before the move to Idle, but if he had not already hired one by 1948 the extra correspondence generated by *Loudspeakers* ensured that he soon did so. 'Miss Isles' (Edith), who became well known through acknowledgements in later books, was the first of only two secretaries that Gilbert employed from then until his death in 1978! About the same time a proper accountant was also taken on, Fred Mason. A photograph of the 'office staff' from about this time has survived. It was almost certainly taken to mark the retirement of Gilbert's sister, Claris, who reached the age of 60 in 1949 (she helped out part-time for a while after 1949, but she died in 1954). The unique importance of this picture is that it includes Gilbert's technical guru Ernest Price, who must have been on a consultation visit at the time. My research turned up two versions of this photograph which show that in the one illustrated (6.18), Bill Escott has been 'added' to the original. Apparently, he arrived too late for the picture-taking and his presence was photographically faked.

It had been one of Gilbert's aims to provide his workforce with a canteen. The buildings at Idle provided the space and as the workforce increased towards the end of the 1940's the critical size for this to be an economic proposition was reached. A cook, Nellie Oats, was taken on and the operation became another of Dorothy Stevens' responsibilities. Its establishment gave Gilbert a great sense of satisfaction and the Christmas lunch, held in the canteen, at which he made witty acknowledgement of the contribution of every individual employee, became a legendary event.

Figure 6.18 Office staff, Idle, about 1949. Back row L to R: NI, Gilbert Briggs, Fred Mason, Bill Escott, Ernest Price. Front row: NI (telephone receptionist), Dorothy Stevens, Claris Briggs, Edith Isles, NI (typist). (NI = not identified). (Photo courtesy of the Escott family.)

Several group photographs, taken on the post-war annual works outings to the seaside survive from the late 1940's and early 1950's. The best of these is shown in (6.19). Gilbert always went along and he may have taken the photograph. The total workforce was around 40 by 1950.

Figure 6.19 Works outing, around 1950. Identifiable employees are: Arnold Hatton (left) with Fred Mason, standing above the group; Bill Escott and Nellie Oats, standing extreme left; Ezra Broadley, Dorothy Stevens and Edith Isles, sitting second, fourth and fifth from the left. (Photo courtesy of Dorothy Stevens.)

Innovations of 1950

The technological innovation of 1950 was the introduction of aluminium voice coils into two drive units, denoted Super 8/CS/AL and Super 12/CS/AL. The advantages of replacing the usual copper wire with aluminium are the reduction in weight by 50% or more and the significant reduction in inductance for a given resistance which, together, extend the high frequency response (say between 8 kHz and 16 kHz) by about 10 dB, in effect adding an octave. The drawbacks, from the production viewpoint at that time, centred on the fragility of the aluminium wire and the difficulties of tinning and soldering it. The latter inevitably led to some joints going open circuit through oxidation some time after manufacture, resulting in loudspeaker failure. Gilbert felt that two percent of such potential returns, from the total factory production, was a reasonable price to pay for the improved performance of these units, which were very well received in the marketplace. The aluminium voice coils commanded an extra 5 s in price. Another change was the introduction of potentiometer type volume controls, replacing the 'Truqual' controls, into the smaller extension speakers and baffles.

Also in early 1950 the nine cu ft corner assembly, whose design had been offered to enthusiasts the previous year, was turned into a product which replaced the two-speaker corner unit of 1947. To complement the non-resonant walls, to which the assembly was fixed, the front panel holding the bass unit was a hollow structure with plywood skins and back-filled with sand. The top was a ¾" thick piece of solid wood. The complete assembly, with separate treble unit (6.20) sold for £26 10 s. The non-resonant sand-filled panel concept eventually became famously associated with Gilbert and Wharfedale, but the idea was not his. He had picked it up from some experiments in progress during one of his visits to the BBC Research Department and quietly filed it away until it could be exploited.[11] Later in the year, the Microgroove Equaliser (or loudspeaker tone control unit) was introduced. This was a tapped inductance with a six position switch giving a series of attenuation rates with increasing frequency. It could be used for matching the characteristics of microgroove recordings (significantly different from those of standard groove 78's) or as a scratch filter and sold for 50 s.

The Wharfedale Catalogue Design

The design and layout of the Wharfedale catalogue had been slowly evolving since 1946 and by 1950 it seems Gilbert was finally satisfied. Thereafter, apart from a change to the basic colour scheme every time it was reissued, it remained the same until 1957. The front cover for 1950 is reproduced in (6.21) and the two introductory pages in (6.22). In the Introduction, we see how

Figure 6.20 Sand-filled corner assembly, 1950 catalogue.

Gilbert cleverly brought together the books and the audio products in his subtle marketing; the Wharfedale brand was imbued with a technical quality derived from his authority as author of the books.

The page layout of the catalogues was done by Gilbert using pictures of the products derived from photographs taken by C.H. Wood's of Bradford, who were specialists in industrial and aerial photography. As with the books (see Chapter 12) he needed peace and quiet in order to concentrate fully on the task. In order to achieve this he would take himself off to a hotel by the sea, such as the Grand in Scarborough or the Midland in Morecombe, as described in the next chapter.

More Books and an Encounter with Raymond Cooke

It must have been in 1951 that Gilbert first encountered Raymond Edgar Cooke, who was later to found KEF Electronics. The event was of such significance for both of them that it would be nice to pin down the date—neither actually did so. At the time Cooke was 26 and studying for a degree in Electrical Engineering at Battersea Polytechnic (later to be transformed into the University of Surrey). According to his recollection,[12] over 30 years later, it happened like this:

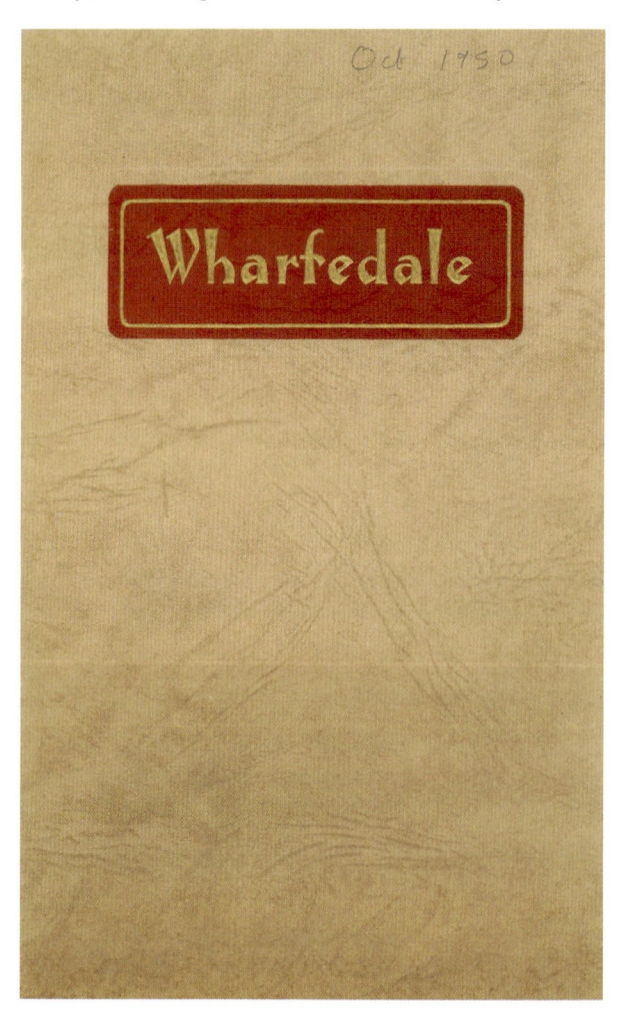

Figure 6.21 Front cover 1950 catalogue.

Chapter Six: 1946–1952

Wharfedale

Introduction —— by G. A. Briggs

IN the following pages will be found full details of the WHARFEDALE range of Loudspeakers, Transformers, Volume and Tone Controls, Frequency Dividers, etc.

All models are designed in accordance with the SOUND principles advocated by the writer in his books *Loudspeakers* and *Sound Reproduction*. First published in May 1948, the book *Loudspeakers* is now in the fourth impression of the third edition. More than 20,000 copies have been sold, including some 8,000 copies in U.S.A. and Canada.

The first edition of 10,000 copies of *Sound Reproduction* was sold out in about six months. The second edition, revised and enlarged, was published in June 1950, and is meeting with a large demand in all countries where English is spoken. A few copies have indeed penetrated the Iron Curtain.

The popularity of these books is mainly due to the fact that problems of high quality reproduction are tackled fairly and squarely without bluff or bias, and the same may truthfully be said about Wharfedale reproducing equipment.

Low resonance and smooth response are the main qualities required for good reproduction—often involving much higher cost of manufacture. A few applications of these principles to Wharfedale Speakers are as follows:

DIE CAST CHASSIS. These cost two to three times as much as pressed steel, but give greater rigidity, less resonance and better air loading. All Wharfedale speakers are now fitted with cast chassis.

CABINET BACKS. Cardboard intensifies resonance, especially on speech. Wharfedale Extension Speakers (except Bantam and Tyny) are now fitted with Expanded Aluminium Backs costing twice as much as cardboard.

Figure 6.22a Introduction from 1950 catalogue (page 1).

Wharfedale

MESH. All cabinets are fitted with Expanded Aluminium Mesh in special non-tarnish Gold finish. This material offers no restriction to sound waves, yet retains its shape indefinitely; its retail value is about 37/6 a square yard.

CLOTH SUSPENSION. The ever-increasing demand for this type of unit proves that the resulting low resonance and refined quality amply justify the higher cost of production. The C.S. range now covers units of 8", 10", 12" and 15" diameter.

All Wharfedale C.S. units cost 10/- more than the corrugated cone.

ALUMINIUM SPEECH COILS. A recent Wharfedale development, which has been highly praised by all quality enthusiasts who have had speakers for test, is the fitting of aluminium coils for extending the H.F. response without introducing peaks in the upper middle register. The light aluminium coil has the effect of increasing the output between 8,000 and 16,000 cycles by about 5 db, at an extra cost of 5/-.

The fragility of the wire imposes limitations on the available impedance as follows:

8" and 10" Units	2 to 10 ohms
12" Units	8 to 15 ohms

It is quite satisfactory to use a 10 ohm treble unit with a 15 ohm bass speaker and a Wharfedale Separator.

CHOICE OF SPEAKER. The best single speaker is the Super 12/C.S./A.L. This unit is extremely sensitive, with excellent transient response. The next choice would be the W.10/C.S., which gives excellent results even if simply mounted on a 3' baffle. Then would follow the Super 8/C.S, or the 10" Golden which is still being supplied to the B.B.C.

The best results are obtained by using two speakers and a separator, with reflex loading for the bass and small open baffle for the treble unit. The selection in order of response would be as follows:—

BASS - W.15/C.S., W.12/C.S., W.10/C.S., Golden/C.S., Golden.

TREBLE - W.10/C.S.B., Super 8/C.S./A.L., Super 8/C.S., W.10/C.S.

PRICES. It is regretted that the prices of most of the better loudspeakers have to be advanced. This is due to a sharp rise in the cost of die-castings and Alcomax magnets.

Page 2

Figure 6.22b Introduction from 1950 catalogue (page 2).

Chapter Six: 1946–1952

'I met Gilbert Briggs. . . . I think in the late 1940s at one of the old BSRA [British Sound Recording Association] Conventions at the Waldorf hotel in London. I had been reading his early book 'Loudspeakers', and at the same time taking an interest in sound reproduction, which had been a hobby of mine since the early days of the Second World War. I found myself at odds with some of the conclusions in the book, particularly in the chapter on reflex enclosures. So I took along to the Waldorf some work that I'd been doing at the University in my spare time, and I challenged him. I said "I think your book is very interesting, but I think some of your conclusions are a bit hasty and wide of the mark. I've got some material here which throws a different light on the matter." So Gilbert said "You'd better come and have a cup of tea with me." We went on to the terrace and talked for about an hour, during which time I think I convinced him that, whereas he had correctly observed what happens in a reflex enclosure, he had drawn the wrong conclusions. So he begged leave to take away my work on the subject, and wrote to me a week or so later, returning the papers and asking me to help with a revision of the book which had sold out. I was still at University, and needed the money, so I gladly helped.'

Whilst the gist of this account is undoubtedly correct, the specific details are not, which is unfortunate because subsequent pieces and obituaries about Cooke all seem to draw on this article! The first BSRA exhibition was held at St Ermin's Hotel in May 1948, when Gilbert was in the USA. The second and third, held in the Waldorf Hotel, were in May 1950 and 1951. Gilbert may have been there in 1950, but he was certainly present in 1951 (see page 259). Cooke's reference to *Loudspeakers* seems mistaken; this book was reprinted many times between 1948 and 1955 (including four editions involving only minor changes) and not updated until 1958. As noted earlier, *Sound Reproduction*, published in July 1949, contained the results of many experiments with reflex cabinets (vented enclosures) and this would be the more likely object of Cooke's disagreement. The second edition, published in May 1950, was greatly expanded, but the section in question was little altered.

All 12,000 copies of the second edition of *Sound Reproduction* sold out 'within a few months of publication' (so well before May 1951), and when *Amplifiers* appeared in April 1952 its cover announced that a revised third edition of *Sound Reproduction* was planned for that autumn. This evidence all points to the meeting being in May 1951* and the collaboration getting underway shortly afterwards, with the aim of revising the Vented Enclosures chapter for *Sound Reproduction 3*.

The collaboration lasted for about a year, with Gilbert sending cabinets to Cooke in London, on which each made different experiments. Most of the discussion was probably carried out by letter and the book contains hints of its nature, as described in Chapter 12. We will hear much more about Raymond Cooke in the following chapter.

Books were taking up an increasing proportion of Gilbert's time—he was hooked. During late 1949 and early 1950 he had been working on the major revision of *Sound Reproduction* and the success of his first two publications had next encouraged him to start work on the book about pianos he had always wanted to write. When Gilbert met Cooke in May 1951 *Pianos, Pianists and Sonics* was in press (to appear in June) and he had recently talked H.H. Garner into being co-author of what was to become *Amplifiers*.

As far as the product range was concerned the only addition during 1951 was a highly efficient 5" tweeter (then referred to as a treble) the Super 5/CS/AL, but the extension speaker range was

---

* In fact, in the programme notes for a joint EMI-KEF concert-demonstration in 1978, Cooke simply states[13] that he first met Gilbert in 1951.

significantly revised. Gilbert usually responded swiftly to perceived changes in the marketplace; the cheapest models, 'Bantam' and 'Tyny', were dropped and a generic style of baffle, like the separate treble unit for the corner enclosure, was used to house the Super 5/CS/AL and Super 8/CS/AL drive units, thus moving the whole range upmarket.

Running out of Space, Again

An insight into the standing of Wharfedale speakers around this time comes from a document in the BBC Written Archive.[14] A problem which arose at one of the BBC studios prompted an audit of loudspeaker types in use at other studios/recording units. It is not clear if the listing in this 1951 report is comprehensive, but, of the 13 locations mentioned, seven had Wharfedales, with a total of 28 speakers (presumably used as monitors). This was the highest of all speaker makes, the next highest being RK with 26 and BTH with 23.

1952 was a year of consolidation with only minor changes to the product line and the publication of *Amplifiers* in March. The focus was on meeting demand. Exports were becoming a significant part of the business, although how quickly these had built up following the USA breakthrough in 1948 is unknown. Apart from re-establishing an agency in Holland (the sole importers before the War), and possibly adding one in France by this time, Gilbert had ignored Europe in favour of Commonwealth countries such as Canada, Australia, New Zealand and South Africa. New agents in far-flung lands were being added all the time, usually with different agents for the audio products and the books, the latter selling in their tens of thousands per year. Bill Escott had taken responsibility for exports at a fairly early stage and, six years after joining Wharfedale, he had become some-

Figure 6.23 Gilbert in his Bradford Road office, about 1952. (Photo courtesy of the Briggs family.)

thing of a 'right hand man' to Gilbert. As 1953 approached it was becoming clear that the production facilities would not be able to cope with much more expansion and the potential for massive growth in the USA alone was very apparent. There was room on the Bradford Road site to increase production; the questions to be addressed were when and how? Maybe Gilbert was contemplating this when the picture of him in his Bradford Road office was taken! (6.23)

Part 2. Away from Wharfedale

Music and Reading

With Ninetta now at University, Valerie, who was 14 in February 1946, was required to take on the role of the listener 'with young ears'. She recalled some 15 years later that:[15]

> *'My introduction to Audio when I was in my teens took place at home in the Music Room; a glorious place bursting with bits and pieces of equipment and littered with wires, and chosen by father for all his tests at home as in his opinion it is an excellent room for listening to any new model he is developing. I accepted as perfectly normal the way in which the house slowly filled up with loudspeakers in various stages of dress, from black cotton bags to plain plywood cabinets smelling pleasantly of sawdust.*
>
> *It was always exciting when Papa brought home a new model for us to listen to. It would start by the front door being flung open, and then "Val, have you a minute?" I knew then that a whopper had arrived. (He carried the small ones in by himself.) On looking back I think we would have done well to invest in a bogie. I thoroughly approve of good quality solid enclosures, but they are not the easiest things to manipulate off the back seat of a car. We would stagger into the room and father was then ready to indulge in his favourite pastimes of joining wires. They hung together for all they were worth, but the music invariably came to an abrupt halt as he crossed the room to a new listening position and wrapped the wires round his feet. There then followed a long drawn out d-a-m-n as things were righted.*
>
> *We would try the speakers facing in all directions, including the ceiling, and in various relationships to the wall. Father told us what he was particularly listening for in each model and I found it very fascinating learning really to use my ears. If there was appropriate music or speech on the BBC this was always chosen in preference to a record as in seemed difficult to better the BBC quality. I distinctly remember listening with rapt attention to applause and learning all about transients.*
>
> *These listening tests were really a series of comparisons, the control eventually being a marble corner cabinet. I do not think this unit was ever bettered or indeed has yet been equalled.*
>
> *Definitely no cheating was allowed at these sessions, so one of us controlled the switch box while the others listened with closed eyes, or their backs to the speakers. It was quite a job trying to identify maybe half-a-dozen models, and I was frequently wrong, but hard as I might try I found it very difficult to fool GAB, and I often tried very hard indeed.'*

This was written in 1961 and a picture of the Music Room taken in about 1958 is shown in (6.24). In the corner is the marble corner cabinet which Valerie mentioned. Two were made as an experiment with marble replacing the sand-filled panels—it was an excellent material but too brittle to be practically transportable. This one was dropped and broke into three pieces following a demonstration to Jim Rogers (manufacturer of amplifiers etc.). Gilbert installed it at home having taped

A Pair of Wharfedales: The Story of Gilbert Briggs and his Loudspeakers

Figure 6.24 Music Room at Woodville, about 1958. Reproduced from *Audio Biographies*.

the pieces together. The other he sold to a friend. On the left of the picture is his Marshall and Rose upright piano, purchased in 1921. Identifiable commercial systems are the SFB/3, next to the piano, and the AF10, second from the right.

It was Gilbert's rule not to work after 7 pm or at weekends, so it seems unlikely that he regarded speaker listening as work! It was around 1946 that he finally stopped having piano lessons from Edgar Knight, because he felt that there was no possibility of further improvement, but he continued his regime of daily playing as before. It was a great calmer and after a frustrating day at work he would go straight to the piano and lose himself in the music. Apart from his other music-related activities, described previously, Gilbert relaxed by playing bridge and the occasional round of golf, and reading. Of course, he devoured the relevant audio journals and magazines, but he was also fond of *Punch*. As far as books were concerned, largely read in bed, he enjoyed biographies—in particular those with a musical connection—the P.G. Wodehouse comedies and the Perry Mason novels by Erle Stanley Gardner. He had a habit of giving books a rating, which he noted at the front. Re-reading them several years later he often found his opinion had changed, so he tried to explain this in additional notes!

Professor Briggs

Gilbert was a great letter writer, an activity prized by his and earlier generations and which he had had to take up early in life. Despite the fact that correspondence occupied a significant

Chapter Six: 1946–1952

fraction of his time at work, he managed to stay in touch with a large network of friends and maintain the friendships for decades. One such friendship, which dated back to at least 1913 and Gilbert's first Indian trip, was with Arthur Spink. Spink was another member of the Bradford textile export merchants' fraternity and is seen standing next to Gilbert (on his left) in the 'Bradford Travellers' football team picture shown in Chapter 3 (3.5). He was two years older than Gilbert, who greatly respected his judgement. Pictures exist of him with Gilbert and Edna on the voyage to India following their honeymoon in 1924, so he and Gilbert both resumed their merchant adventuring where they had left off when the First World War intervened. When Gilbert stopped these India trips his advice to his successor was to arrange his schedules and the all-important timing of when to arrive in each major city according to Mr Spink's movements. In September 1949 Spink's daughter Cecilia was married in the London Oratory and Gilbert and Edna were invited. The occasion produced a rare picture of the couple together, taken by the Oratory steps in Brompton Road (6.25). On the back of the original photograph,

Figure 6.25 Gilbert and Edna at the wedding of Cecilia Spink, London Oratory, September 1949. (Photo courtesy of the Briggs family.)

in Gilbert's handwriting, is 'Professor G.A. Briggs, 29 September 1949'. Since his second book, *Sound Reproduction*, had been published two months earlier, he had probably been jokingly introduced by someone in the know—maybe by Spink himself—as 'The Professor'. This would have gone down very well; one of Gilbert's regrets in life was that he had not had the opportunity to go to University.

As described in Chapter 12, one of the features of *Sound Reproduction* was the collection of photomicrographs, illustrating wear of disc recordings and styli, taken by the incomparable Cecil Watts. Most of the data was the result of research commissioned by Gilbert especially for the book and its publication not only added greatly to the book's appeal, but also ensured that the naturally reticent Watts gained wide recognition for his pioneering work. Since Gilbert had made several visits during the collaboration to Darby House in Sunbury-on-Thames, where Watts lived and worked, he invited Cecil, with his wife Agnes and daughter Susan, to Ilkley for a celebration following publication. By this time Edna had expanded into pig keeping at Woodville and Agnes Watts had this recollection of the visit:[16]

'Apparently the Briggs family are nearly as mad as the Watts. When we visited their home the next day, Mrs Briggs had just come in from seeing to her animals. She remarked that it was a pity we had not come earlier as Susan could have helped her milk the goats. Susan was only eight years old and looked very disappointed at the missed opportunity. So Mrs Briggs took her outside.

We soon heard riots of laughter and Susan suddenly appeared in the drawing room with a baby piglet in her arms. It soon escaped and, to Mr Briggs' horror, careered round and round the room, squealing like mad. Susan and Mrs Briggs were in hot pursuit until they finally collapsed with laughter.

Before the visit with the piggie, they had been out in the garden shed with a goat, which Mrs Briggs had brought in from the paddock. She gave Susan a lesson in milking, then allowed her to try her hand. Being an opportunist, Susan pretended she could not see where she was directing the jet of milk and swamped Mrs Briggs; it does not need much imagination to picture the state that both of them were in. Mrs Briggs won our affection immediately for the understanding she displayed in making Susan's visit as enjoyable as our own; children are often overlooked when business people meet.'

In the summer of 1950 Ninetta completed her degree and qualified as a veterinary surgeon; she was totally exhausted by the final examinations. Gilbert decided that she needed a holiday and dropped everything to take her on a tour of some of the European capitals he knew. It was on this trip that she first fully realised his facility with foreign languages—he seemed to speak about seven. Gilbert was very modest about his abilities, claiming that he only knew enough to get by with his foreign agents in sales discussions, but she was not totally convinced. He also revealed his knowledge of architectural styles—he had a sketcher's eye for detail and had probably sketched buildings during his early visits to these cities before the First World War—but now photography was his preferred means of recording them.

High Fidelity

In mid-1951 the first issue of a new magazine, *High Fidelity*, appeared in the USA. Its founder and editor, Charles Fowler (6.26), had been greatly impressed by the content and style of

Loudspeakers and *Sound Reproduction*, both of which had already given rise to a devoted following there. The reason for this he later analysed as follows:[17]

> 'This man could talk to his readers and explain complex matters in a fashion lucid even to the most thick-skulled neophyte. There was always a twinkle in his literary eye, always a feeling of warmth and of a genuine, friendly desire to help the other fellow acquire some of his own broad knowledge.'

Fowler was convinced that early articles by Gilbert would help establish the right tone for his magazine and ensure its success: 'I needed this author, with his particular ability.' He wrote to him only to receive a polite but firm 'Sorry, too busy'. He wrote again, more pleadingly, and made the breakthrough. A series of four articles was agreed upon and the first, entitled 'The Loudspeaker and the Ear', appeared in the third issue towards the end of the year. The second, on 'Response Curves', was in the following issue in spring 1952 and here Gilbert was in good company—also published was a monumental discography of all Beethoven LPs by C.G. Burke. Fowler felt that in this issue he had embodied the approach to both the technical and musical aspects of the magazine which he desired. In the end Gilbert wrote eight articles, the last appearing in March 1954 (full details are given in Appendix 3). As Fowler had foreseen they were a great success.

As Gilbert was contemplating these articles, Valerie left school and Ninetta's marriage to Leslie Theobald, a fellow vet, took place. They soon set up a practice together and, since they had no experience of running a business, Gilbert offered his advice. When he saw the operation at first hand he was appalled to find that everything was done 'on account' and even the most trivial pieces of work were billed. Chasing up these small accounts involved a ridiculous amount of time and effort, but that was the way the whole veterinary profession worked. Ever the one to 'tell it straight', Gilbert told them this had to stop and all such work should only be carried out for on-the-spot cash payment. He even wrote out notices for the waiting room explaining the revolutionary new regime, which Ninetta thought were almost incomprehensible. However, the approach was adopted and their practice, they believe, was the first to break the mould. Gilbert's first grandchild arrived in late 1952.

Figure 6.26 Charles Fowler. Reproduced from *Audio Biographies*.

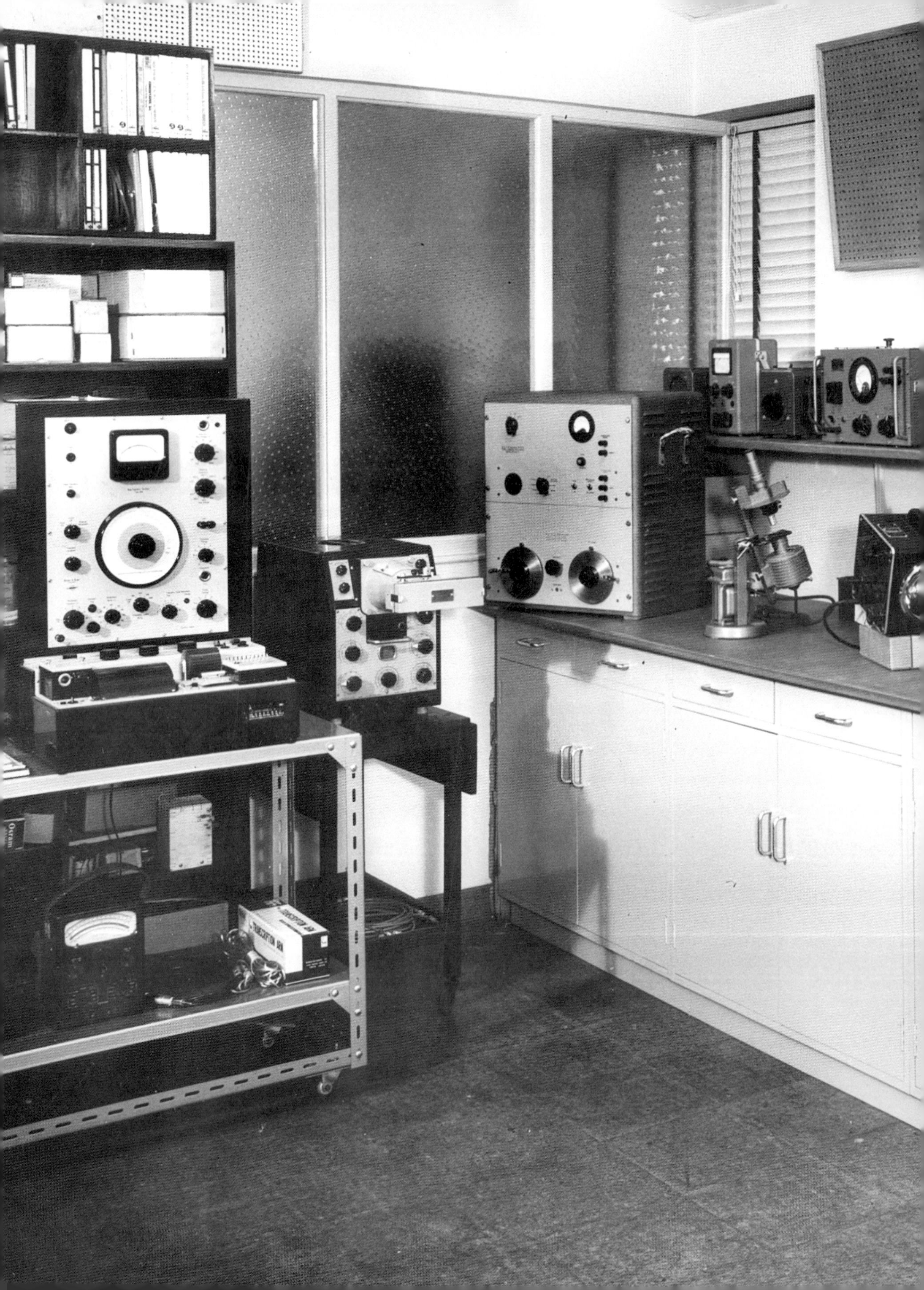
Osram
TRANSCRIPTION ARM

CHAPTER SEVEN: 1953–1958

Part 1. Wharfedale Wireless Works Ltd, Idle

Expansion

Twenty years after Gilbert first started the firm as his sideline in 1932, Wharfedale Wireless Works was still solely owned by Edna. Redevelopment of the Bradford Road site—a necessity for any significant expansion—would have had serious financial implications for the firm, the consideration of which probably caused the situation to be changed. The upshot was that in late 1953 Gilbert took over ownership before the firm became incorporated as a Limited Company, around February 1954. The directors were Gilbert, Edna and Fred Mason, who was secretary. Exactly when the redevelopment started is not known, but the change of telephone number to Idle 1235/6 during September 1953 is indicative of the fact that expansion was underway. A picture of the site taken across Bradford Road, shortly after its completion in 1960, is shown in (7.1). Further up Bradford Road, off to the right from this picture, is the multiple road junction known as Five Lane Ends and the area shown is now the site of a supermarket. The huge expanse of earth, which looks to have been recently bulldozed, is on the edge of the Springfield Works site, which had been the Jowett Cars factory. Car production ceased in the summer of 1954 and the site was sold to International Harvesters, the tractor manufacturers, who began operations in 1955.

The two buildings in the foreground constitute most of the new development. The single-storey new building, to the left of the entrance gate, was built first and originally housed a purpose-designed laboratory, technical office and listening room, as well as offices for Gilbert and his secretary. It was certainly functioning in 1957 as such. By the time of this photo it had Gilbert's office on the right, the accountant's office in the centre and the larger general office (two windows) on the left. The longer building, finally fitted out by 1960, started, on the visible side, with the shipping office (two windows, extreme right), then the laboratory, then the technical office. The rest of the building, under the roof lights, was the cabinet shop. On the opposite side, the first office was Bill Escott's and opposite the lab was the listening room. The transformer department was in a cellar below the end of the cabinet shop, where the ground seems to fall away. A flat-roofed building, which is just visible by the extraction vent pipe, may also be newly erected; this was probably the stores on the ground floor and the canteen above. The collection of buildings near the entrance, behind a section of wall, was inherited in 1946, along with a single-storey building almost hidden behind the new office block. These, together with the more modern building next to it, with three roof lights, were mainly used for loudspeaker production, assembly/trimming of loudspeaker systems (cabinets with drive units) and packing.

Figure 7.1 Bradford Road site, probably end 1960. (Photo courtesy of the Escott family.)

Figure 7.2 Idle factory yard early 1950s. (Photo courtesy of the Escott family.)

Chapter Seven: 1953–1958

Figure 7.3 Idle factory yard, late 1950s. (Photo courtesy of the Escott family).

The older buildings on the right of the picture bounded a yard. One view of this (7.2) predates the development. The Citroën car belonged to Gilbert and originally he had his office on the upper floor of the building on the left, later taken over by Ezra Broadley as Works Manager. Just on-picture on the left is the single-storey building in which the loudspeakers were assembled. The crates are awaiting shipment to C.F. Kauderer, the agents for The Netherlands, in Muiden. Another, different, view (7.3), taken at a later time, shows the works van being loaded. The top of the derrick can just be made out in (7.1) above the roof line of the new office block. The crates are destined for British Industries Corporation in New York and the The Radio People in Hong Kong.

Maybe because the export market was very much in mind at the time, Gilbert first published a list of overseas agents, for the audio products, in the catalogue of May 1953 — reproduced in (7.4). Note that TRP in Hong Kong are not on this list; they were appointed later (but before 1956).

The First Three-Speaker System

Since the introduction of the first crossover unit in 1947 there had been a gradual expansion of the range. By 1951 there were eight variants of the 'quarter section type separator' with permutations of crossover frequency (1 kHz or 3 kHz), impedance (2–6 ohms or 7–17 ohms) and connections (terminals or soldering tags). By 1953 'half section type' three-way separators (second order circuits) had been introduced with crossovers at 800 Hz and 5 kHz or 400 Hz and 5 kHz. This paved the way for the first three-speaker system to be marketed, based on the sand-filled corner panel and including a horizontal cabinet with treble and tweeter units (7.5). It was described as 'omni-directional' and Gilbert characterised the assembly as 'the most natural reproduction so far achieved by Wharfedale'.

Overseas Agents

U.S.A.	British Industries Corporation, 164 Duane Street, NEW YORK, 13, N.Y.
CANADA	J. B. Smyth, 2063 Victoria Street, MONTREAL, Prov. Que.
AUSTRALIA	Stooke Electrical and Radio Services, 6 Balfour Street, BRIGHTON, Victoria.
NEW ZEALAND	Alan R. Seccombe, 2 Mount Royal Avenue, MOUNT ALBERT, S.W.2, Auckland.
SOUTH AFRICA	L. Lloyd (Pty) Ltd., 30/34 Rosettenville Road, Village Main, JOHANNESBURG.
FRANCE	Setton & Co., 55 Avenue Hoche, 55, PARIS, 8.
HOLLAND	Geo. C. F. Kauderer, MUIDEN, near Amsterdam.
MALAYA	Eastland Trading Co., P.O. Box 1128, 1 Prince Street, SINGAPORE.
JAMAICA	Stanley Motta Ltd., 109 Harbour Road, KINGSTON.
TRINIDAD	Investments and Agencies Ltd., 2 Charlotte Street, PORT OF SPAIN.
BERMUDA	Masters, Ltd., HAMILTON.
MEXICO	Ultra-Fidelidad, S.A., MEXICO 10, D.F.
JAPAN	Kawamura Electrical Laboratory, No. 92, 2-Chome Koenji, Suginami-Ku, TOKYO.
CEYLON	Siedles Cineradio, COLOMBO.

Figure 7.4 Overseas agents listed in 1953 catalogue.

Chapter Seven: 1953–1958

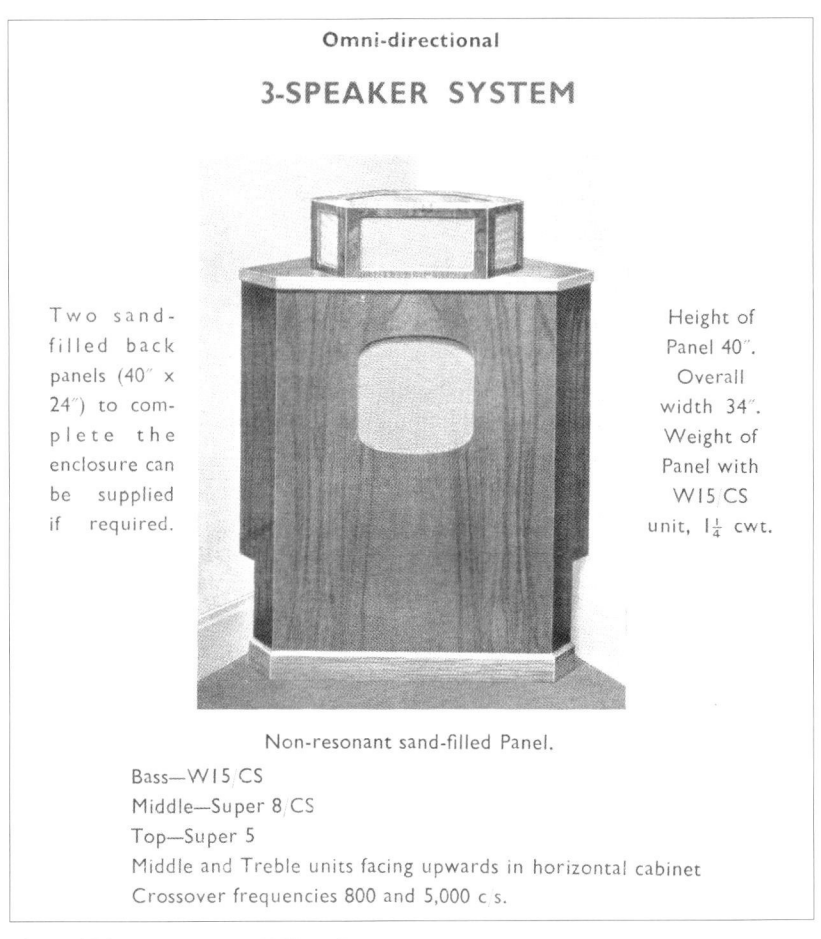

> **Omni-directional**
>
> **3-SPEAKER SYSTEM**
>
> Two sand-filled back panels (40" x 24") to complete the enclosure can be supplied if required.
>
> Height of Panel 40". Overall width 34". Weight of Panel with W15/CS unit, 1¼ cwt.
>
> Non-resonant sand-filled Panel.
> Bass—W15/CS
> Middle—Super 8/CS
> Top—Super 5
> Middle and Treble units facing upwards in horizontal cabinet
> Crossover frequencies 800 and 5,000 c/s.

Figure 7.5 3-speaker system, 1953 catalogue.

This top-of-the-range product, which Gilbert later used in all his concert-demonstrations, was available for demonstration at carefully selected retailers some months after it was first launched in the spring of 1953. The first batch was announced when the product was first advertised in *Wireless World* in September 1953: Scarborough, Leeds, Bath, Birmingham and four in the London area. In subsequent months others were added in London, Manchester, Salisbury and Stoke-on-Trent.

The new horizontal cabinet was not sold separately until the following year, when it was called the 'Twin Treble Cabinet'. The vagaries of the purchase tax regime in force at the time meant that there was a significant advantage in buying the corner assemblies; they qualified as 'kits' since they had to be assembled from 'parts' and were therefore free of the 33% tax. Thus the price of the whole assembly came to £73 10 s (about £1500 in 2010) which included the horizontal top unit with its two drivers at £18. However, if the latter was bought separately, being complete in itself, its price was £24 due to the addition of £6 purchase tax! A pair of sand-filled back panels to make a complete, stand-alone enclosure was an extra £12. The high level of purchase tax which audio products attracted was a great source of irritation to Gilbert and he complained about it bitterly, in print, at any opportunity. The 'kit' avoidance measure, though fortuitous, would have given him a crumb of comfort.

141

Introducing Oscillograms

Another innovation of the 1953 catalogue was the inclusion of 'oscillograms' to illustrate aspects of loudspeaker performance. When Gilbert was working on *Pianos, Pianists and Sonics* in 1950/51, he embarked on a series of experiments, with the unflagging support of Ernest Price, to harness oscilloscope techniques in the cause of demystifying several aspects of piano performance and behaviour. Not least of these was the recording of time-dependant behaviour by capturing the oscilloscope trace on moving film (such as the decay of a piano note with or without pedal application). Gilbert knew that the reproduction of solo piano was one of the sternest tests of a loudspeaker and it was but a short step from these experiments to the use of similar techniques to generate oscillograms for the investigation of loudspeaker behaviour per se. He and Price produced a prodigious amount of such data for inclusion in *Sound Reproduction 3*, published May 1953, and examples found their way into the catalogue of that year (7.6), and of subsequent years. Whether or not the reader understood what was being demonstrated (and it is certainly not always clear!), it all looked impressive and reinforced the feeling that Wharfedale, and Gilbert in particular, was at the cutting edge of loudspeaker development. Chapter 12 covers this in more detail.

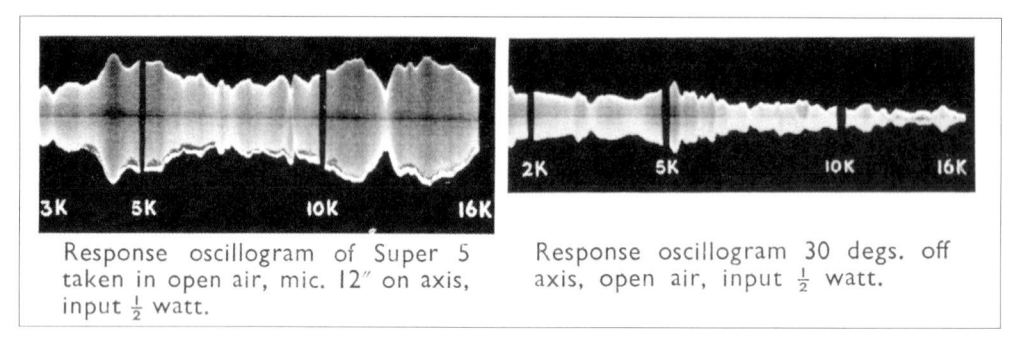

Figure 7.6a Equivalent of the response curve measured by feeding the microphone output to an oscilloscope whilst scanning the frequency range (in sections). 1953 catalogue.

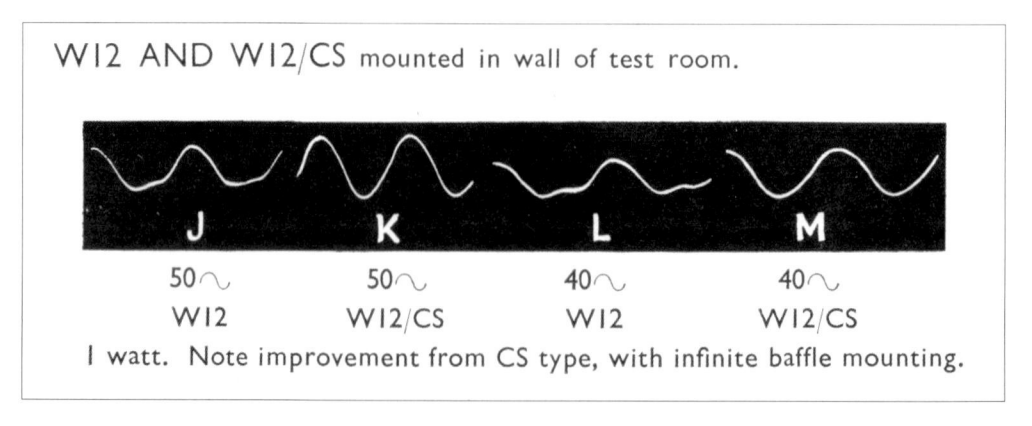

Figure 7.6b Waveform of speaker output for a fixed input frequency. A pure sine wave indicates zero distortion. 1953 catalogue.

Chapter Seven: 1953–1958

To the USA with Harold Leak

It was also in 1953 that Gilbert returned to the concert platform for the first time since his illustrated lectures on composers following the end of the First World War, although, as noted in Chapter 4, giving lectures about, and demonstrations of, loudspeakers had been going on for 20 years. It seems that his first invitation to give a lecture-demonstration on the subject of 'Sound Reproduction', in which the playing of records was an important feature, led him to Torquay. No more is known, but a strong possibility is that it was at a National Federation of Gramophone Societies meeting during the summer. He gave another in the Craiglands hotel in Ilkley, which he organised for invited friends, so this could have been a rehearsal for Torquay. On the strength of this he was booked to give a performance to the Ilkley Gramophone Society in February 1954 (their September–April programme was arranged during the summer months). Gilbert had arranged to attend his first Audio Fair in New York in October 1953 and, probably through an initiative of his Canadian agent Barney Smyth, he was invited to give another lecture-demonstration to the Society of Music Enthusiasts at Toronto University on 28 October, before returning home (see also Chapter 11).

Gilbert's trip to the States in October 1953 was his first since the adventure of 1948. In the meantime exports to that country had become very significant and it seems probable that he was initially persuaded to go by his agent, British Industries Corporation, so that he could spend time on their stand in the exhibition area and personally demonstrate Wharfedale products, which was done in an allocated room in the New Yorker Hotel in which the event took place. Gilbert flew there for the first time and travelled with Harold J. Leak. Leak was, in fact, the first British manufacturer to take part in an American Audio Fair. This was in 1949; he read about the event in a newspaper and decided to go. According to Gilbert,[1] Leak:

> '... applied to the Bank of England for the necessary dollar allowance, which was granted. He immediately cabled the organiser of the Fair as follows; "Will take room 60. Dollars available." The message was, however, received and interpreted as: "Will take room. 60 Dollars available." Harry Reizes, the organiser, concluded that here was another mad Englishman and promptly threw the cable in the waste paper basket.'

Leak, receiving no answer, made a transatlantic phone call to Reizes and eventually twigged what had happened. It all ended happily. Before the Fair, Leak had put a full-page advert in *Audio Engineering* in which he added that he would be seeking a suitable US agency for his company's products during the Fair. So it was that BIC became agents for Leak amplifiers and now Harold Leak was attending the Fair for the same reason as Gilbert. In fact, there was a plan for a joint demonstration at the Fair and Gilbert wanted to use a BBC recording as part of this. His written request to the BBC (reproduced in Appendix 1) was the first of what was to become a stream of letters on this theme, related mainly to his concert-demonstrations, which went on until 1964.

Flying would not have been Gilbert's choice for getting to America; he preferred the relaxed and convivial five days at sea with the time changed by an hour each day. Harold Leak always flew but presumably Gilbert decided to travel with him in order to get to know him better. At that time, Leak would only have been a passing acquaintance from UK industry events and they were likely to be together for much of the Fair (7.7). Flight BA 521/121 was a tiring overnight route from London to New York via Prestwick (Scotland), Keflavick (Iceland), Gander (Newfoundland) and Boston on a

A Pair of Wharfedales: The Story of Gilbert Briggs and his Loudspeakers

Figure 7.7 BIC stand, 1953 New York Audio Fair. Standing left to right: Harold Leak, Gilbert Briggs, Leonard Carduner of BIC. The person sitting is thought to be Edward W. Kellogg. (Reproduced from *Audio Engineering*, 1953.)

Lockheed Constellation aircraft. It is no surprise that Gilbert returned to Southampton from Quebec on board the *Sythia*!

New Friends and Fans

The day after he arrived in New York Gilbert received a telegram which, since he was used to such messages being extremely terse, was of such a remarkable length that he counted the words—70. It was from Milton David Kramer, of whom he had never heard, inviting him to visit his apartment and listen to his speaker set-up which involved no fewer than 12 Wharfedale units. A drawing of the cabinet construction he found there is shown in (7.8). At the time Kramer was Vice-President for Public Relations and Personnel with Associated Transport, but he would later become Public Relations Director of Hertz. He was an amateur hi-fi enthusiast who built systems for himself and friends. What Gilbert heard, and was mightily impressed by, involved a two-channel (mono) amplifier with an electronic crossover at about 200 Hz with 40 watts output in the treble and 100 watts in the bass—all built by Kramer himself. As Gilbert put it:[2]

> '… and so sprang up one of those valued friendships which make life worth living.'

On this trip Gilbert also got to meet his fans Charles Fowler and John Conly of *High Fidelity* magazine. The visit to Great Barrington, home of the publication, was organised by Leonard Carduner and Arthur Gasman from BIC. Of the drive from New York, Gilbert tells the following story which perfectly illustrates his self-deprecating wit:[3]

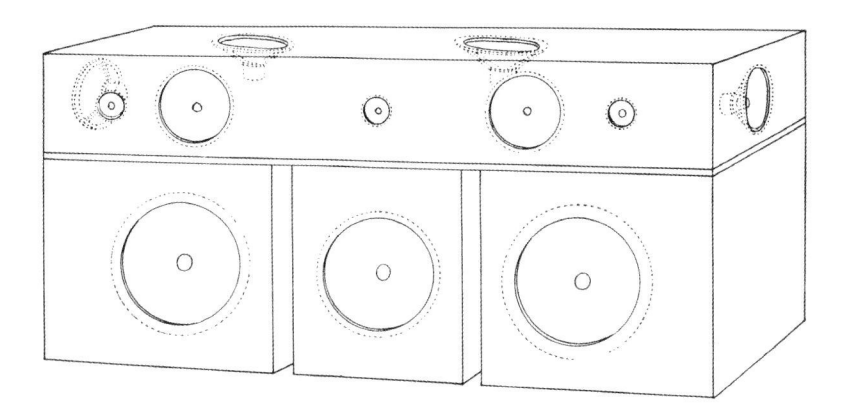

Figure 7.8 Kramer loudspeaker system. Reproduced from *Loudspeakers* 5th Edition.

'I was naturally keenly interested in the topography, which was all new to me, but when I asked if the river on our left was the Mississippi and the hills on our right were the Rockies our driver-host, Arthur Gasman, refused to answer any further questions and concentrated on his Cadillac. First he pressed a button which automatically opened a window and let some fresh air into the car. Then he pressed another which caused a metal rod to rise outside to act as an aerial, and a third button to switch on the radio. "Which station is that?" I enquired. "Crystal Palace" replied Mr Gasman innocently.'

The R-J Enclosure

Another development in this eventful year, also with a strong American connection, was a new loudspeaker system of rather smaller size than hitherto contemplated. It was based on the 'R-J' enclosure design of William Joseph and Franklin Robbins (US patent application 2694463, filed in April 1952) which they discussed in two articles in *Audio Engineering*.[4,5] The enclosure, of which more below, was for a single wide-range speaker and Wharfedale's 8" units with aluminium voice coils, the Super 8/CS/AL and 8" Bronze/AL, were ideally suited. Sales of these chassis for fitting into R-J enclosures, for which BIC was also an agent, took off in the States and led Gilbert to experiment with the design at Idle. The result was a license to manufacture and sell R-J enclosures in the UK and the launch of the 'R-J Cabinet' during in the summer of 1954 (7.9).

The cabinet design effectively involved two resonant chambers with the speaker mounted on an internal baffle, set back from the cabinet front, and with slightly smaller dimensions. The asymmetric cut-out, with maximum dimension slightly smaller than the speaker, acted as both port and disperser. The larger rear chamber was largely filled with padding. Tuning was rather critical but Gilbert reported that:[6]

'This type of loading improves the waveform at frequencies around and below the normal cone resonance, and smoothes the response in the upper middle register where resonant peaks are often troublesome. As the HF output from the back of the cone is mostly absorbed, it is essential to use a speaker unit with a very good HF response, such as one fitted with aluminium voice coil.'

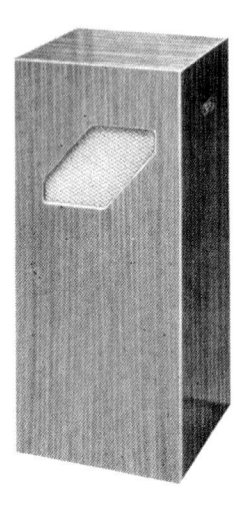

Figure 7.9 R-J cabinet, 1954 catalogue.

New Ideas

The Toronto lecture-demonstration, mentioned above, inspired Gilbert to put on the first of his 'concert' demonstrations with 'live versus recorded' items at St. George's Hall in Bradford on 25 March 1954. The full story of this concert and the many that followed during the next six years, for which he became justly famous, is told in Chapter 11. After the St. George's Hall concert Gilbert's daughter Valerie invited him to perform at Studley Horticultural College, where she was in the second year of her course. Of this event, in early summer 1954, Gilbert recalled:[7]

'I know of nothing more exhilarating than the spontaneous enthusiasm of young people, and my lecture was popular—not because I was giving it—but because of the novelty of having a lecture in the college illustrated by records. (The usual topics were soil pests, insecticides and manures.) One of the most popular records of that period was 'O Mein Papa' played by Eddie Calvert,[8] and the mere announcement that this would be included in the programme ensured a full house'

The other introduction of 1954 was another sand-filled panel/corner assembly variant. This had a capacity of 4.75 cu ft, with reflex openings which also allowed skirting board clearance, and it was specifically designed, for a single wide-range speaker, around the Golden CSB unit.

By now there were owners of Wharfedale speakers and readers of Gilbert's books all around the world. His 'fan mail' was considerable and he started to use some of the testimonial letters from far-flung lands in advertising. A typical example of the genre, from *Wireless World* of February 1954, is reproduced in (7.10).

Cocktails in New York

That autumn Gilbert was once again doing his stint at the New York Audio Fair, but also making a bit more of a holiday out of the trip. He took along Edna and Valerie, sailing on 25 September from Liverpool to Quebec (on the *Franconia*) in order to visit his Canadian agent first, eventually ending up in New York. Harold Leak was there again, also with his wife, Muriel. Two senior representatives of Garrard's, BIC's first clients, were at the event to promote the now legendary Garrard 301 transcription turntable, launched that year: Hector V. Slade, the chief executive and Edmund W. Mortimer, the chief designer. To mark their collective presence BIC organised a cocktail party, the invitation card for which is reproduced in (7.11). A picture of the Briggses with the Leaks taken at the gathering is shown in (7.12). Edna and Valerie left on 20 October, for Southampton on the *Queen Mary*, without Gilbert; he was obliged to fly home because of the need to fit in several more rehearsals for his first Royal Festival Hall concert on 1 November (see Chapter 11).

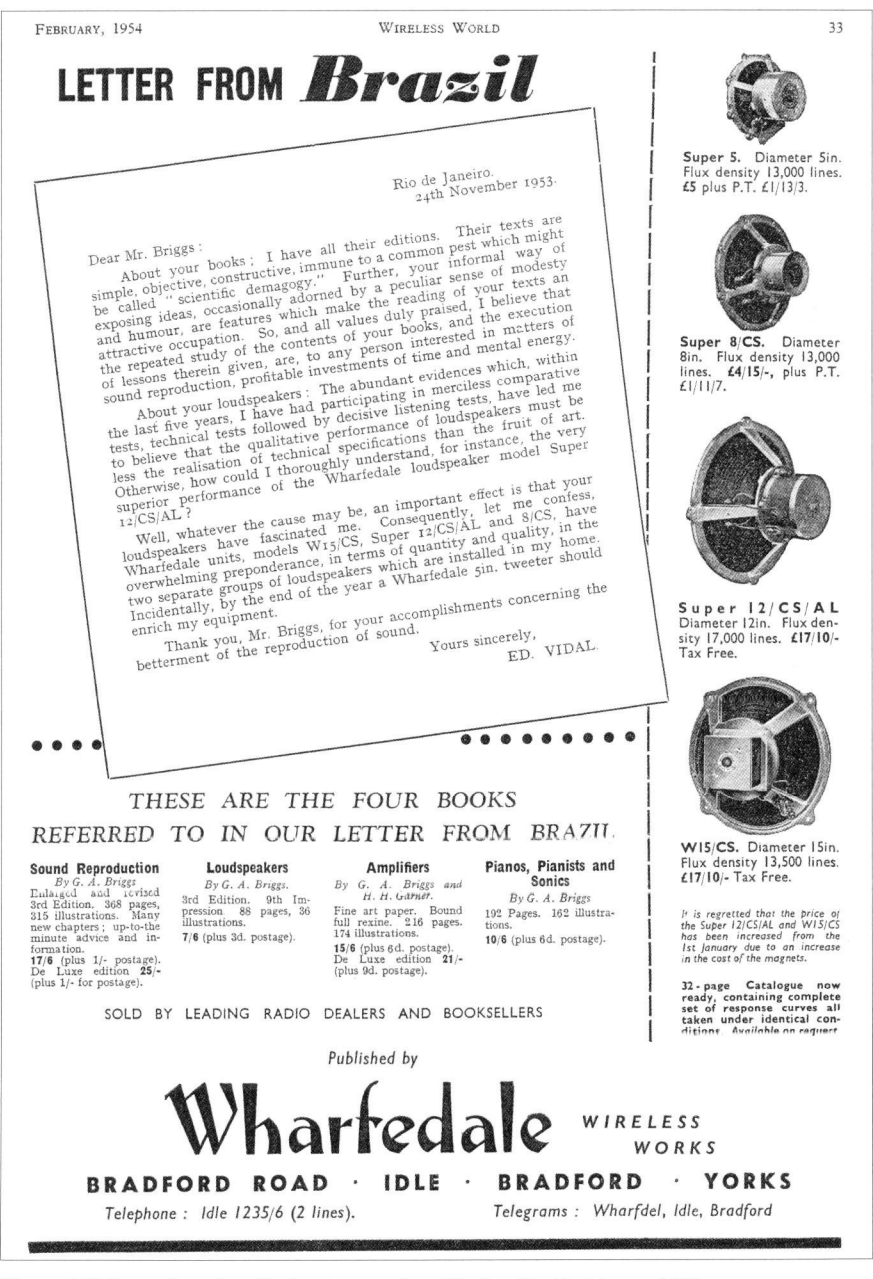

Figure 7.10 'Letter from Brazil' advertisement from *Wireless World*, February 1954.

A Pair of Wharfedales: The Story of Gilbert Briggs and his Loudspeakers

The Management and Staff of British Industries Corporation requests the pleasure of your company for cocktails in honor of

Mr. and Mrs. Gilbert A. Briggs *Mr. Hector V. Slade*
Miss Valerie Edna Briggs *Mr. E. W. Mortimer*
Mr. and Mrs. Harold J. Leak

on Wednesday, the thirteenth of October at five-thirty at the Hotel New Yorker New York

R.S.V.P.

Figure 7.11 BIC cocktails invitation. (Scan courtesy of the Briggs family.)

Figure 7.12 Gilbert with Harold Leak, Muriel Leak (left) and Edna (right). New Yorker Hotel, 1954. (Photo courtesy of the Leak family.)

Chapter Seven: 1953–1958

The Arrival of Raymond Cooke

In the summer of 1955 Gilbert invited Raymond Cooke to join Wharfedale as Technical Manager. When and why he decided to do this has not been recorded, but a number of factors were involved. He was now in his 65th year with an expanding company to run and two related activities which were demanding: writing/publishing and lecture-demonstrating. The importance of the US market meant accepting that travelling there on a regular basis would continue. His health was not great; some aspects have been dealt with in earlier chapters but additionally he was a migraine sufferer—attacks would follow bursts of intense concentration like night follows day—and there was a legacy of years of heavy cigarette smoking. Not long after the end of the War his GP changed and on his first check-up with the new doctor he was told that if he did not give up smoking he would be dead within a couple of years. Gilbert stopped immediately but continued to carry a silver cigarette case to prove that he had chosen this path! Several years later he wrote that he had enjoyed his audio life:[9]

> *'tremendously, apart from many illnesses and migraine attacks brought on by strain and overwork. Recovered good health at the age of 65 by deciding to put health first, work second and money last.'*

So, looking after his health was clearly one factor and handing over some responsibility for the technical developments at Wharfedale would certainly ease the workload. The marketplace was changing rapidly and it had probably been clear for some time that new initiatives were required to come up with the compact systems which were being demanded, without sacrificing the reproduction quality so carefully established over the years. New, younger blood and a complementary technical experience were called for.

Comparing him with Gilbert, in many respects Raymond Edgar Cooke was a 'chip off the old block'. Born in 1925, he had started playing the violin aged eight, was passionate about classical music and an avid concert-goer. On leaving Doncaster Grammar School during the early years of the War he had worked as an analytical chemist with the London and North Eastern Railway (LNER) and become interested in sound reproduction. He served as a radio operator in the Fleet Air Arm, on the aircraft carrier HMS *Hunter*, from 1943 to 1946 and then resumed his career in chemistry, but soon decided that electronics and audio was what he wanted to do. In 1948 he started a degree course in electrical engineering at Battersea Polytechnic, London and he was in his final year of this when he first met Gilbert in 1951, as described in the previous chapter. After graduating he worked for Philips–Mullard in television tube production and then, in 1954, moved to the BBC Engineering Designs Department and worked alongside such notables as D.E.L. Shorter and Dudley Harwood (founder of the Harbeth Company) on the development of record and tape reproducers for broadcasting. His spare-time work on loudspeaker enclosure behaviour, started during his degree, had carried on in collaboration with Gilbert during 1952 and beyond. They would have been in communication throughout the production of *Sound Reproduction 3*, into 1953. Whilst at the BBC, but seemingly not as part of his work there, Cooke was developing some revolutionary ideas about bass reproduction and required chassis diameter, described later, and he claims[10] to have made Gilbert aware of these before he joined Wharfedale. This may well have been the moment when Gilbert realised he had been presented with a golden opportunity. He invited Cooke to come and visit Wharfedale and offered him the job of Technical Manager over lunch. (As noted in

Chapter 12, the context of the initial discussions about bass reproduction may have been Gilbert's new book project—*High Fidelity*—which he was contemplating in mid-1955.)* Gilbert did not want Cooke to be under illusions about his role at Wharfedale and told him, with his characteristic frankness, that whilst he would give him every opportunity to develop his own ideas, he would not get the credit for them because he (Gilbert) was not only the boss but also, in effect, an integral part of the Wharfedale image. Cooke decided that it was better to become a big fish in this small pond, rather than remain a small fish in the huge BBC pond, and accepted the job! He moved to Idle in August 1955.

Foam Surrounds

Cooke soon made his presence felt. It would have been one of his early priorities to get to know Wharfedale's most important suppliers and Gilbert had the following recollection about the result of one of these encounters:[11]

> 'Towards the end of 1955 . . . Mr R.E. Cooke . . . returned one day from London with glowing reports about smooth response from a foam surround cone assembly shown to him by our cone makers. Our own tests confirmed the results, and we found that the foam plastic covers sold for ladies' coat hangers were big enough to be cut into segments, and the density and thickness were at least suitable for experimental purposes.'

Once again the Ilkley shops were raided so that enough material could be procured for experimentation! The results eventually enabled specifications for variations of the foam, to suit the different chassis, to be provided for the manufacturers. However, a major concern was the longevity of the foam, especially under the cycles of temperature and humidity experienced in some of the export markets, and tests to provide the necessary reassurance on this score went on for several months.

A Memorable Year

Earlier in 1955, before Cooke's arrival, there were several additions to the product line. The 10" Bronze chassis was modified with the bakelised cone and cloth suspension treatment to produce the /CSB variant—the most affordable, wide-range driver with this performance then available. A new, very sensitive, tweeter—the Super 3—for use at frequencies above 3 kHz had a small cone with no centring device and an aluminium voice coil, the magnet providing a flux density of 13,000 oersteds (the same as the Super 5). Its response was very similar to a ribbon device. Reaction to the appearance of the R-J cabinet—the asymmetric aperture was not to everyone's taste—led to the RJ-2 in which the front of the cabinet was covered in anodised aluminium mesh. Finally, in response to demand for a system like the three-way corner design, from potential customers who did not have a suitable corner space available, came the FS/3-way—the free-standing version. The rectangular box base unit was 'acoustically treated'

---

* In *The Gramophone* interview[10], mentioned also in the previous chapter, Cooke's recollection was that he joined the BBC in 1955 and Wharfedale in late 1956. These dates, subsequently copied elsewhere, are incorrect. The definitive information, from the BBC staff cessation card archive, gives his appointment date as 17 May 1954 and resignation (leaving) date as 2 August 1955.

Chapter Seven: 1953–1958

to give an acceptable performance from the W15/CS whilst the treble (Super 8/CS) and tweeter (Super 3) units in the horizontal cabinet on top, of course, gave identical performance to the corner unit (7.13).

'Live versus recorded' concert-demonstrations were given in St. George's Hall, Bradford (1 April), The Royal Festival Hall, London (21 May) and Carnegie Hall, New York (9 October). The RFH event was scheduled to coincide with the BSRA Exhibition, memorable because Peter J. Walker of the Acoustical Manufacturing Co. ('Quad') demonstrated his prototype full-range electrostatic speaker there for the first time. It sent shock waves through the audio world. As Gilbert humorously recalled just over three years later:[12]

'Few developments during the last twenty-five years have made a greater impact on the science of sound reproduction than the introduction of push-pull electrostatic speakers some three or four years ago. Nearly all our technical writers soared high into the realms of adulation and praise; one British critic writing in an American journal actually foretold the doom of the moving coil. . . .

I must admit that I was considerably impressed myself, for I remember meeting Stanley Kelly (a maker of ribbon speakers), at the first demonstration of a full range ESL at the Waldorf in 1955, when we solemnly agreed to change into black and meet in due course at the workhouse.'

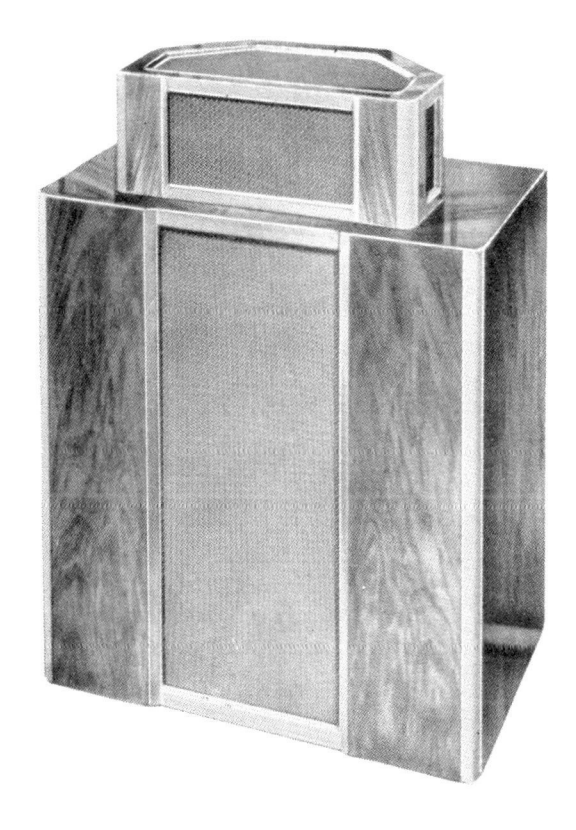

Figure 7.13 FS/3-way, 1955 catalogue.
Fitted W15/CS, Super 8/CS, Super 3 and HS/CR3/2 Crossover.

When Gilbert gave the usual 15 minute communal demonstration of what was new from Wharfedale, later in the exhibition, he opened by playing part of the Chopin Funeral March as a mark of respect for his doomed moving coil speakers! Since Peter Walker was, by now, his trusted collaborator in the concert-demonstrations (see Chapter 11) this development created an interesting situation. Although Gilbert made light of the electrostatic speaker threat in later years it was a real cause for concern at the time and, in his writings, he had to make objective comparisons between the two approaches to transduction—electrostatic and moving coil—whilst putting these fears to one side.

The Carnegie Hall concert was similarly scheduled to precede the start of the New York Audio Fair. Gilbert reprised his first trip to the USA by travelling first class from Southampton to New York on board the *Queen Elizabeth* (this time without assistance from his Swift–Levick friends !) and staying in the Barbizon Plaza hotel. Arnold Hatton, in charge of loudspeaker setting up and switching at countless lecture-demonstrations over the years also went out to perform this vital role, taking the same flight as Peter Walker. This gave him a short taste of life in the USA and it is likely that during this visit the idea that he help to establish a Wharfedale service facility at BIC in Port Washington was first mooted.

Reflex Cabinets and Early Stereo

By April 1956 Raymond Cooke's work on small enclosures bore fruit in the form of two new reflex cabinets. Both incorporated the 'Wharfedale Acoustic Filter' for which a patent application was made—a horizontal partition placed beneath the driver into which a series of parallel 1/16" slits were cut, the number and length depending on the size of cabinet and driver. The function of the filter was to obviate standing waves and reduce panel resonance. The first of these, the Bronze Reflex Cabinet, was designed around the 10" Bronze/CSB speaker; it looked almost identical to the RJ-2 cabinet whilst about 20% larger (28 × 14 × 12.25"). The second was the FS/10 Reflex cabinet, also designed for a 10" speaker—in particular the W10/CSB. This unusual design is shown in (7.14) and the accompanying description from the catalogue dated April 1956 tells the story.

This catalogue was timed to coincide with the first London Audio Fair, held in the Washington hotel during 13–15 April. This first commercial venture, run along the lines of the New York Fair, was a success and attracted about 24,000 attendees. A picture of Raymond Cooke taken during the event is shown in (7.15). The identical systems to his left and right are Bronze Reflex Cabinets with Super 3 tweeter cabinets on top, the tweeter cabinet also being a new introduction. They were being used to make an early stereo demonstration, described by Gilbert in a 'stop press' supplement at the end of *High Fidelity*:[13]

> 'We were demonstrating stereophonic reproduction in our rather small room at the Audio Fair, using two-channel tapes at 7½ in/sec and a Ferrograph machine with stacked heads. It is difficult to obtain true stereophonic effects in a small room, especially when packed with human beings; but my impression was that the two-channel working gave sweeter treble and rounder base than is possible with single-channel working. These tonal qualities are actually more important than receiving a true impression of the position of the artists.'

The April 1956 catalogue went beyond the audio product range to underline the other strengths of the Wharfedale brand. Pictures of the live versus recorded concerts at the Royal Festival Hall in

FS/10 REFLEX CABINET

Size 29½″ × 26½″ × 16¼″
Weight 60 lb. (less unit)

Primarily intended for use with the W10/CSB unit, this elegant cabinet has been designed to give the best possible reproduction from a single wide range speaker in an enclosure of reasonable size. It can, however, be used with any good 10″ loudspeaker having an open baffle resonance below 45 c/s.

The unusual appearance of this model results from the application of fundamental acoustic principles. The broken line of the front face avoids the worst effects of acoustic diffraction, and the inclined angle of the loudspeaker axis directs the high frequencies upwards into the room. The edge of the speaker aperture is radiussed to minimise cavity resonances, and the irregular shape of the sides reduced the effect of panel resonances. The interior is acoustically treated and incorporates the new Wharfedale Acoustic Filter. *** When fitted with a W10/CSB the result is clean and well balanced sound, free from the "chestiness" associated with many wooden enclosures.

The cabinet is strongly constructed from high-grade ¾″ plywood and is rigidly cross braced where necessary. Available in figured walnut veneer (oak or mahogany veneer to order, at the same price).

*** *Patent App. No. 4483/56*

Figure 7.14 FS/10, catalogue April 1956.

Figure 7.15 Raymond Cooke at the London Audio Fair, Washington hotel, April 1956. (Photo courtesy of the Escott family.)

May and Carnegie Hall in October of the previous year were included, the books currently on offer were fully described and the forthcoming publication (in May) of *High Fidelity: The Why and How for Amateurs* trumpeted. Gilbert also updated his list of overseas agents, now totalling 21. The additions since 1953 were Cuba, Irish Republic, Greece, Hong Kong, Italy, Panama, Portugal/Spain and Venezuela, with Bermuda no longer included.

Re-emergence of the Flat Panel Baffle
The new Audio Fair had not supplanted the BSRA Exhibition so in May Gilbert was back at the Waldorf Hotel as usual. Peter Walker of Quad demonstrated his full-range electrostatic speaker again, but this time it was in a commercial form and with an indicative price tag. Once again the comments were very favourable, particularly in respect of its handy size, box-free performance and appearance. In the year following the prediction of the demise of the moving coil driver, Gilbert had often reflected on the similarities and differences of the two approaches to transduction. He was now inspired to believe that similar results must be possible with a flat

panel baffle and moving coil units, despite the fact that baffle mounting had long since been abandoned—mainly because of the performance of the early drive units, in particular their high cone resonance.

At exactly the same time enough data had been gathered on the performance of the foam surround material under adverse conditions to discuss its future. The decision was made to replace cloth suspension on all models with the foam (so /CS became /FS). Fortunately, this provided units with just the required characteristics to reinvestigate baffle mounting and for the next three months Gilbert and Raymond Cooke experimented continuously with baffle materials and driver combinations. What emerged as the commercial proposition in late 1956 was the SFB/3. The 'good old sand-filled panel' gave the best compromise (weight, cost, rigidity, ease of manufacture, reproduction tone) for the baffle construction, whilst a parallel combination of two wide-range units (10" Bronze/FSB plus W10/FSB) and a Super 3 tweeter gave the best performance. Thus 'SFB/3' for Sand Filled Baffle with three drive units. The front and back views of the construction are shown in 7.16 and 7.17 respectively.

Writing later Gilbert made the following observations:[14]

'Apart from convenience in use . . . [i.e. portability] . . . flat baffles are better than is generally believed for the following reasons. They are free from cabinet resonance. Two speakers in parallel on one baffle give a 3 dB gain at low frequency, and double the power handling capacity so that bass lift can safely be used in the amplifier, resulting in the possibility of four times the LF output of a single unit. The floor also improves bass reflection and walls can be harnessed in the same good cause. Baffles are efficient because sound from both sides of the cone is used and I am inclined to the belief that equal air loading front and back is a good thing. Also, by parallel working, an almost level impedance characteristic from 60 to 20000 c/s provides a good match to the amplifier.'

The SFB/3 was originally priced at £37 10 s and, writing in a 1957 brochure not many months after its introduction, Gilbert claimed its success was unprecedented. Perhaps this is why the price was rapidly increased by £2! A leaflet was issued in which Gilbert discussed at length the use of listening tests to find the best position for the baffle in a typical living room—taking advantage of the fact that its portability facilitated this procedure and also allowed it to be stored against a wall when not actually in use. He also gave the layout of his own listening room at Woodville, which had an unusual shape, and described the effects of different placement positions within it. In addition, he made the point that the speaker would do full justice to FM radio transmissions, now available in most parts of the country and having a wider frequency range than AM, in a way which a typical radio set, with its small internal speaker, could not. To facilitate connection to radio sets, in particular, a new auto-transformer, the WMT-1, was produced. It was designed to match any combination of load and output impedances in the range 2–16 ohms; it would handle 15 watts and had a response level within 1 dB from 20 Hz to 15 kHz. Percy Wilson, reviewing the new speaker in *The Gramophone* started his enthusiastic piece thus:[15]

'The SFB/3 system is a reversion to an old idea, but with a difference—and what a difference! I surmise that it was suggested, as an experiment, by the characteristics of Mr P.J. Walker's Electrostatic Doublet, last spring. But whatever the genesis, there is no doubt about the result; it is so successful as to be a revelation. So there you have it from alpha to omega.'

A Pair of Wharfedales: The Story of Gilbert Briggs and his Loudspeakers

Figure 7.16 SFB/3 taken from the first leaflet in late 1956.

Figure 7.17 Rear view of the SFB/3. Reproduced from *Loudspeakers*, 5th edition.

Chapter Seven: 1953–1958

Developments at BIC

Good customer support had always been a key part of the Wharfedale ethos; Gilbert placed great emphasis on the 12 months guarantee attached to all the products and to honouring the pledges. He would often 'go the extra mile' repairing very old speakers and those damaged by their owners—often in the course of DIY installation. As sales to the USA became an increasingly significant fraction of total output, the after-sales service situation there became a more pressing issue. Gilbert was lucky to have on his staff a husband and wife combination, in Arnold and Winifred Hatton, who were willing to go to the States and establish an after-sales presence. As one of Wharfedale's longest serving employees and, over time, a considerable but understated source of practical input into the firm's progress, Arnold was the ideal person to start the operation. The Hattons moved to Port Washington, Long Island, on 16 September 1956—initially on a six-month secondment to BIC. By late 1955, BIC appear, from an advert in *High Fidelity* featuring Garrard, Wharfedale and Leak items (November 1955), to have married together Wharfedale speakers with R-J cabinets, selling models designed for floor and shelf mounting. In March 1956 *High Fidelity* carried a full-page advert solely for Wharfedale, which may have been the first to establish the theme in which the reputation of the product was allied to that of Gilbert himself. The words, accompanying a picture of the great man, included:

> '*Wharfedale loudspeakers . . . respected throughout the world because of the personality and principles which guide their manufacture . . . designed and built under the personal supervision of G.A. Briggs.*'

By 1956 BIC were also agents for River Edge 'customised cabinets' and this may have been indicative of the desire to sell cabinet systems with Wharfedale components but locally sourced and styled enclosures. The subsidiary Wharfedale Audio Products was set up by BIC as the vehicle to achieve this. Looked after by Hatton it sourced cabinets, made to basic Wharfedale designs and for which a 5% royalty was paid, and assembled systems using drive units, crossovers etc. from Idle. The completed systems were then sold to the parent company. Arnold's title later became Service and Production Manager and he branched out into cabinet design for the DIY enthusiasts wanting to use Wharfedale drive units. Presumably this started out as the customisation of basic designs which Wharfedale had made available in the 'Cabinet Construction' leaflets from about the mid-50's; Issue 4 from 1957 had non-resonant reflex cabinet designs for 8", 10", 12" and 15" units with foam surrounds, including designs with the patented acoustic filter, made freely available for DIY use (see page 308).

Hatton had just settled into his new surroundings when Gilbert arrived in New York for the second Carnegie Hall concert of 3 October. For the last time he was able to fulfil his responsibility for checking and connecting up the speakers and switching gear and doing the actual switching during the performance. This was a role he had been performing with Gilbert for at least 20 years and which had taken him from small halls in Bradford to the premier concert venues of London and New York!

All the publicity in the USA over these two years seems to have created the problem of success. A full-page BIC advert in *High Fidelity* in November 1956 posed two questions: 'Why are Wharfedale speakers so hard to get?' and 'Are they worth ordering and even waiting for?' The answers, of course, were a variation on Gilbert's 'quality not quantity'. Early in 1957, in an attempt to overcome the supply shortage, BIC started assembling W10 and W15 speakers from UK components (supplied by or via Wharfedale) in a small factory called the 'Wharfedale Room', supervised by Hatton. Previously BIC had only imported complete units, so this was a further development in the collaboration.

Later in 1957 BIC moved up another gear by producing their own versions of the SFB/3, first advertised in *High Fidelity* in October. The 'Warwick Custom' model was similar in appearance to the UK model but less austere, whilst the 'Windsor Deluxe' version appears to have the baffle contained in an ornate, open-backed cabinet. The two versions sold for $199 and $249 respectively. The following month another full-page advert announced the Super 12/FS/AL and the AF-12 cabinet. There was no UK equivalent of this cabinet (a larger, 'Americanised' version of the AF-10) so at this point the US operation was beginning to go its own way. Sometime later, they started to sell a system which married the speaker and cabinet together, called the W/AF/1 and followed this with a two-way system, the W/AF/2, which added a Super 3 tweeter and balance control.

An insight into the relationship between the two companies, or more accurately between Gilbert and the Carduner brothers, is contained in a rare handwritten draft reply to a letter from BIC, dated 11 February 1957, which ended up in Raymond Cooke's possession. The main subject is the SFB/3 and BIC's desire to sell their own variants on the design. Gilbert writes:

'. . . But first we should see your [SFB/3] baffle. This is so important that I must emphasis the fact that alterations made without my approval could have extremely serious results. Under no conditions should I allow my name to be used on any design which did not satisfy me in every detail. Whatever the cost in dollars to you, to me or to Wharfedale, this principle would apply, and my co-directors* are of the same opinion. It applies to SFB/3 and any other model. I think it is in both your interests and ours that the position should be clearly understood.

Now that you are making Wharfedale speakers the importance of conforming technically—not necessarily furniture-wise—to our designs is far greater than it was when you bought our units and made your own cabinets. The impact which the SFB/3 has had on our home market has been fantastic, especially in appearance, so please don't alter anything before you realise fully the significance of what you are doing. In other words, don't alter the shape of the Johnny Walker whisky bottle, unless you want to stop the sale.'

Gilbert also instructed Edith Isles to send a copy privately to 'A.H.'. Arnold Hatton's unofficial roles at BIC were clearly to be Wharfedale's diplomat and Gilbert's mole.

Anticipating Stereo

1957 was primarily a year of rationalisation as far as the product range was concerned, but also the year in which stereo was addressed for the first time. Loudspeaker production had been gearing up to make the change from cloth to foam surrounds during the latter half of the previous year and this was fully implemented at the beginning of 1957—by October output of speakers with foam surrounds was running at 800 per week. Some drivers were dropped from the range and those remaining are shown in (7.18).

---

* Raymond Cooke, Bill Escott and Albert Smith became directors on 4 January 1957.

DATA SHEET

MODEL	CHASSIS DIA. ins.	BASS RESONANCE c/s	FLUX DENSITY	TOTAL FLUX	PEAK INPUT watts	WEIGHT lbs.	BAFFLE OPENING ins.	DEPTH ins.	FIXING HOLES P.C. DIA. ins.
Super 3	3⅝	—	13,000	54,000	Treble only	2⅞	3	2⅝	3 15/16
8″ Bronze	8	75/85	10,000	39,500	5	2¼	7	3¾	7¾
8″ Bronze/FS/AL	8	45/55	10,000	39,500	4	2¼	7	3¾	7¾
Super 8	8	70/80	13,000	54,000	6	3¼	7	4	7¾
Super 8/FS	8	45/55	13,000	54,000	5	3¼	7	4	7¾
Super 8/FS/AL	8	55/65	13,000	54,000	4	3¼	7	4	7¾
10″ Bronze/FSB	10¼	30/38	10,000	39,500	6	4½	8½	4¼	9⅞
Golden/FSB	10¼	30/38	13,000	54,000	8	5⅜	8½	4¾	9⅞
W10/FSB	10¼	30/38	14,000	74,000	10	9	8½	5¼	9⅞
W12/FS	12½	30/35	13,000	145,000	12	12½	10⅞	6½	12⅜
Su 12/FS/AL	12½	30/38	17,000	190,000	12	18¼	10⅞	7	12⅜
W15/FS	14½	25/30	13,500	180,000	15	17	12¾	7½	14½

AL = Aluminium Speech Coil   FS = Foam Surround   B = Bakelised Apex

Figure 7.18 Drive units available in early 1957. Flux densities in oersteds, total flux in maxwells.

No extension speakers were listed in the 1957 catalogue and it was probably the intension to stop making them because demand was falling. Four models: the Reflex Baffle with 6" chassis; Bijou, Bronze Baffle and Bronzian—all with 8" chassis—were still being offered in 1956. However, the Reflex and Bronze baffles did re-appear a couple of years later so there must have been a residual demand. The SFB/3 replaced all the large systems except the 9 cu ft three-speaker corner unit, which was still regarded as providing the best reproduction available. Raymond Cooke's first two models, the Bronze Reflex and FS/10 Reflex cabinets, were replaced in the April by the AF 10 Reflex Cabinet designed around the new foam surround 10" drivers and the acoustic filter (see below). Later in the year the similarly styled, but smaller R-J 8 Cabinet replaced the R-J and R-J2 models.

Two-channel stereophonic sound reproduction ('stereo') had been creeping up on the audio industry since the development of reel-to-reel tape recorders following the Second World War. The introduction of two-channel recording using staggered, or better stacked, heads from about 1949 eventually led to the marketing of rather expensive, pre-recorded stereo tapes by RCA-Victor in the USA in 1954. Gilbert was fully aware of these innovations and had first used a stereo tape recording, made for him in early 1955 by his colleague Arnold Sugden (in Brighouse), for one of the 'live versus recorded' items in the St. George's Hall concert-demonstration of 1 April (see Chapter 11). He also knew, from his contacts in the recording industry, that there was feverish activity to produce stereo disc recordings. By 1957 the problem was not how to do this so much as establishing which approach to adopt universally, to avoid future incompatibility, and gaining the industry-wide agreement. The record companies had been building up their stereo repertoire on master tapes for at least a couple of years so it was a question of when, rather than if, stereo discs would appear and reduce the cost of stereo reproduction. By addressing the issue for the first time in early 1957 Wharfedale was catering for the existing, limited interest in stereo reproduction from tape but anticipating the arrival of stereo LPs.

Their recommendation for a dual speaker system was for two each of the AF10 Reflex Cabinet with 10" Bronze/FSB and the Super 3 Cabinet with Super 3 tweeter, as shown in (7.19). The total

A Pair of Wharfedales: The Story of Gilbert Briggs and his Loudspeakers

Figure 7.19 Stereo reproduction with two AF10/Super 3 cabinets with 10" Bronze/FSB and Super 3 drivers, photographed in the Listening Room at the Idle factory in 1957. Note the reel-to-reel stereo tapes on the table. Reproduced from a 1957 catalogue.

cost for the dual system of £62–2-4 (roughly £1100 in 2010). A pair of SFB/3s was regarded as 'a more ambitious set-up' at £79 (equivalent to £1400).

Control from a Distance

The letter from Gilbert to BIC quoted above was one of a handful of handwritten missives, kept by Raymond Cooke, which Gilbert sent back to Wharfedale whilst on an extended absence from Yorkshire in early February 1957. He went to London for a week (where 'you learn more in three days than three months in Idle') to spend time with, amongst others, D.E.L. Shorter and his colleagues at the BBC and Ralph West at the Northern Polytechnic, staying at the Rembrandt hotel. He then moved on to the Highcliff hotel in Bournemouth where he was going to concentrate on generating some product leaflets and starting work on a series of magazine articles (see below). It was some days after his arrival there before he could do anything, because he had developed a nasty cold, so his sojourn was longer than anticipated. The letters show that he was receiving information and important letters (by post), from Bill Escott or Raymond Cooke as well as phone calls from Albert Smith, on an almost daily basis. Gilbert's replies, whilst including almost 'picture postcard' details of his health, accommodation and the weather, deal with all the topics raised succinctly and authoritatively and fire off ideas and instructions to all and sundry back in the factory. Any sense that these will not be followed up immediately if so required, sometimes backed-up by recall of past transgressions, leads to underlining and repeats. But there is a 'family' feel to the correspondence which softens the message. He was in complete control.

The Sell-out to Rank

In the summer of 1957 Gilbert was making plans for his, by now, annual autumn trip to the USA for the New York Audio Fair. For some reason, maybe prompted by travel insurance considerations, he began to ponder the financial consequences of sudden death through illness or fatal accident. With most of his assets tied up in Wharfedale, death duties would require a forced sale of the company and be disastrous for both his family and the firm. It seemed unlikely that a take-over bid would be on the cards in the foreseeable future so the only way out of the dilemma would be to seek a buyer. Having come to this conclusion, Gilbert seems not to have wasted any time. Leonard Carduner (representing the American interests) was made a director and, if it had not already happened at the beginning of the year, the directors were given a share allocation in Wharfedale Wireless Works before the end of 1957.

Gilbert's enquiries led to the suggestion that the Rank Organisation might be interested. Rank had grown from small beginnings in the financing of religious films in the 1930s, backed by a powerful flour milling business, to become the dominant film-maker and distributor in the UK by 1949. A financial crisis in that year triggered a diversification of interests, beginning with the acquisition of Bush Radio. The founder, J. Arthur Rank, stepped down as Managing Director in 1952, becoming Chairman, in favour of John H. Davis who continued to consolidate the film production interests and further diversify, including going into partnership with the Haloid Corporation to form Rank Xerox in 1956. At the time of Gilbert's enquiries Rank were looking to move into record production (which led to Rank Records Ltd in 1958) and into television (they became part of the consortium which set up Southern Television—the ITV south franchise—also in 1958). Part of this strategy of expansion into the entertainment field had been the take-over of Thompson, Diamond and Butcher Ltd, who were wholesalers for records and audio equipment, including Wharfedale speakers. Gilbert had friends there who had done well from their sell-out to Rank and they arranged an introduction. Gilbert accordingly made contact with Kenneth Winkles, Assistant MD at Rank, with the result that:[16]

'... suitable terms were easily agreed upon and the transfer was made at the end of September 1958.'

This may be a typical understatement, or it may accurately reflect what happened. It is believed that Gilbert was bought out for £97,000 (roughly £1.6 m in 2010). At the time the turnover was around £250,000 and net profit around £17,000. Dorothy Stevens remembers Gilbert's accountant, Fred Mason, telling her on completion of the deal that 'He gave the Company away!' Raymond Cooke, also there at the time, recalled in 1974:[17]

'Gilbert always put human values above personal gain and glory and the sell-out to Rank in 1959 brought him very little hard cash, but it did allow him to provide modestly for his wife and two daughters, which was all that he required from that particular deal.'

The reference here to 1959 confirms the fact that the formal announcement by the Rank Organisation that it had acquired Wharfedale Wireless Works Ltd was not made until 20 May 1959, nearly eight months after what Gilbert referred to as 'the transfer' (see next chapter). However, John Balls (see below) reckoned that judged solely on the basis of the price/earnings ratio (nearly six) or on the value of the assets, Rank's offer was a fair one.

Not very long after the initial agreement had been made Fred Mason either died of, or became seriously ill with, heart problems and he was replaced, both as accountant and director-secretary, by Gilbert's cousin Luther B. Emsley. Having to find someone suddenly to fill this key position,

Gilbert probably induced him to do the job for a few years in the run-up to retirement, since Emsley was already in his late 50's.

Stereo in Earnest

The consolidation of the previous year, following the burst of innovation in 1956, might have been expected to continue, with a change of ownership going on in the background, but this was far from the case. The record industry had resolved the issues about recording stereo discs by universally adopting the approach first described by Alan Blumlein, working for EMI, as far back as 1931. This involved cutting the two channels into the same groove at 90° to each other, and at 45° to the record surface (the 45/45 solution). These discs finally hit the market in 1958. Development work at Wharfedale, aimed at new products for the stereo market, was intense.

Towards the end of the year two products appeared which showed that the issues raised by stereo were being tackled from several standpoints. The first was the Omni-directional Column Eight loudspeaker. One of Gilbert's local hi-fi colleagues was the inventive engineer Arnold Sugden, of A.R. Sugden and Co. in Brighouse—maker of 'Connoisseur' turntables. Sugden had developed stereo tape recorders and also stereo disc cutting and playback systems, amongst other things, and he was a proponent of column loudspeakers. Gilbert had witnessed a number of his successful demonstrations of such designs and thought their time might now have come. They are, of course, omni-directional and Gilbert's stance on this aspect of stereo reproduction was this:[18]

'There are two schools of thought [on directional effects] *here. Some technicians favour directional loudspeakers to give maximum stereophonic effects for listeners placed at the apex of a triangle. Other experimenters prefer omni-directional speakers which enlarge the available listening area. We consider that the advantages of the wider area far outweigh the arguments for a directional set-up.'*

This said, he went on to say of column speakers:[18]

'This type seems to give the best possible sound in relation to cost and amount of floor space taken up, but the main advantage is probably the fact that the middle and upper registers are distributed around the room at a height which clears obstructions from tables, chairs and the listeners themselves. Having decided to use a column speaker . . . the next step is to get away from the column as much as possible in both appearance and performance. (Column resonance suits the organ but nothing else.)'

The original design of the Column Eight is shown in (7.20), but the horizontal slots of the top portion were soon replaced by a mesh. In this part of the cabinet—which was not supplied with a drive unit—was a metal diffuser which gave 360° high frequency output in the horizontal plane, with little effect on vertical output. The acoustic filter was also incorporated. A special driver, the 8/145 (a modified Super 8/FS/AL), was developed to suit the cabinet and recommended for use therein. The column was 44" tall with a cross-section of 14 × 12". The basic details of the design had actually come from The Netherlands; they we sent to Gilbert by a Mr H. Bach of Trianon

Chapter Seven: 1953–1958

Figure 7.20 Column Eight Cabinet, from the first sales leaflet.

Electric Ltd, distributors of the Ronette stereo pickup, the first of its type used by Wharfedale for demonstration purposes.

The second approach was to significantly reduce the weight of a cabinet so that it could be shelf mounted. The PST/8 cabinet was 24 × 12 × 12" with large sections of the side panels being made from expanded polystyrene, which had recently become available in sheet form. As well as reducing the weight, panel resonance was much reduced by the acoustically dead expanded polystyrene. The cabinet was intended to be mounted horizontally, with the loudspeaker facing upwards to reduce directionality, if required. Any good 8" driver was suitable. This product proved to be very popular in use with tape recorders.

The Word from Paris

Towards the end of January 1958 Gilbert was in Paris for a combined business and pleasure trip (he took Valerie with him). As usual he was in touch with the Works by handwritten letter and one gem, dated 25 January, was kept by Raymond Cooke. It includes this passage:

'Film et Radio. I spent yesterday afternoon there. They tell me the Ionophone is now withdrawn as they all broke down after large numbers had been sold. They say Mr Klein is tickled to death at having sold the rights for cash in England and America. I am glad I turned it down when Plessey offered it to us. The ELS [electrostatic loudspeaker] could easily suffer a similar fate but I am not saying it will. Let caution be your watchword.'

and finishes as follows:

'Test Cards. Mr Carduner came out with an idea which I like very much. This is to take a snapshot of a waveform of each speaker at a low frequency—say 25c/s for a W15—and attach it to each unit as a guarantee of LF performance. This idea must be tried out at once. There is a small scope which I bought from Sugden a few years ago. If this works get it rigged up in Mr Broadley's test room and we will see what happens. If it does not work get the Cossor across as I require these tests in Mr Broadley's room and not in the Lab. A cheap crystal mike is suitable. If Mr Cooke is busy for his London visit get Stewart on the job. The last time I suggested this routine nothing happened! Quoth the raven, never more.
We will then report at once to Mr Carduner on the feasibility of his quite brilliant idea.
Bright sunny day here.
G.A.B.'

Literary Collaboration

October 1958 saw the publication of a new and very much enlarged 5th Edition of *Loudspeakers*. The previous editions, with only minor alterations, had gone through a total of 16 impressions in 10 years and sales were still high, but so much had changed that Gilbert decided that a new book should be attempted. During 1957 he and Raymond Cooke had collaborated on a series of eight articles published in the American journal *Radio and TV News* and parts of these were reworked to form the basis of about a quarter of the chapters in the book. As described in more detail in Chapter 12, Gilbert also invited others to contribute specialist chapters, and he included two chapters of reminiscences about the early years of Wharfedale and the book publishing activities. Of particular interest here is what he had to say in his 'Looking Back' chapter about 'technical knowledge'—one of his 'four essentials for success as a manufacturer in a new venture':[19]

'Technical knowledge is a great asset, but is the least essential because it is usually possible to acquire the services of a good man. If you engage a genius it is most important to start by limiting his activities to a narrow circle, because he will disrupt your production by brilliant ideas and land you in Queer Street if you give him the run of the place'

One wonders what Raymond Cooke would have made of this! As Technical Editor of, and significant contributor to, the book he was intimately involved with all the material except the two 'non-technical chapters' which were sub-edited by F. Keir Dawson. So he may only have read this on publication. Looking back himself, he claimed[10] that Gilbert gave him a free hand, whilst retaining a keen interest in developments, but the evidence suggests this to have been the case only after he had been at Wharfedale for at least a couple of years. However, since he was the first technical

Chapter Seven: 1953–1958

specialist actually on the staff of the company, as opposed to the largely unsung 'consultant' Ernest Price, only he could be the 'genius' Gilbert had in mind.

As with *Sound Reproduction 3*, a great deal of work went into generating new data for inclusion in *Loudspeakers 5*—most of it produced by Cooke and his assistant, Stewart Horne. Gilbert was at pains to acknowledge this effort at the end of the book. A picture of Cooke in the Wharfedale laboratory taken at the time is shown in (7.21), whilst a slightly earlier picture of the laboratory, of larger view, is given in (7.22).

Another key employee included in a picture taken for the book was Ezra Broadley, the Works Manager. Figure (7.23) shows him operating the equipment for magnetising loudspeaker permanent magnets. The saturating magnetic field was produced by a direct current energised coil, the d.c. itself coming from a motor-generator running off the a.c. mains supply. When Wharfedale first established itself on the Idle site there was a mains d.c supply which could be used directly and this also powered the electric trolley buses which ran on Bradford Road alongside the Works (known locally as the 'Trackless' to distinguish the system from the trams running on tracks); the supply wires can actually be seen, strung above the road, in (7.1). Ezra liked to tell the story that whenever the switch was thrown to power the electromagnet, any trolley bus which happened to be passing was brought to an abrupt halt!

Figure 7.21 Raymond E Cooke in the Wharfedale Laboratory, Idle, March 1958. SFB/3 speaker in the foreground. On the left is a Brüel & Kjar instrument for recording response curves. (Copyright—Bradford Museums and Galleries. From the C. H. Wood photographic archive negative number W32266, photo courtesy of the Escott family.)

A Pair of Wharfedales: The Story of Gilbert Briggs and his Loudspeakers

Figure 7.22 Wharfedale Laboratory, Idle, early 1957. (Photo courtesy of the Escott family.)

Figure 7.23 Ezra Broadley operating the magnetising equipment at Idle, 1958. Reproduced from *Loudspeakers* 5th Edition.

Chapter Seven: 1953–1958

Catching the Big Boys

The end of 1958 witnessed a small but significant change in Wharfedale's advertising in *Wireless World*. Commenting on the situation in the early days, when the adverts were rarely larger than quarter-page, Edna writes:[20]

'I used to be nearly ill when rival firms extolled their goods in adverts four times as big as ours.'

Even after the War, when advertising in the monthly issues was a fixture, this was still the case. From the mid-1950s half-page adverts for the audio products became the norm, with additional small adverts for the books, but in October 1958 the leap to regular full-page adverts for the audio products was made. Finally, after 25 years, Wharfedale commanded as much advertising space as its larger, long-time competitors such as Goodmans and Rola–Celestion.

Part 2. Away from Wharfedale

Births and Deaths

As far as Gilbert's close family are concerned this was a period in which one generation was replaced by another. A few years earlier his aunt Jinny, who had been part of his extended family at 5 Clayton Lane for a time, went to live with his sisters Claris and Mabel. In between she had achieved a late marriage, aged 60, and been widowed. Visitors to the three ladies at Hollybank Grove were greeted by a pet budgerigar which cried 'Wharfedale loudspeakers are the best!' Aunt Jinny died in 1952 to be followed by Claris in 1954 and Mabel in 1957. Gilbert had spent a lot of time with his sisters during the War and worked with Claris for a decade so their battles with cancer must have been very distressing—and an additional strain during a particularly busy period in his Wharfedale life. During the same period Ninetta had three children, girls in 1952 and 1956 and a boy in 1954.

Grandchildren became an important part of Gilbert's life (eventually there were seven) and although the trip to Staffordshire to see the Theobalds was not an easy one he went as often as possible. Whilst their first three children were still very young, Leslie and Ninetta were living in a property which was literally being rebuilt around them. One of Gilbert's trips coincided with a building inspector's visit to inspect a new sceptic tank and associated drainage. He caused consternation by advising the inspector, apparently in all seriousness, not to trust his daughter and son-in-law and told him that the pipes to the tank probably ran uphill!

This is a typical example of the wicked side of Gilbert's sense of humour. He would provoke a reaction by making an unexpected comment or statement and the better he knew someone, or the more fond he was of them, the more outrageous this might be. He sometimes extended this approach into harmless practical jokes. His daughters' favourite recollection is of his way of deflating a golfing partner who could be irritatingly pompous. Gilbert spent some time fashioning a fake golf-ball from a bar of soap. He somehow managed to switch his partner's real ball just before he teed off and had the satisfaction of witnessing the man's shocked disbelief when the 'ball' disintegrated on impact.

The Concerts Phenomenon

Between March 1954 and the end of 1958 Gilbert was involved almost continuously in putting on concert-demonstrations which featured, to an ever greater and more sophisticated extent, 'live versus recorded' items. There were at least a dozen such events during this period (described in detail in Chapter 11) which were influential way beyond the halls in which they took place, and on both sides of the Atlantic. They led to a great deal of discussion, almost philosophical, about the way in which recordings were made—and the choices available to the engineers—and the relationship between this, the nature of the space in which the sound was reproduced, and the listeners experience. Gilbert wrote several articles for audio journals about various aspects of the demonstrations and what he learned from them himself, as well as responding, in press, to letters and comments which resulted from the articles and the events themselves.

The effect on one individual of attending the first Royal Festival Hall concert in 1954 was quite dramatic. R.V. de Carvalho (7.24) was a young man in his early 20's who was in charge of recordings at his uncle's firm of Valentim de Carvalho Ltd in Lisbon. As part of his training he had spent time at the EMI studios in London and on a return visit, to see friends he had made during his stay, he was taken to Gilbert's concert-demonstration. He was astonished at the results and decided that his firm must get into the high fidelity business. He succeeded in becoming agent for Wharfedale speakers, Acoustical (Quad) amplifiers and Ortofon pickups. Once the business was fully established he decided the best publicity must be a demonstration à la G.A.B. in a fashionable Lisbon theatre. He wrote to Gilbert to tell him of his plan and was pleased and surprised to receive a reply in which Gilbert offered to help, attend and even address the audience. The event took place on November 17 1956 using four three-way corner enclosures, two of the normal sand-filled panel type but the other two were brick enclosures and in total they weighed about a ton. Since the theatre was being used as a cinema at the time, they had to be installed after midnight for rehearsals and removed afterwards, every day for a week. Gilbert was impressed. The demonstration included a live

Figure 7.24 R.V. de Carvalho. Reproduced from *Audio Biographies.*

versus recorded item with solo piano and was a great success. A second show was put on in Porto a year later.[21] Carvalho's also became agents for Spain and today they are a large organisation.

Other agents around the world realised the marketing power of these events, despite the effort required to stage them, and they were subsequently copied elsewhere. For example, Wilfred Proctor, the South African agent, attended rehearsals for the third RFH concert in May 1956 (and was photographed in the process, see Fig 11.20) to learn 'from the master' before staging an event in the Cape Town City Hall that December. He recalled, ruefully, how the acoustics of that venue were seriously affected by the presence of the audience. Since the recordings were made in the empty hall, the 'live versus recorded' items turned out to be rather disappointing.[22]

In Demand

As 1956 drew to a close Gilbert received an invitation from the editor of the American magazine *Radio* and *TV News* to write a series of articles under the general title 'All about Audio and Hi-fi'. This turned out to include nine parts, written by Gilbert but with significant contributions from Raymond Cooke, published almost contiguously in monthly instalments between May 1957 and March 1958 (for details see Appendix 3). The style adopted was even more conversational than in the books with Gilbert disclosing in his preamble, for instance, that his early visits to the USA had quickly converted him to the use of teabags! An Editor's Note to the first instalment started as follows:[23]

'We take great pleasure in welcoming to these pages one whose wide experience and knowledge truly entitle him to be called a "noted authority" in the hi-fi field. Mr G.A. Briggs' pre-eminence stems not from a theoretical, ivory-tower approach to the subject, but rather it is the result of endless experimentation, a well-developed sense of inquiry and a serious (although good-humoured) interest in good audio reproduction. Besides all this Mr Briggs has the peculiar ability to tell of his experiences in a crystal-clear, personal, down-to-earth manner which is a pleasure to read.'

Around the beginning of 1958 Gilbert's standing in the high fidelity world was further recognised in a rather different way. By now G.A. Briggs the author was known world-wide not least since *Loudspeakers* and *Sound Reproduction* alone, in their various editions to date, had notched up sales in excess of 90,000 copies. A researcher by the name of K.J. Spencer had put together a bibliography, drawn from the international literature, on all aspects of sound reproduction from the subject's beginnings up to June 1957. He invited Gilbert to write the Foreword, which he typically began with a quotation from Samuel Johnson: 'Knowledge is of two kinds: we know a subject ourselves, or we know where we can find information upon it.' As the literature was now rather extensive (the bibliography contained 2600 references) Gilbert was in no doubt that here indeed was a valuable contribution to the second kind. The book, published in 1958 by Iota Services Ltd, was entitled *High Fidelity: a bibliography of sound reproduction*.

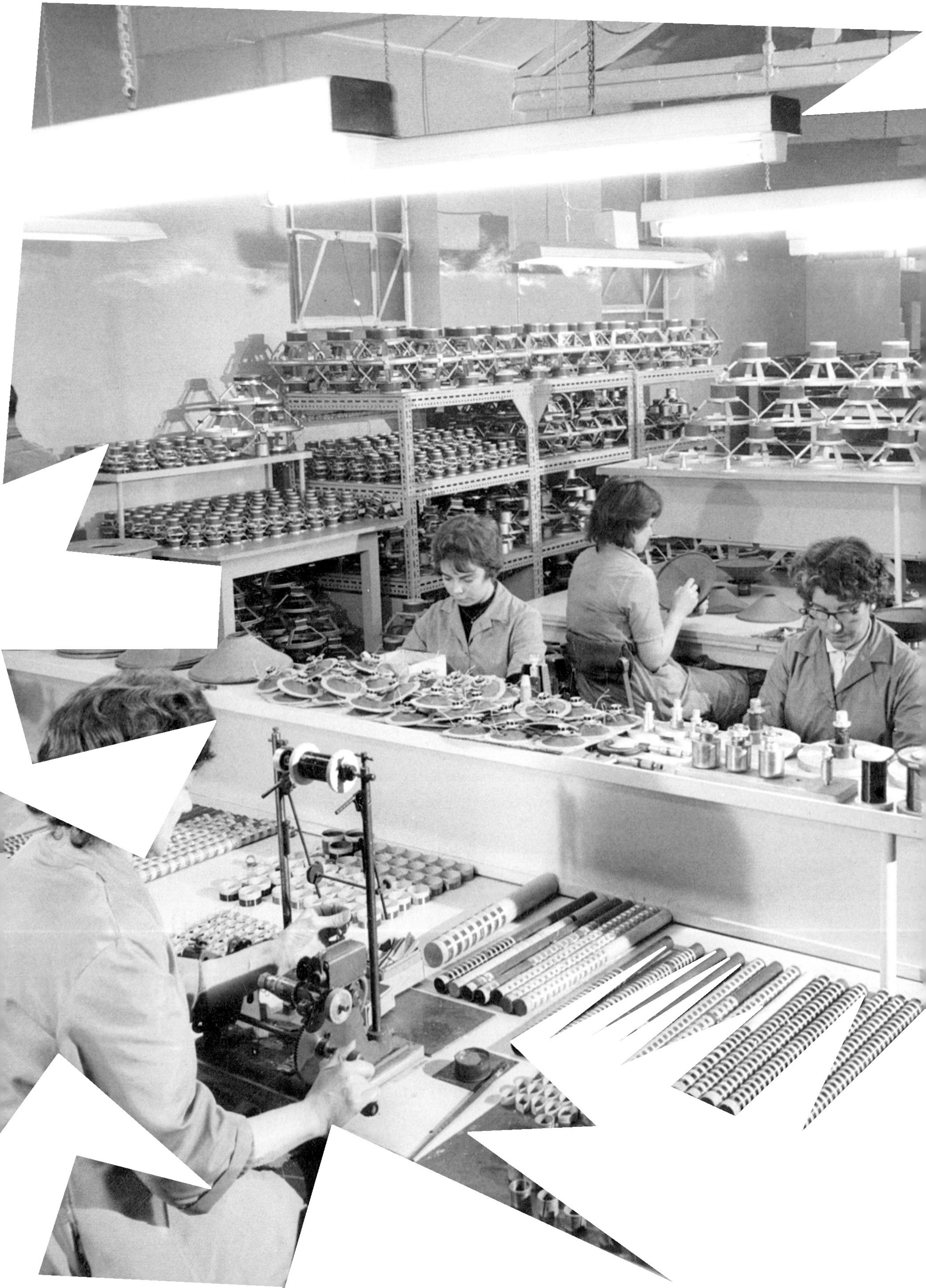

CHAPTER EIGHT: 1959–1965

Wharfedale Wireless Works Ltd (within Rank Organisation Ltd), Idle

In late January 1959 Gilbert was back in the USA, the main reason being 'The International Hi-Fi Festival' in Washington D.C., held during the first week of February. This was the largest and, in terms of participating companies, the most international Audio Fair so far organised in the USA (and by this time there had been about 40 across the country visited by some half-a-million people). It was billed as the venue where the enthusiast could question the President or CEO of a company directly on its stand. The buzz generated by the arrival of stereo discs was considerable and even led a member of the House of Representatives, Philip J. Philbin of Massachusetts, to exhort his House colleagues to attend in order to share the exciting experience of hearing stereophonic sound for the first time! [1]

Wither BIC-Wharfedale?

Gilbert had started out in New York where the future direction of BIC's Wharfedale Audio Products Division was under serious discussion. The occasion gave rise to a picture, which Gilbert subsequently had framed for hanging in his office, of himself with Leonard Carduner (8.1). The North American market had performed poorly in 1958, despite the move to stereo, with retailers reporting a large drop in sales compared with the previous year. Against this backdrop, Gilbert still had nagging doubts about BIC's ambitions for their 'Wharfedale' products and he did not really like the arrangement, started two years earlier, whereby they assembled certain drive units from components shipped from the UK. In addition there was a clear difference in the US customers' perception of 'good sound output' from cabinet systems, compared with the UK, which essentially amounted to a requirement for accentuated bass. Gilbert had wanted to meet this need with systems incorporating W15s, which needed large cabinets, but this was counter to the trend for 'bookshelf' speakers as exemplified by Acoustic Research's AR-1 and AR-2 and the early KLH models. There was much to discuss.

It should be noted here that Edgar M. Villchur had come up with his 'acoustic suspension' idea, which delivered exceptional bass from a very small enclosure, in 1952 (US patent 2775309 in 1953). His company, Acoustic Research, was incorporated in 1954 to produce the AR-1 speaker because no established company wanted to take it up. A 12" driver with a very low resonance (10–15 Hz) in a small (1.7 cu ft) virtually airtight enclosure, filled with absorbent material, utilised the air in the enclosure to provide a linear restoring force to the diaphragm. The result was hitherto unattainably low distortion at low frequencies coupled with the major reduction in enclosure size. Acoustic Research Inc. also received a UK patent and over time the AR models achieved a very high market share in the USA. The 'bookshelf' cabinets were particularly apposite when stereo arrived. Gilbert was dramatically introduced to the AR-1 by Villchur himself at the New York Audio Fair in 1954, when Edgar challenged him to a 'dual' between his 'baby' and the Wharfedale 9 cu ft corner unit.[2] Gilbert's only consolation—he wondered why on earth he had not thought of the idea—was that the AR-1 speaker efficiency was very low in comparison with his flagship and required about ten times the input power. He also disliked the inevitable 'boxiness', which he strove throughout his audio life to avoid, and the directional properties of a small area source.[3] However, he was always fulsome in his praise for this invention, and several other pioneering developments by Edgar Villchur—ranking him alongside Paul Voigt. Villchur may have learned from Gilbert, however. In the early 1960s Acoustic Research gave around 75 live versus recorded demonstrations of their speakers, across the States, in which the Fine Arts Quartet gave the live performances. Henry Kloss, who co-founded Acoustic Research with Villchur, formed his own company (KLH) with two former AR investors in 1957.

At BIC two key decisions were quickly made. The first was to end drive unit assembly in the USA and concentrate on assembling cabinet systems. Gilbert felt the expanded capacity at Idle should now handle not only this requirement but also the speakers for an expanded range of models, which BIC had ambitions to market. As an incentive he offered an extra discount of 5% on all drive units in excess of sales worth £35,000 and this was accepted. The second was to develop a series of products which could seriously take on the US competition. Gilbert was finally convinced by the Carduners, with private confirmation from Arnold Hatton, that BIC would not do anything without his approval. The starting point was an intensive series of listening tests over two days in which Gilbert and Leonard Carduner ('who is not a bad judge of sound') assessed the AR1 and 2, the KLH 4 and 6, a new Electrovoice model and a prototype W3. (This model, see below, was almost ready for launch in the UK but the new 12" unit in it had not survived shipping unscathed; the cone had sunk slightly and it rattled. Gilbert was alarmed and wrote home on 22 January to suspend all advance publicity work, in case there was a systemic unreliability problem.) They also tried the new Coaxial 12 driver (see below) in a modified W/AF/1 cabinet.

In this letter home Gilbert described the AR2 as 'shocking' and had this to say about one of the KLH models:

'The KLH which is knocking Villchur [Acoustic Research] *has a 10" unit without die-cast chassis and the resonance with open back is 28 c/s. It bumps out the bass but the quality is poor, but by gum it sells!'*

It appears that a small two-way system, with 8" bass and 4" tweeter, was settled on as the immediate response to the bookshelf competition (it became the WS/2, see below). Gilbert doubted whether the W3 had enough bass to satisfy the American customer and experiments on alternative three-way approaches were started. Out of this intensive interaction came ideas for Raymond Cooke to follow up in detail once Gilbert returned home.

Whilst in New York, Gilbert had his customary get-together with his friend Milton D. Kramer and presented him with a copy of the recently published *Loudspeakers 5*; this too was recorded (8.2).

Chapter Eight: 1959–1965

Figure 8.1 Gilbert Briggs, holding a Super 3 tweeter, with Leonard Carduner at BIC, New York, January 1959. (Photo courtesy of the Escott family.)

Figure 8.2 Gilbert Briggs with Milton D. Kramer, New York, January 1959. (Photo courtesy of the Briggs family.)

173

Rank Take-over—Official

In May, the announcement that the Rank Organisation had acquired Wharfedale Wireless Works was headline news in Bradford, as Gilbert indicated on a copy of the Rank press release which he sent to his daughter Valerie (8.3). What lay behind the delay between Gilbert's 'transfer' and the formal announcement is not known, but it is probable that he had signed a 'Heads of Agreement' in September 1958 and the working through of all the details went on until May 1959. One of these was the setting up of a pension scheme—something Gilbert had wanted for his employees but which Wharfedale Wireless Works Ltd did not have the resources to fund. He took the opportunity to persuade Rank to establish a Wharfedale fund, which was replaced in a couple of years by the general Rank Organisation scheme. John Davis and Kenneth Winkles, MD and Assistant MD of the Rank Organisation, respectively, joined the Wharfedale Board at this point and Edna's anomalous directorship finally came to an end. Gilbert had indicated his intention to retire fairly soon and in June Rank appointed John Balls, an engineer working for ICI in Sussex, as MD-designate to understudy him. For contractual reasons Balls could not move to Idle until November, but Gilbert kept him abreast of what was happening there. When he did move to Yorkshire, Gilbert almost immediately sent him off on a 'world tour' to get to know the overseas agents.

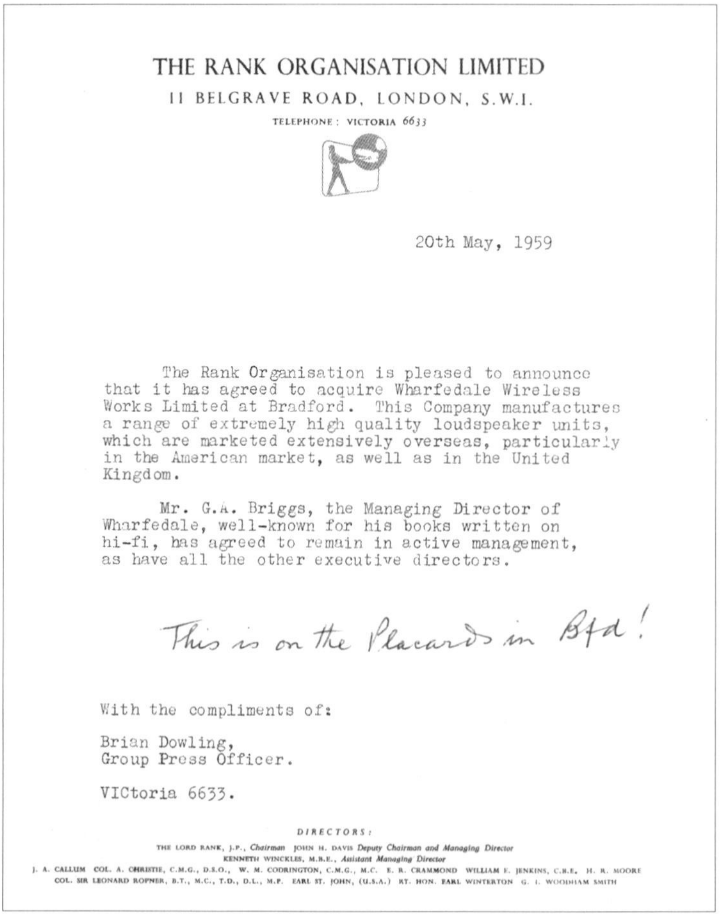
THE RANK ORGANISATION LIMITED
11 BELGRAVE ROAD, LONDON, S.W.1.
TELEPHONE: VICTORIA 6633

20th May, 1959

The Rank Organisation is pleased to announce that it has agreed to acquire Wharfedale Wireless Works Limited at Bradford. This Company manufactures a range of extremely high quality loudspeaker units, which are marketed extensively overseas, particularly in the American market, as well as in the United Kingdom.

Mr. G.A. Briggs, the Managing Director of Wharfedale, well-known for his books written on hi-fi, has agreed to remain in active management, as have all the other executive directors.

This is on the Placards in Bfd!

With the compliments of:

Brian Dowling,
Group Press Officer.

VICtoria 6633.

DIRECTORS:
THE LORD RANK, J.P., *Chairman* JOHN H. DAVIS *Deputy Chairman and Managing Director*
KENNETH WINCKLES, M.B.E., *Assistant Managing Director*
J. A. CALLUM COL. A. CHRISTIE, C.M.G., D.S.O., W. M. CODRINGTON, C.M.G., M.C. E. R. CRAMMOND WILLIAM F. JENKINS, C.B.E. H. R. MOORE
COL. SIR LEONARD ROPNER, B.T., M.C., T.D., D.L., M.P. EARL ST. JOHN, (U.S.A.) RT. HON. EARL WINTERTON G. I. WOODHAM SMITH

Figure 8.3 Rank press release. (Scan courtesy of the Briggs family.)

For several years an outside observer would have been hard pressed to detect any difference overall. The company name remained unaltered and there was no mention of the parent company in brochures or advertising. Rank had learned the hard way, from previous take-overs, that an immediate, strong association of their name with a well-known brand led to a drop in sales, so they kept a very low profile. Eventually, the Rank logo (as on the letterhead in (8.3)) was added to the side of company vans and to the Wharfedale letterhead, but this only happened at the time of Gilbert's 're-tirement' (January 1964). Gilbert reported to Kenneth Winkles, with whom major policy decisions had to be agreed, and he attended certain Rank Board meetings chaired by John Davis. Wharfedale was well-run and reasonably profitable, but a relatively small enterprise within the Rank Organisation so, at least initially, Gilbert was left to run the business more-or-less as before.

Not long after the acquisition, Rank summoned the Wharfedale 'Personnel Officer' to the London Head Office for a meeting of all such representatives from within the Organisation. This posed a slight problem since there was no such person. Dorothy Stevens had already got 16 years service under her belt and, having been there since the intimate days of the War when there were only about 20 employees (by 1959 the number was around 60) there wasn't much she didn't know about the operation. Gilbert asked her to go into the lions' den. Her novel presentation went down well and the head of Personnel expressed an interest in visiting the factory; Dorothy said he would be most welcome and she would arrange it with Gilbert. When he came, she was dispatched to meet him at Leeds railway station—driving Bill Escott's Jaguar, which doubled as the company car for such purposes. It had a habit of stalling when idling and could be difficult to restart, so she was instructed to keep the engine running at all costs. As she approached the station she was brought to a halt at traffic lights. She was so scared that the engine would die that she revved it like a racing car and never forgot the various indications of admiration she received from other (male) motorists! The Personnel chief's verdict was that Wharfedale was a very happy place to work.

Driven by Stereo

The intensive development work during 1958, driven by the arrival of stereo discs and the ensuing hype about the glories of stereo sound reproduction, led to a raft of new products from the end of that year; the Column 8 and PST/8 cabinets have already been described. The thinking behind these was two-fold: firstly, how best to cater for those who already possessed good quality single-channel equipment and who would be unwilling to start again from scratch and secondly, how to interest those buying into hi fi for the first time and wanting stereo. In both cases there were considerations of budget, performance and footprint (occupied floor space), but in the former the added issue of how best to match existing speakers. As always with any significant new development, Gilbert's 1959 sales leaflets for the new products described the logic behind the designs, allowing the potential purchaser to choose their optimum solution.

The spring 1959 offerings included three cabinet systems, the W2, W3 and W4 (all based around a new 12" bass chassis, the WLS/12) and the AF12 reflex cabinet designed for the 12" units with foam surround but especially for a completely new twin-cone chassis, the Coaxial 12. The three cabinet systems were arguably the first products with a modern feel and the first thought-out range or 'family' of loudspeakers in which performance for a chosen retail price was optimised whilst striving to minimise the size, particularly the occupied floor space. They were, as the names might imply, respectively, two-, three- and four-way systems. Raymond Cooke, convinced well before he joined Wharfedale that a 15" speaker in a large enclosure was not essential for really good base reproduction (the received wisdom at that time) provided some loss of efficiency was accepted, had the incentive to prove his theory. He was, however, assisted by the fact that speaker efficiency was becoming less of an issue; as amplifiers improved, speakers did not have to do their best to preserve every precious watt of output so efficiency was becoming a tradable commodity.

The common base unit in all three new Wharfedale systems was the WLS/12, a 12" chassis with a heavy cone, high compliance suspension (resin-impregnated, moulded fabric treated with synthetic rubber to make it airtight) to give improved linearity and an expanded polystyrene diaphragm, known as a 'bung', which fitted inside the cone to provide extra stiffening and a degree of acoustic filtering. Although not described as such, the suspension was effectively 'roll surround', a precursor of things to come. The result was a bass resonance in all three cabinets of 30 Hz, undistorted at 4 watts output. The systems are shown in (8.4).

The W2 had a modified Super 5 treble unit (crossover at 1 kHz) and, with dimensions of 23.5 × 12 × 14", a volume of 2.3 cu ft. The W3 had a special 5" treble unit and a Super 3 tweeter

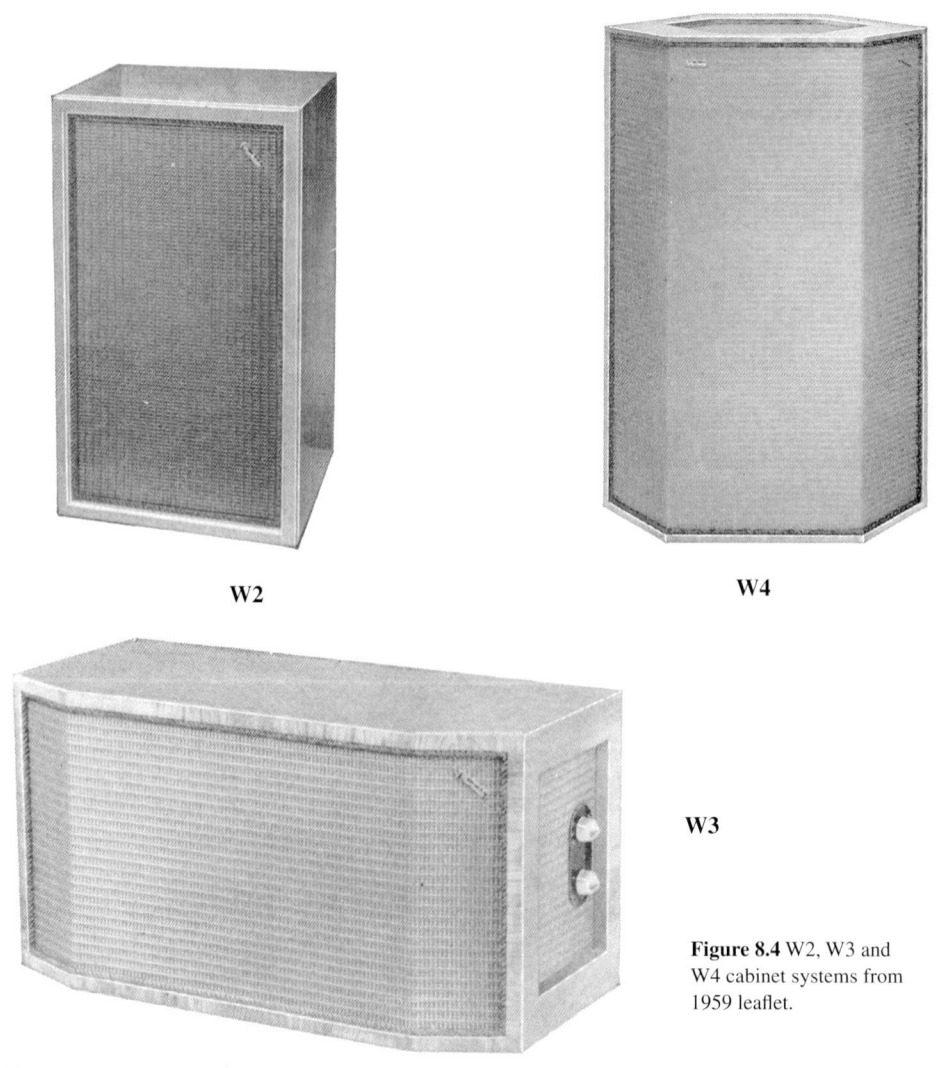

W2

W4

W3

Figure 8.4 W2, W3 and W4 cabinet systems from 1959 leaflet.

(crossovers at 1 kHz and 5 kHz) and a volume of 2.7 cu ft (28 × 14 × 12"). The hexagonal W4 had two 5" treble units and a Super 3 tweeter (crossovers at 400 Hz and 5 kHz) and a volume of 5.8 cu ft (35 × 24 × 12"). All had balance controls. The veneered systems carried price tags of just under £30, £40 and £50 respectively. The WLS/12 and the special 5" driver were specifically designed for these cabinet systems and not sold separately, although the Super 5 was reintroduced into the range.

Commenting on speakers for stereo reproduction Gilbert wrote:[4]

'Whilst many will prefer to use similar models in each channel, specially matched pairs are not required because production tolerances are closely controlled, and any slight variations in sensitivity and response will in any case be much less than the unbalance due to room acoustics, records and variations in output from each side of the pickup/amplifier channels on stereo. (Such variations in output level are far greater than those associated with loudspeakers.)

Where a speaker is required to work in conjunction with an existing model it is advisable to ensure that the directional properties are similar at high frequencies. Thus the W2 should be used with a forward facing speaker, and the W3 is an ideal partner for the SFB/3. The W4 is a natural choice for use with the 9 cu ft corner 3-speaker system for optimum results. Many compromises are possible by suitable orientation of speakers and use of balance control.'

One of the very few pictures which have come to light showing a cabinet system being assembled features the W3. This is reproduced in (8.5)

Figure 8.5 W3 assembly. (Photo courtesy of the Escott family.)

The Coaxial 12

The Coaxial 12 was a novel design for a twin-cone driver which utilised a patented magnet design with two concentric gaps in the same plane. The 12" low frequency unit had a heavy cone, with foam surround, and was attached to a copper voice coil moving in a gap of 1¾" diameter. The high frequency unit had a steep-sided 2" paper cone, also with foam surround, attached to an aluminium voice coil moving in a gap of 1" diameter, and fitted with an aluminium dome. The tweeter chassis was specially shaped to avoid cancellation effects, which would otherwise have produced a response dip in the 7–8 kHz range. Overall a level response from 40 Hz to 5000 Hz was achieved with a loss of only 10 dB over the range 30–20,000 Hz, coupled with very high efficiency. The AF12 reflex cabinet, sold separately, was a larger version of the AF10 (shown in 7.19) and the Coaxial 12 was the recommended driver, although any of the foam surround 12" drivers could be used. It is interesting that the AF12 cabinet was only introduced at this point, the US model having been introduced two years earlier to accommodate the Super 12/FS/AL. In May 1959 BIC-Wharfedale made a big splash about their new two-way model, the WS/2.

Speakers for Schools Revisited

It was also during 1959 that the range of speakers for school use was augmented and Raymond Cooke was probably mainly responsible. Speakers designed especially for school use—which aimed to mitigate some of the worse effects of acoustically problematic spaces, such as large halls or classrooms, and provide good reproduction of speech in particular—had been a steady source of sales since their introduction in 1949/50. The various models for table and wall mounting or portable use had remained essentially unchanged for a decade. Around 5000 WM8 units, for example, had been sold to Educational Authorities during the period. An interesting footnote relates to the uncharacteristically named 'Essex Grey' portable model of 1951 (8.6). Gilbert must have spent some time liaising with the Essex Educational Authority on the design of this model, presumably for a large order, for it to be so christened. The Technical Officer of the Education Committee, at the time, was Major H.H. Garner—Gilbert's co-author of *Amplifiers* published in 1952.

Raymond Cooke's arrival at Wharfedale in 1955 was an opportunity for Gilbert to either offload some of his lecturing to music/record groups and societies or, if he already cut down

ESSEX GREY

A portable cabinet speaker fitted with 10" Bronze unit and carrying handle. Finished in grey cellulose.

Jack socket, "Truqual" volume control and transformer can be fitted if required.

Size: 18" x 16" x 6".
Weight: 12 lb. complete.
Impedance: 2/3 or 12/15 ohms.
Power handling capacity: 8 watts.

Figure 8.6 Essex Grey portable school speaker, 1952 catalogue.

LINE SOURCE SPEAKER TYPE LS/7

INSTALLATION

The LS/7 should be mounted vertically, thus giving the widest horizontal coverage and a space of at least 3" should be left clear at the rear of the cabinet.

Size: 60" x 9½" x 5½".
Weight: 37 lb. complete.
Impedance: 15 ohms and 70 v line.
   Other impedances available to order.
Power handling capacity: 15 watts.

The LS/7 is fitted with seven special 8" foam surround units.

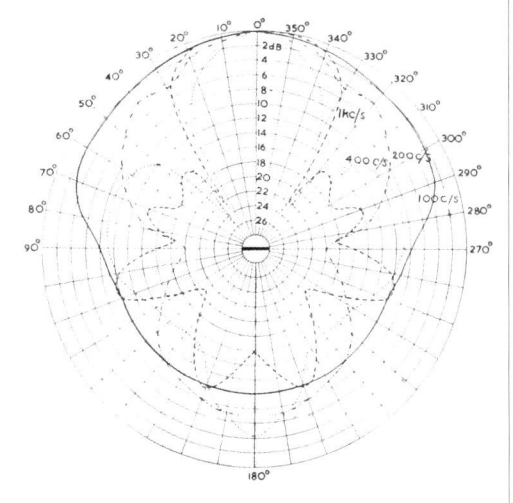

This has been specially designed for use in school halls and other auditoria having difficult acoustic conditions. The radiation pattern is highly directional to reduce reflections from adjacent walls, floor and ceiling, thereby improving intelligibility. The loudspeakers are loaded at the rear by means of slit type acoustical resistors which attenuate the response at the rear and sides down to below 100 c/s. as shown by the accompanying polar diagrams.

The normal response gives good quality reproduction of speech and music. A bass cut switch is provided to attentuate output below 200 c/s. for use on boomy speech transmissions and recordings.

Figure 8.7 LS/7 Line Source Speaker, 1959 catalogue.

on this activity (which seems likely), to encourage Cooke to take up more invitations on behalf of the company. Cooke was no stranger to giving lectures, indeed, like Gilbert, he had started young. During his wartime radar training in the Fleet Air Arm he had met Dobson and Young, a famous duo who visited the services and gave musical lectures, which had stimulated him to put together his own show based on classical recordings to entertain his fellow servicemen. So, he was soon established on the circuit in his own right and landed an invitation to give a lecture on sound reproduction in schools, at the Annual Conference of the National Committee for Visual Aids in Education (Bedford College, London, July 1957). Schools broadcasting had become an important feature of education, with improved quality following the introduction of VHF transmissions, but often the reproduction was compromised by the acoustic characteristics of the listening space. Lecturing and demonstrating in similar environments meant Cooke had recent first-hand experience of the problems and he put a lot of effort into researching and preparing this talk. It cannot be a coincidence that some 18 months later Wharfedale launched the LS/7 line source speaker for use in school halls, churches etc. and a new portable unit for classrooms—the P8 (which was smaller than previous models). The entry for the LS/7 in the new catalogue provides all the essential information, as would be expected (8.7).

The Wharfedale 60

After the hectic activity of the previous two years, 1960 appeared—at least from the new products viewpoint—to be much calmer; only one new item appeared in the UK. This was the SM-1 stereo mixing (1:1 ratio) transformer, designed to assist with the uprating of mono equipment to stereo. The advertised applications were threefold: combining the bass from two stereo channels into a common woofer; adding a third full-range speaker to reduce the 'hole-in-the-middle' effect; combining the bass output from both channels in a full-range speaker on one channel so a small unit operating above 300 Hz could be used on the other channel. This followed the publication, at the end of the previous year, of *Stereo Handbook* which, as described in Chapter 12, aimed to provide some clarity in a confused world. Meanwhile the transformer product range had been reduced to include just the OP3, P-Type, GP8, W12, W15 and WMT1.

However, across the Atlantic, in Port Washington, Wharfedale Audio Products was gearing up to launch a novel new product onto the North American market. This was the Wharfedale 60, designed in the UK and named after the year, as indicated in the flyer reproduced in (8.8). It was a two-way system with a 12" bass unit (W12/FS) and 5" tweeter (a variant of the original Super 5) and with dimensions of 14¼ × 13 × 24" (just over 2½ cu ft volume) it was marketed as a potential bookshelf speaker. Its unique feature was the use of sand-filled panels to produce a completely non-resonant cabinet. The fully-veneered version sold for $105. All the components were supplied from Idle, but the furniture-styled cabinets were locally sourced and the systems assembled at Port Washington. The drive units used were specially produced for the system and were not included in the list of units for sale through BIC.

The appearance of this product was another logical step in the progression from putting Wharfedale speakers into other people's cabinets, started some four years earlier, followed by the production of 'Americanised' versions of UK cabinet systems of increasing differentiation. The Wharfedale 60 (later simply W60) had no UK equivalent and was created for the US market in response to BIC's particular requirements. They also sold a kit for a DIY two-way system, based on this design, with the standard US 12" and 4" units (WNA/12 and WNA/4), crossover components and cabinet construction details. The origin of the image of Gilbert in (8.8) is interesting. Another of the framed photographs which he had hanging on his office wall, at least after he retired from Wharfedale (see Chapter 10), is shown below (8.9).

It was probably taken in mid-1956 and shows Gilbert with Arnold Hatton (not long before he left for the USA) and an unidentified lady discussing a new device from Mullard for tuned ultrasonic tinning of aluminium voice-coil wire. Hatton would surely have taken this picture with him to the

Figure 8.8 BIC Flyer for the Wharfedale 60, USA 1960. (Scan courtesy of the Briggs family.)

A Pair of Wharfedales: The Story of Gilbert Briggs and his Loudspeakers

Figure 8.9 Gilbert with Arnold Hatton and an unidentified lady demonstrating ultrasonic tinning of aluminium voice-coil wire*. Idle, probably mid-1956. (Photo courtesy of the Escott family.)

USA and probably had it on display in his office. His BIC marketing colleagues did not have to look far for a suitable image of Gilbert to extract for their flyer.

The W60 was very well received in the USA but when John Balls first visited BIC in May 1960 he received a cool reception. The Carduners were very frustrated because, once again, they could not get a reliable and large enough supply of speakers from the UK and they were confident that they could expand the product range if this problem could be solved. They were even threatening to drop Wharfedale and take speakers from Leak. Balls played the Rank card and promised to meet their requirements. This changed the atmosphere and Leonard Carduner outlined his ambitions. They discussed plans for a cheaper companion to the W60 which BIC contemplated selling in the thousands per year. They also expressed their interest in pursuing the three-way system, first talked about during Gilbert's visit the previous year, as a potential W70 addition to the sand-filled cabinet family.

---

* When Cooke joined Wharfedale from the BBC in 1955 he brought news of this development and was subsequently involved in the research which perfected the method. In the picture the aluminium voice-coil is being held so that the end of the wire to be tinned is dipped into a small heated bath containing liquid solder, which is ultrasonically agitated. This mechanically removes the oxide layer from the wire surface allowing effective tinning before re-oxidation can occur. The ultrasonic frequency can be adjusted ('tuned') to optimise the results under any given circumstances. This technique finally solved the problem of tinning which had been a potential source of joint failure ever since aluminium voice-coils had been introduced in 1950.

Altogether this represented a potentially huge increase in USA business and when John Balls returned from his visit, Gilbert decided they should celebrate. He took Balls and the Wharfedale directors out to dinner at the Midland hotel, in Bradford, and what happened there became legendary. During the meal, Gilbert invited Balls to give a summary of his trip for the benefit of the rest of the party. Having talked about BIC, Balls turned to Barney Smyth, the Canadian agent, whom he had also visited in Montreal. He said Smyth wanted to build systems using Canadian-made cabinets, so why not have an agreement for this to be done under licence, as with BIC? Gilbert, who had been dealing with Smyth for almost a decade and had cause to question his administrative capabilities, dismissed this by saying: "A leopard never changes its spots." Balls, perhaps not recognising the finality of this response, asked again. Gilbert, growing very red in the face replied, "I said, a leopard never changes its spots!" Then, suddenly boiling over with anger, he walked out. Raymond Cooke, one of the witnesses, used to tell this story as an illustration of how Gilbert's loss of temper could be quite frightening in its speed and intensity. It made those who worked with him somewhat wary in social situations because, no matter how good a mood he seemed to be in, this might flip in an instant.

Raymond Cooke then produced a new version of their standard WNA/10 unit, named the WLS/10, and designs for a system to incorporate this, with the WNA/4 tweeter, into another cabinet with sand-filled panels. The dimensions were the same as the W60 and this cheaper model was dubbed the W50. BIC confirmed an initial order for 2500 sets of components in June.

A New Technical Assistant

It was probably in late 1959, just before Balls arrived, that Cooke acquired a new assistant, a 30-year-old Scot with an interesting background in audio. William A. (Bill) Jamieson first met Gilbert when he was working for EMI, on stereo tapes, in 1956. EMI gave the first completely stereo recorded presentation at the Royal Festival Hall on 26 April and Bill was very involved in all the secret rehearsals. Only two weeks later Gilbert was putting on his third concert-demonstration there and, as described in Chapter 11, stereo recordings were demonstrated through two commercial stereo tapes and in several 'live versus recorded' items. All of these involved EMI and Bill Jamieson's recent experience proved valuable. He was again on hand to help and advise at the next major concert in Liverpool in July 1957. He left the UK for adventures in Africa in early 1958 but had formed a sufficient attachment to Gilbert to write to him several times from there. When the company he was working for folded, Gilbert offered him the lifeline of a job at Idle.[5] He appears in a fine photograph with his colleagues (8.10), which may relate to preparations for a concert-demonstration given at Leeds University in July 1960.

The 10th Wharfedale book emerged in November 1960: *A to Z in Audio*. Although Raymond Cooke was again credited as the Technical Editor, it is certain that he had a great deal of input and could deservedly have been named as co-author. He kept the complete original handwritten manuscript until his death in 1995 (see Chapter 12). Cooke was also busy during the year working on new products for the following year and ideas for the future, the latter eventually leading to dramatic events as we shall see.

As described in the previous chapter the major new building, housing the cabinet shop amongst other departments, was completed in 1960. A picture of most of the Wharfedale products on offer about this time was taken in the new demonstration/listening room there (8.11). Shown from left-right are: top shelf—Reflex and Bronze Baffle extension speakers; bottom shelf—PST/8 cabinet,

Figure 8.10 Clockwise from bottom left: Bill Jamieson, Raymond Cooke, Gilbert Briggs, Bill Escott. Probably mid-1960. (Photo courtesy of the Escott family.)

Figure 8.11 Product line-up around 1960 displayed in the new demonstration/listening room. (Photo courtesy of the Escott family.)

W3 and W2 systems; floor standing—drainpipe column speaker (demonstration only), three-speaker corner system, W3, SFB/3, W4, Column 8 and LS/7; table and floor—chassis, transformers, crossovers, volume controls and air-cored inductor.

Airedales, Ceramic Magnets and Double Diaphragms

The W2, W3 and W4 had proved very successful in the two years since their launch, the Column 8 model less so. This, along with the AF10, was dropped from the range for 1961, although the cabinet construction details for both were made available to DIY enthusiasts. The new system introduced early that year was the 'Airedale', a free-standing version of the three-way, omni-directional corner unit with styling similar to the W4 (8.11). One of it functions was to provide a stereo partner for an existing corner unit where another corner was unavailable or too far away. The bass chamber was reduced to 4 cu ft and to provide low distortion at low frequency under airtight conditions a new version of the W15 with roll surround—giving a large linear excursion—was produced. The W15/RS was also fitted with a Feroba II ceramic (barium ferrite) magnet which increased the flux density to 14,000 oersteds (from 13,000 in the W15/FS) whilst reducing the size significantly. This is believed to be the first time a ceramic magnet was used on a high fidelity speaker in the UK.

When *Loudspeakers 5* appeared in 1958 small isotropic ceramic magnets, such as Feroba I, were already in widespread use for transistor radio speakers, cathode ray tubes, door catches etc. and anisotropic types, such as Ferobar II, were starting to be used in the USA for large loudspeaker magnets. In the UK demand limitations and patent obscurities prevented investment for a time, but by 1961 Gilbert's friends at Swift–Levick were in full production.[6] There was no fanfare; for Gilbert this was just another step along the well-trodden path of incorporating the most efficient magnet available, without increasing the overall price.

As shown in (8.12) the Airdale had a Super 8/FS mid-range driver and Super 3 tweeter which were mounted facing upwards with a second order separator providing crossovers at 400 Hz and 5 kHz. The front panel was sand-filled and the rear side panels were made non-resonant by attaching ceramic tiles.

A joke which Gilbert was fond of repeating related to his earliest large hall demonstration in Toronto. An ex-Yorkshireman in the audience asked why the speakers were called 'Wharfedales', when Bradford was actually in Airedale. Gilbert replied that if the speakers had been no good they would have been called 'dirty dogs', at which point the questioner's wife piped up 'I suppose that's why you call them woofers!' Despite, or even because of, this history the name was used—the first in what became a tradition of using the names of Yorkshire Dales and associated villages for cabinet systems.

Raymond Cooke had spent some of 1960 revisiting the 'double diaphragm' approach for increasing the range of single speakers. This went back to the Voigt patent which Gilbert had licensed for a short period in 1936/7 to produce the 'Twin-Cone Auditorium'. Whilst very effective in increasing the high frequency response from typically 4 kHz to 15 kHz in a 12" unit, the problem with the inner trumpet-like cone was the introduction of pronounced resonances which were difficult to damp. Cooke overcame this by giving the inner 'whizzer' cone a foam surround which he was able to bond to the main cone. The whizzer cone itself had a design akin to that used on the Super 3 tweeter with an integral dome in a single moulding. Two new drivers using this patented method were produced, also incorporating roll surrounds—the RS12/DD with a 14,000 oersteds ceramic

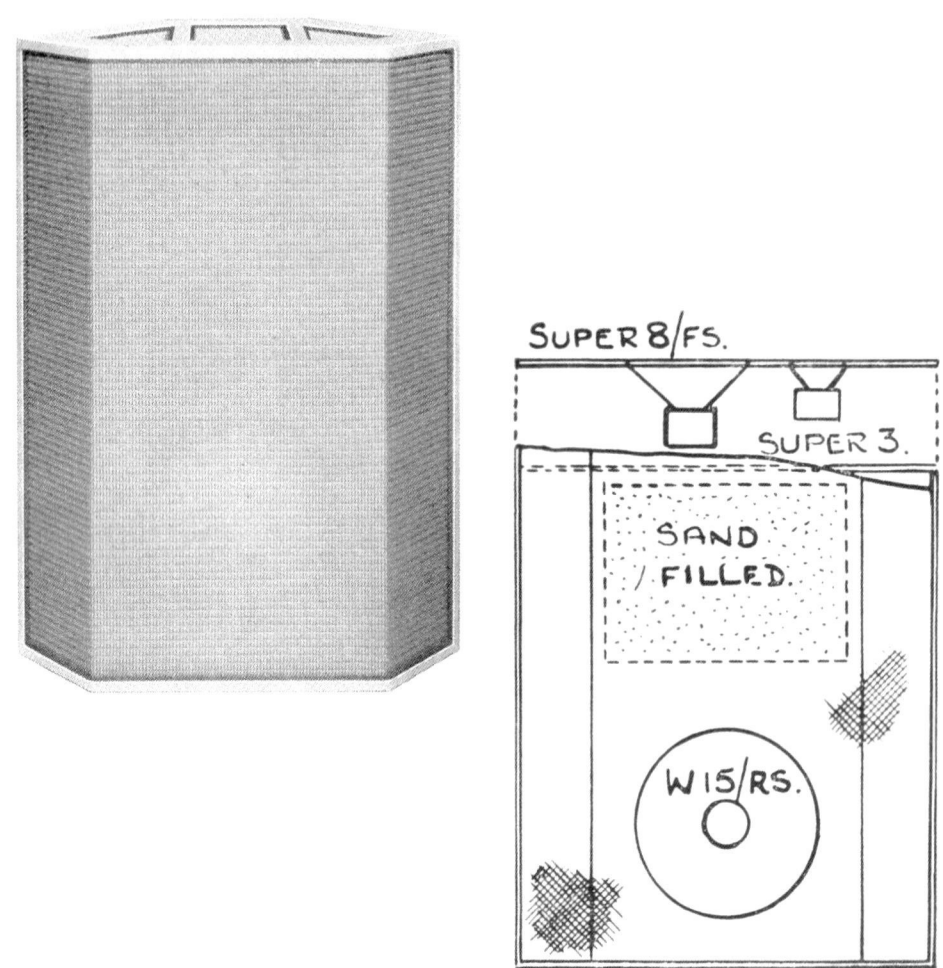

Figure 8.12 Airedale, 1961 catalogue.

magnet and the Super 12/RS/DD with a 17,000 oersteds Alcomax III magnet. They both had a quoted frequency range of 25 Hz–15 kHz, which compared with the Coaxial 12 unit's range of 25 Hz–20 kHz (converted to roll surround at the same time). All the roll surrounds were produced in-house using pneumatically operated and timed moulding presses.

Sudden Departures

These new products had all been introduced by June 1961 and not long after Gilbert faced the prospect of losing two of his closest colleagues. The first may not have been a big surprise: Edith Isles, his secretary who was now 63, decided she would like to retire. This was going to be a wrench since she had been with Gilbert for about 12 years. It was agreed that Edith would take two weeks holiday in July, during which time she would organise a temporary replacement from an agency. If the 'temp' turned out to be satisfactory then she could be offered the job full-time. The second loss must have come as a shock: Raymond Cooke was going to set up his own loudspeaker business. The reason was recalled by Cooke, himself, some 20 years later[7]:

Chapter Eight: 1959–1965

'I had developed a number of ideas, mainly about materials for diaphragms. I was seeking to liberate the loudspeaker design from the need always to use a very simple conical paper cone. It seemed to me that there were really three objections to paper cone technology at that time. One was the enormous expense in producing a set of tools for moulding and drying even to try out a new profile. Another problem was the long time it took to produce the tools—up to six months. Then one was limited to very simple shapes which could be turned on a lathe. Grading the thickness of the pulp, for instance was extremely difficult. So I thought that we needed a process that was cheap—at least at the experimental stage—and quick to produce. We also needed to be able to vary the characteristics of the material. Now it so happened that a chap called Bob Appleyard, who previously had been a famous cricketer, worked at Waddingtons the playing card people in Leeds and used to drop in to chat to Bill Escott and me. One day he brought with him one of the very first moulded plastic cups. He described the very simple process of taking a flat sheet of the material, warming it up and vacuum-forming it over a shaped tool. Prototypes could be produced on wooden moulds or plaster of Paris. So this started me thinking.* I also looked at expanded polystyrene which some people in the business were already starting to use. . . . Other people had tried such materials but had made the fatal mistake of thinking that, if it feels stiff to the touch under steady pressure conditions, then it will probably be stiff dynamically—which it usually isn't. Well, I could see that the Rank Organisation was not very interested at that time in diversifying to a completely new technique. So I decided that the only way was for me to risk everything and start on my own.'

In fact, Cooke's account tells less than half the story. In March 1961, John Balls, MD in waiting, left Wharfedale. In the first year that he had been at Idle and, in large part through the strong working partnership that he had established with Cooke (8.13), the turnover at Wharfedale had increased by some 40% and net profit by over 50%. Gilbert's reaction was to forget about retiring; indeed he persuaded Rank that his presence was required to 'steady the ship'. Balls was wondering where this left him when he had a surprise visit from Bob Pearch, MD of Kent Engineering and Foundry Ltd (KE&F) in Maidstone, whom Balls knew from his previous job with ICI. The company was a metalworking business that principally manufactured crop sprayers and associated equipment. Pearch's father, Leonard, the owner of the business, wanted to retire and he insisted that his son find a working partner, otherwise the business would have to be sold. Bob Pearch proposed that Balls be that partner and, in view of the uncertainty of his position at Wharfedale, he agreed. He kept in touch with Cooke who was, in turn, concerned about the future for Wharfedale when Balls was not replaced. All that happened was that Rank appointed another director, Leonard Chapman, to the Wharfedale Board, but he would have this as an additional responsibility and not be moving to Idle. So, what was Rank's likely strategy—a potentially damaging merger? All the original Wharfedale directors had been given three year service contracts, back-dated to October 1958 when Gilbert negotiated the sell-out, and they would run out in September 1961—an additional uncertainty.

Once at KE&F, Balls realised that all the facilities were in place to make loudspeakers, so he

---

*Appleyard played for Yorkshire and England during the 1950s and is most famous for taking 200 first-class wickets in his first full season (1951)—a unique achievement. Escott was a keen cricketer and met Appleyard through the game; as teenagers they both played in the Bradford Central League, before they were caught up in the War. Waddington's, though best known for its board games, was a diversified printing and packaging company. At the time of this story Appleyard had senior responsibilities in the drinks packaging business and was working on improving waxed-cardboard drinks cartons, hence his interest in new plastics technology[8].

Figure 8.13 Raymond Cooke (extreme left), Bill Escott (centre) and John Balls (second from right) on a visit to Mullards in 1960. (Photo courtesy of the Escott family.)

proposed to Cooke that they form a new company, with Bob Pearch as third director. This was around May 1961. Cooke immediately started working on possible designs and even tested out ideas in the marketplace, all the while still working at Wharfedale, before deciding whether he would take the plunge. He had more-or-less made up his mind to leave by mid-June when he sent Balls the draft of a resignation letter to Gilbert. He had no one else to turn to for advice and he was clearly uncertain about the best approach to adopt. On the one hand he would like to tell the truth about his frustrations with Wharfedale, and Gilbert in particular, but on the other hand he did not wish to cause the 'old man' undue hurt. Like everyone else who came under the G.A.B. influence for any length of time he still had a fondness for him. In the end, it was not until the end of July that he actually resigned. Cooke liked the idea of using the initials K.E.F. as a word, KEF, because he did not particularly want to use his own surname or a combination of surnames in the new company's name. He researched the word KEF to make sure it did not have any negative connotations in other languages and decided it was good to use. So it was that KEF Electronics came into being on 5 September 1961.

The rest of that story is told elsewhere[9] but germane to our story is Cooke's recollection[7] that, when considering this bold move, he only had a vague idea in mind for a first product. Given that he also says that operations in Maidstone started in early October and the first product—a three-way speaker system in which all the drivers were unconventional (see below)—was being shipped in late November, this claim cannot be taken too seriously. The extent of any discussions with Gilbert, which might have led to Cooke's conclusion that his ideas were unlikely to be adopted, is not known. However, he told Malcolm Jones, who joined him at KEF (as his technical assistant) in early October 1961, that Gilbert threw his first attempt at a moulded plastic cone into the waste paper bin! It is the case that Gilbert became increasingly cautious about investment in his last two to three years before stepping down as MD (in January 1964), almost certainly because Rank was exercising more financial control and demanding higher profitability.

The Parting of the Ways

In July Gilbert finished compiling a new and rather different book, *Audio Biographies*, in which 64 of his friends and acquaintances in the audio business wrote about their experiences (see Chapter 12). The Introduction is very precise about how the final number of contributions was derived from the original invitations, numbering just over 100. He records that three were omitted by mutual agreement and one of these must have been Cooke's. Bill Jamieson told Ken Russell (Cooke's successor, see below), rather dramatically, that his contribution 'was expunged'. It would have been out of character for Gilbert to have done this in anger or for revenge, although he set great store by loyalty and felt badly let down. More likely is that it was a matter of timing. By August Gilbert had written a Conclusion to fill up a couple of blank pages revealed when the proofs had been pasted up and paginated. A copy of the galley proof of the first part of Cooke's entry (his education and positions held), with Gilbert's introduction, has survived. It is stamped 21 July 1961, the date Gilbert received it from the typesetter, perhaps only days before Cooke resigned, so the alterations necessitated by his decision (see Chapter 12) could not be made. The biographical section has been lost. History is the loser. Raymond Cooke and KEF Electronics both became well known and highly respected and insights into Cooke's pre-Wharfedale days are scarce.

The anecdotal evidence is that Gilbert was livid and Cooke was probably the recipient of some very harsh words. In addition he was essentially ordered to 'clear his desk' and leave the premises. Their relationship was complex because it operated on several levels. At Wharfedale, Gilbert had given Cooke a great deal of freedom to pursue his own development ideas but at the end of the day it was he, as boss and embodiment of the firm to the outside world, who took the credit for their success. The situation was slightly different in terms of collaboration on books, magazine articles and the later concerts, where Cooke's contribution was scrupulously recognised, but he was always going to be in the shadow. However, through his own lecturing and demonstrating programme, which Gilbert had also encouraged, Cooke had built a reputation which gave him a taste of doing things largely for himself. They often worked together at Cooke's home, particularly carrying out listening tests. His living room was typical of modern houses, unlike Gilbert's, which made it much more appropriate for testing stereo pairs in a domestic setting. This brought Gilbert into contact with Cooke's wife, Marjorie, and two young children. Cooke was exactly the age that Gilbert's son Peter would have been, had he lived, and the children were the same age as his own grandchildren. Marjorie and the children were charmed and Gilbert became a sort of grandfather figure. Cooke's departure would leave a hole not just in Gilbert's firm. The real tension was probably created when Gilbert became aware of the forward planning which had been going on, under his nose, and the fact that Cooke's first product would be a direct competitor to one under consideration, if not active development, within Wharfedale (see below).

Gilbert might well have reflected on his own drawn-out and indecisive struggle with Lund—an older, more cautious partner—when he was Cooke's age and recognised that his ambitious colleague had what it took to make the break and be in control of his own destiny. When the dust had settled, relations resumed and Cooke came to realise, and admit, the debt that he owed 'the old man'. They became, and remained, close friends until Gilbert died (see Chapter 10). When that happened Cooke wrote to his son, Martin, in Canada:

'G.A.B. is a loss and will be missed. I had a tremendous regard for him and learnt much from him by example. He was so upright in everything that he did, like Herbert Batchelor and Leon Goossens. Three out of the same 19th century mould. I never felt easy with him though, he was restless, mercurial and frequently very bad tempered, but the sun eventually shone through and at his best he was marvellous company.'

The New Arrivals

The temporary secretary who stood in for Edith Isles, whilst she was on holiday, was Mrs Dorothy P. Dawson who, at 28, was about the same age as Gilbert's youngest daughter. This represented quite a change but Gilbert had no hesitation in offering her the full-time job. He seemed to her to be 'a nice man to work for' so she accepted—and stayed with him as his 'second best secretary' for the rest of his life. Her arrival at Wharfedale and Raymond Cooke's departure almost coincided; the overlap was only a couple of weeks or so.

Not long before these summer manoeuvrings Gilbert, almost certainly with Cooke alongside him, had given one of his last large lecture-demonstrations to the Audio Society of the Automatic Telephone Manufacturing Co. Ltd (ATMC) in Liverpool. At the time the ATMC had around 10,000 employees and its Audio Society members could fill its lecture theatre which had a seating capacity of about 500. In casting his net to find a new Technical Manager, Gilbert wrote to the President of this society to enquire whether he thought any of its members might be interested in the position. A notice was duly posted and seen by Kenneth F. Russell, an electrical engineer with ATMC who had wound up with responsibilities in the patents field. A DIY hi-fi enthusiast, whose interest had been nurtured by Gilbert's books and cabinet construction sheets, he thought this was an opportunity he could not afford to pass up. His details found their way to Idle and in due course he was paid a surprise visit at work by Ralph West, lecturer in radio at the Northern Polytechnic in London and technical writer on hi-fi. West had known Gilbert for years, contributed to his books and collaborated on his Bristol concert-demonstration in October 1959—his judgement was trusted. An unconventional approach to assessing a candidate maybe, but his feedback ensured that Russell was invited to visit Idle. There he was shown round the factory by Bill Escott before Gilbert took him to his favourite watering hole, the Midland Hotel in Bradford, for lunch. There was no technical interview, but Gilbert would have been pleased to discover a fellow music lover—in his view a key requisite for working in loudspeaker development—and shortly after he offered Russell the job with a start date of 1 January 1962.

He arrived to find that winter in Bradford was somewhat harsher than on the Wirral coast he had left behind. Snow a foot deep covered the pavements and it stayed there until well into March. This was not the only surprise. Raymond Cooke had left no records of development projects or any indications of work in progress. According to Bill Jamieson the work he had been doing latterly was clearly, with hindsight, directed towards what soon appeared in the first KEF offering. What Cooke did bequeath was an accumulation of technical correspondence, from customers and dealers, which needed to be answered and a list of speaking engagements at venues all over the country, which would have to be honoured. A further surprise was when Boyd Crabtree, in charge of transformers

Chapter Eight: 1959–1965

(who had only recently taken over from Frank Mann), handed Russell a specification for a transformer and asked when the design would be ready; Wharfedale made a fairly open offer in its sales brochures that transformers with ratios and impedances different from the standard models could be supplied on request. Ezra Broadley determined that here was a fresh mind which could address some of his technical problems in Production. One of these was rather important: Gilbert had invested in a modern, German, semi-automatic coil-winding machine which had never really worked properly and was sitting idle in the factory. One unique aspect of the machine (shown in 8.14) was that it automatically dispensed the adhesive which ultimately held a fragile voice-coil together—in manual coil-winding the adhesive was applied using a finger. In sorting out the problems with this machine, Russell realised the importance of adhesives, as materials, in the behaviour of loudspeakers and embarked on a programme to investigate the products then available and to optimise the materials for their specific roles. Fundamental work on loudspeaker development was going to have to wait and in retrospect Gilbert's description of the post in his letter to ATMC as 'an interesting job connected with loudspeakers' was seen to be the literal truth.

Figure 8.14 Aumann semi-automatic coil-winding machine operated by Winnie Kimber, about 1963. Note the pendulum clock which hung above the coil-winding bench in Brighouse, seen in (5.10) (Photo courtesy of the Escott family.)

First Impressions

Ken Russell's impressions of the firm at this time are very interesting. The number of employees had risen to over 80 and day-to-day operations, in the capable hands of people who had worked together for many years, seemed to run on clockwork. Bill Escott, as deputy MD, was essentially in overall charge. Ezra Broadly looked after production—a burly man who 'called a spade a spade' he nevertheless had an instinctive understanding of his large contingent of female employees which ensured harmonious labour relations. Apart from coil-winding and roll-surround production, loudspeaker and transformer manufacture was assembly work with all the other components coming from outside suppliers (8.15). The cabinets were different; these were all produced from the basic raw materials (plywood, wood veneers etc.) on site (8.16) under the supervision of Ralph White who had also been at Wharfedale a long time (over 10 years). Eric Barker ran the general office, in charge of purchasing, stores and so on and Dorothy Stevens looked after shipping. Albert Smith was still the only sales representative outside Bradford. Luther Emsley, the accountant who had taken over from Fred Mason, had stayed less than three years, complaining to Dorothy Dawson that the incessant demand for budget projections and the like from Rank was 'getting him down'. He also retired before the end of 1961. This supports the idea, suggested earlier, that Wharfedale was coming under much tighter financial control when Raymond Cooke was contemplating radical new developments. By the time Russell arrived the accountant was Harold Reynard.

Russell found that Gilbert himself was something of an enigma. Tucked away in his corner of the Office Block, accessed only via his secretary's office, and usually arriving there in mid-morning he could have seemed rather remote. If he wanted to talk to anyone they would be summoned to his office. Matters needing his attention seemed to be filtered up to him via Ezra Broadley and Bill Escott and it was expected that anything beyond the day-to-day operations required his executive decision. However, on arrival at work Gilbert would usually go round the different production areas and wish his employees 'Good Morning'. Most of the key employees had grown up under him from the days when the firm was very small and his presence was keenly felt and appreciated, so this was enough to ensure they knew he was still 'the boss'. Equally, he might appear in the yard at lunchtime and be unable to resist turning his arm over if the lads were playing cricket. At least once a week, he liked to go for a leisurely lunch at one of his favourite watering holes, with someone from the firm if a visitor did not provide the opportunity. Everybody who was a potential companion knew that Gilbert would not make any allowance for the lost working time and, though tempted by a free lunch, they usually declined—but more often than not pleaded with Dorothy Stevens to go and keep him company. Since she didn't mind taking unfinished work home with her she was often treated in this way. Gilbert started off with a 'gin and It' and always asked for a cherry with it, which he gave to Dorothy. He had three topics of conversation during the meal: progress or otherwise on his current book project; the performance of his share portfolio and the vagaries of the stock market; and Perry Mason!

Much of the time when he was not in his office Gilbert would be in the listening room adding to the thousands of hours he had spent assessing new speaker designs, but he would also disappear into the small building where his books were stored and even wrap up a few orders! In meetings, discussions had to be kept to the point; Gilbert would never gainsay anyone with specialist knowledge but he hated digression and this made him irascible. His normal facial expression was faintly mischievous with a twinkle in the eye, in keeping with his 'impish humour' as Raymond Cooke put

Chapter Eight: 1959–1965

Figure 8.15 Coil-winding and loudspeaker assembly, about 1963. (Photo courtesy of the Escott family.)

Figure 8.16 Cabinet shop, about 1963. (Photo courtesy of the Escott family.)

it, but when he was losing his temper the smile faded and the twinkle turned to flint. Such losses could be dramatic but were short-lived and Gilbert's inherent charm ensured that everyone left smiling in the end. He knew that committee work would drive him crazy and so he consistently refused such invitations to become involved (for instance to be Chairman of the BSRA).

More Books

Raymond Cooke implied that at least one reason why technical developments were left largely to him was that Gilbert 'spent a lot of time on the books'.[7] Although Gilbert produced his first drafts at home in Ilkley, working in the early morning, everything else was done at Idle and this involved a great deal of corresponding and collating material. No sooner had he compiled *Audio Biographies* in July 1961 (published in November) than he started work on *Cabinet Handbook* which appeared in March 1962. As described in Chapter 12 this included a large number of photographs taken of activities in the cabinet shop; a collective shot of the cabinet makers taken at the same time but not included in shown in (8.17).

It was fortunate for Ken Russell that he arrived during the gestation of *Cabinet Handbook* because it did not really impact on him. The next project, *More about Loudspeakers*, published in March 1963, did. He became used to receiving a phone call on arrival at work, from Gilbert at home, which went along these lines: "I'm working on a chapter of my book and I need to make some tests." The 'tests' would be described and they could easily involve several days work in the lab. Gilbert would end the instructions with: "I'll be in about 10.30, when do you think you will have the results?" This work, continued for the next book '*Audio and Acoustics*', published in November 1963, occupied a lot of Technical Department time.

Figure 8.17 Cabinet makers, Idle, September 1961. Sitting, fifth from left is Ralph White (head of the cabinet shop). Standing, second from left is Frank Dawson (chief machinist) and fifth from left is Wilfred Malton (specialist in hand veneering). (Copyright—Bradford Museums and Galleries. From the C. H. Wood photographic archive negative number 45476, photo courtesy of the Escott family.)

Chapter Eight: 1959–1965

The Housewife's Prayer

The major new product offerings of 1962 were driven by two influences which would have barely registered ten years earlier: housewives and pop groups! In January the 'Slimline 2' two-way cabinet system, designed to minimise the occupied space and be suitable for floor, table or wall mounting, was launched. The criteria imposed on the design were a volume of less than 1½ cu ft, depth no greater than 7", an attractive appearance and a price tag appreciably less than £25. Gilbert was looking for 'the answer to the housewife's prayer'. The shallow depth requirement could now be entertained by using a ceramic magnet but the resulting problems of standing waves and the transmission of reflected sound from the rear of the speaker through the cone had to be overcome. Polystyrene came to the rescue and Gilbert published an article on the acoustic principles of using thick expanded polystyrene diaphragms, or bungs, for damping and acoustic filtering within shallow enclosures in the January (1962) issue of *Wireless World*. The 12" bass unit (W12/RS/PST) was new with a smaller magnet (11,500 oersted, 1½" diameter pole) than the W12/RS and having the polystyrene diaphragm (a patent application was made). Also new was the PST 4, a 4" unit with a ¾" pole, centre dome and a similar flux density magnet. The suspension was cloth and the cone was covered with a thin skin of polystyrene to mitigate cone break-up. Front views of the system, with and without the covering, are shown in (8.18). The treble driver response picked up where the bass driver rolled off, at about 700 Hz, to produce a remarkably level response curve from 50 Hz to around 15 kHz.

Raymond Cooke's first product from KEF, the K1 launched at the Russell hotel in London on 28 November 1961, was aimed at the same market, and this reopens the question about when and why he first started work on a slimline concept. The market pull for such a design had been clear for a long time and when Ken Russell went to Idle to be interviewed for Cooke's old job, around

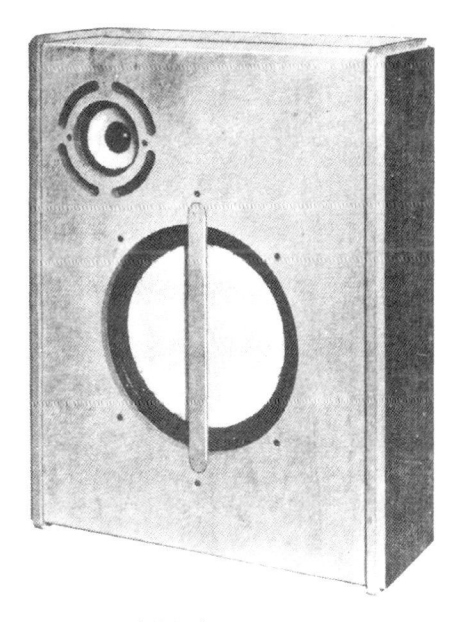

Figure 8.18 Slimline 2, floor-standing (left) and with front exposed (right). 1962 leaflet.

October 1961, he was asked to give his opinion on the performance of a prototype Slimline 2 in the Listening Room. It obviously needed more development. Gilbert's article in *Wireless World* would also have been submitted around this time. This suggests the project must have been underway for some time beforehand, but it would have been possible to knock up a prototype in only a few weeks. As described earlier, Russell took up his position at the beginning of 1962 to find that Cooke had left no evidence whatsoever of the work he had been doing in the months leading up to his departure, which was because, his assistant Bill Jamieson explained, it had been done 'under the table'. The following quotation, from KEF, might sum up Cooke's thinking at an early stage in a 'slimline' project at Wharfedale,[10] if there was one:

'The chief motivation behind the K1 series speakers was the desire to achieve outstanding high-fidelity reproduction from a slimline enclosure. To do this with existing techniques and materials was to invite failure; a complete rethink was called for.'

His completely unconventional three-way system,[11] packed into an enclosure only 26½ × 16½ × 4¾", must have occupied most of Cooke's waking hours, in and out of Wharfedale, between May and August 1961 and John Balls was gearing up for production in Kent even before KEF Electronics was officially incorporated in September. By that time Cooke had produced a K1 prototype, constructed at home. They had a head start, although Cooke discovered that Jensen in the USA had stolen a march on both of them with a similar offering. However, Gilbert was galvanised and put in 'two-three months of intense research' to finalise his product by January 1962. The name he gave to it, 'Slimline 2', seems to have defined the genre, since the K1 was not advertised as the 'K1 Slimline' until two months later, in March 1962.

Electric Guitars

The pop groups of the early 1960s, exemplified by The Beatles, with their line-up of lead, rhythm and bass guitarists plus drums had evolved from skiffle and rock 'n roll groups of the mid-late 1950s. Coupled with the increasing output available from amplifiers, they engendered a new market for speakers capable of working at the sustained high input levels from amplified electric guitars. Gilbert, prejudiced against this form of music,[12] probably wanted to keep Wharfedale at arm's length but Albert Smith, at the sharp end of dealer's requirements, and Bill Escott, wanting to expand order books, must have prevailed. In 1962, generally regarded as the year when the pop scene exploded (The Beatles 'Love me Do' made the mainstream pop charts), saw a crash programme to develop suitable products. Until relatively recently the larger drivers (12" and 15") had been rated at about 12 watts and 15 watts **peak** input respectively, but the introduction of roll surrounds had more than doubled these ratings, which provided a platform for development. The modifications were all aimed at making the drivers more robust. At the same time cabinets were developed with the appropriate size and weight for portability. All this was done in collaboration with 'various guitar groups', and presumably led by Bill Escott. The end result, in October 1962, was in the form of two drivers and associated cabinets, which could be purchased separately. The W12/EG, intended for use with lead guitars, was derived from the RS12/DD and had a frequency range of 40 Hz to 17 kHz with a maximum peak input of 30 watts. The W15/EG, derived from the W15/RS, was for bass guitars; this had a frequency range of 35 Hz to 5 kHz and a maximum peak input of

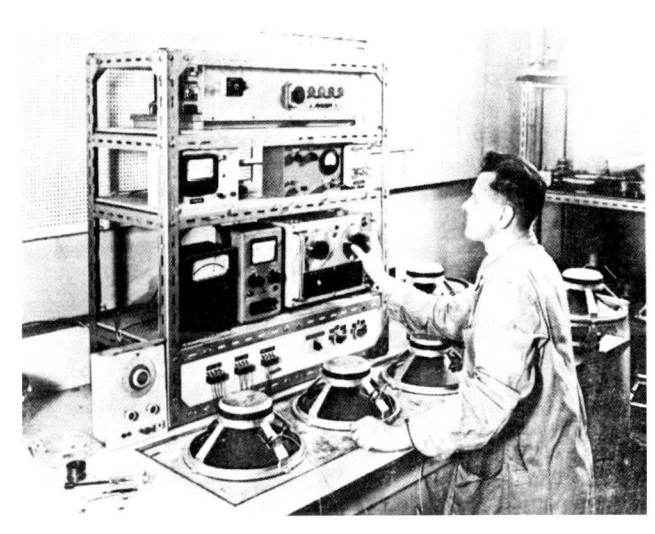

Figure 8.19 Testing drive units, from a brochure of 1963.

40 watts. The corresponding cabinets were named the EG12 and EG15. Their appearance was no different from the cabinets destined for the living room, showing that form had yet to be sacrificed to function in this rather different market. An 'Electric Guitars' leaflet gave the full constructional details of these cabinets for the DIY enthusiast along with a design for a larger 15" enclosure for those preferring extra bass over mobility! Despite the efforts to make these drivers much more robust, the pop groups found many ways to damage them and Gilbert was rueful of the marked increase in units returned for repair. A picture of drive units being tested, taken about this time is shown in (8.19).

RS/DD Units and Columns

Early in 1962 the 8" and 10" Bronze speakers were given roll surrounds and three more drive units were updated by incorporation of roll surrounds and double diaphragms; these were now the Super 8/RS/DD, the Golden 10/RS/DD and the Super 10/RS/DD. When he submitted the first of these to the magazine *Audio and Record Review* for possible review, Gilbert expressed the view that 'in his humble opinion' the Super 8/RS/DD was easily the best 8" model Wharfedale had ever produced. The reviewers (Donald Aldous, David Phillips and Frank Roberts) wholeheartedly agreed and asked him for a second unit so that they could carry out extra stereo tests. The conclusion to their rave review was:[13] 'a stereo pair in small enclosures gives sound quality that will come as a revelation to many listeners wedded to massive enclosures'.

Although the Column 8 speaker cabinet had been dropped, Gilbert had not given up on columns. Judging by the continuing volume of requests for the cabinet construction details of this model in particular, it was clearly a DIY favourite. Pursuing the logic of this observation led to two products which probably only he could have dreamt up (or had the courage to market). These were kits of parts for creating a column speaker from a 9" or 11" (internal diameter) concrete drain pipe! The components are illustrated in (8.20), made to fit into either type of pipe made by Stanton and Stavely. The former required a Super 8/RS/DD and the latter a Super 10/RS/DD unit to complete the speaker. The estimated price for the completed 8" system was under £12 which compares with the price of the Column 8 plus 8/145 system of over £29, when available. As Gilbert opined in the sales leaflet:

A Pair of Wharfedales: The Story of Gilbert Briggs and his Loudspeakers

'For the floor space used and the very low financial outlay, it is impossible to equal the impressive sound that comes from this enclosure—and you can build it yourself in next to no time. The concrete column is an unusual shape for a loudspeaker enclosure. Cabinets which are tall compared with their cross-sectional area are normally unsatisfactory because they tend to behave like organ pipes, i.e. they resonate at a fundamental frequency and at a series of harmonics. By the use of a specially designed filter and absorbent pads of bonded acetate fibre, all carefully positioned, this has been overcome in the Wharfedale column, and the result is a remarkable freedom from colouration for an enclosure of this type.'

The kits proved to be products with longevity. Although dropped in by Wharfedale in 1968, the parts were sold by Eaton Audio Fitments after that date and the two drivers were still available from Wharfedale until 1973.

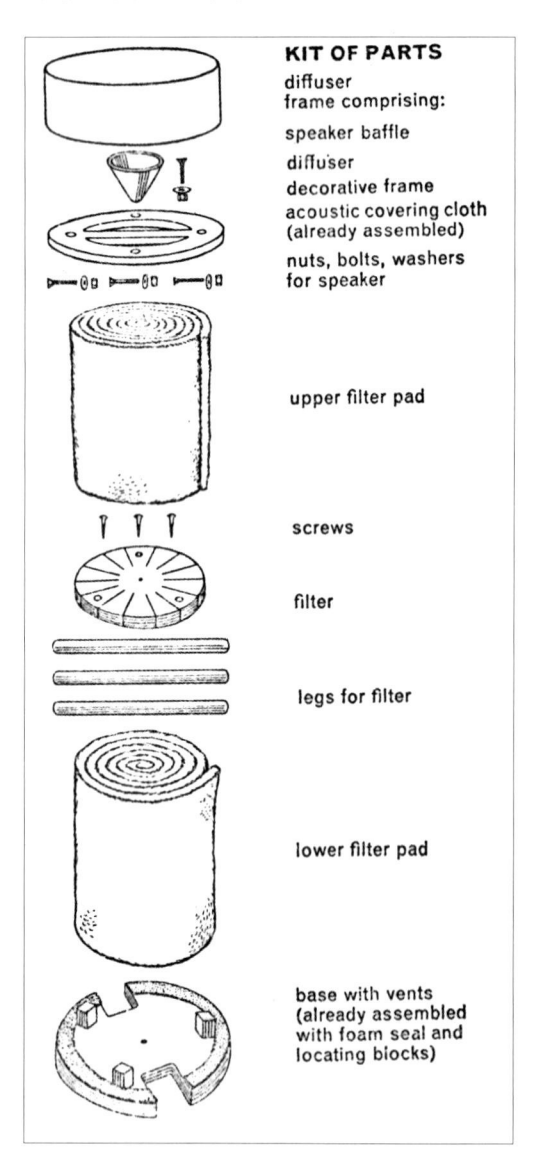

Figure 8.20 Exploded view of parts in the kit for a concrete pipe enclosure (speaker unit not included). 1962 leaflet.

KIT OF PARTS
diffuser
frame comprising:
speaker baffle
diffuser
decorative frame
acoustic covering cloth
(already assembled)
nuts, bolts, washers
for speaker

upper filter pad

screws

filter

legs for filter

lower filter pad

base with vents
(already assembled
with foam seal and
locating blocks)

Chapter Eight: 1959–1965

Later in the year a further foray was made into the 'school' speakers field (a shorthand description for spaces with unfavourable acoustic conditions). This was the combination of a bass-only version of the W2 and a six-driver version of the line source speaker (LS/6) together with a 400 Hz crossover unit. The combination, referred to as the LS/6B, was aimed at adding faithful music reproduction to the speech intelligibility given by the line source alone.

Bookshelf 2

1963 saw further progress in the pursuit of more compact systems with good base response. The Bookshelf 2 measured 19 × 11 × 6¾" and had a volume of just over 0.8 cu ft. The airtight enclosure was fitted with specially developed, highly efficient 10" and 5" drivers, the former, with roll surround, having a resonance of 25–30 Hz. As with the Slimline 2, a key component was a 1½" thick polystyrene diaphragm fitted to the 10" unit. The frequency range was 50 Hz–16 kHz. In the octaves above and below 1 kHz the units operated in parallel to smooth the response, with the dividing network working outside this range. The speakers were ideal for use with mono or stereo tape recorders. By 1963 transistor radios had been around for several years and their quality, especially on VHF, had been steadily improving. Gilbert, always an early purchaser of new gadgets, owned a 'high class' Bush TR82C (first produced in 1959) and he used this in promoting the Bookshelf 2 as an extension speaker for a transistor radio 'giving astonishing results' (8.21).

Another use for the Bookshelf 2 was as extension speaker to a TV set. Gilbert was scathing in his assessment of music broadcast on TV—'a contradiction in terms'. He could not see the point of watching a performance when the sound reproduction was so appalling. But the broadcasters per-

Figure 8.21 Bookshelf 2 connected to Gilbert's Bush TR82C VHF transistor radio. Sales leaflet, February 1963.

sisted with programmes of this ilk and he thought the situation might be ameliorated by the use of a suitable extension speaker. However, this required a suitable matching transformer and so the WM2 was born. This was similar to the existing WM1 but having separate windings with heavy insulation to provide sufficient isolation.

The only other product of this year was the W12, an updated version of the chassis produced during 1942–1956 for public address (PA) work and reintroduced for the same purpose. This brought the number of drive unit types in production to 18, the highest ever, as detailed in (8.22).

MODEL	BASS RESONANCE c/s	FLUX DENSITY oersteds	TOTAL FLUX maxwells	MAX. INPUT RMS watts	MAX. INPUT PEAK watts	WEIGHT lb.	BAFFLE OPENING in.	CHASSIS DIA. in.	OVER-ALL DEPTH in.	FIXING HOLES P.C.D. in.
Super 3	Treble only	14,500	60,000	6	12	3	3	3 5/8	2 7/8	3 15/16
PST/4	Treble only	11,500	28,800	5	10	1 1/2	3 5/8	4 3/8	2 5/8	4 3/8
Super 5	—	14,500	60,000	6	12	3 1/4	4 1/2	5 1/4	3 1/4	5 3/8
8″ Bronze	70/80	10,500	41,500	5	10	2 1/2	7	8	3 3/4	7 3/4
8″ Bronze/RS	50/60	10,500	41,500	5	10	2 1/2	7	8	3 3/4	7 3/4
8″ Bronze/FS/AL	50/60	10,500	41,500	4	8	2 1/2	7	8	3 3/4	7 3/4
Super 8/FS	50/60	14,500	60,000	5	10	3 1/2	7	8	4 1/8	7 3/4
Super 8/RS/DD	50/60	14,500	60,000	6	12	4 1/2	7	8	3 1/2	7 3/4
10″ Bronze/RS	38/43	10,500	41,500	6	12	4 1/4	8 3/4	9 3/4	4 3/4	9 7/8
10″ Bronze/FSB	38/43	10,500	41,500	6	12	4 3/4	8 3/4	10 1/4	4 1/2	9 7/8
Golden 10/RS/DD	38/43	14,500	60,000	8	16	5 3/4	8 3/4	9 3/4	5 3/4	9 7/8
Super 10/RS/DD	38/43	16,000	85,000	10	20	10 1/2	8 3/4	9 3/4	5 3/4	9 7/8
W12	45/50	14,000	156,000	15	30	12	10 7/8	12 3/8	5 1/4	12 3/8
W12/RS/PST	26/32	14,000	156,000	15	30	12	10 7/8	12 3/8	5 1/4	12 3/8
RS12/DD	26/32	14,000	156,000	15	30	12	10 7/8	12 3/8	5 1/4	12 3/8
Super 12/RS/DD	26/32	17,000	190,000	20	40	18 3/4	10 7/8	12 3/8	7	12 3/8
Coaxial 12	26/32	LF 14,000	LF 155,000	15	30	19	10 7/8	12 3/8	7	12 3/8
		HF 13,200	HF 44,000							
W15/RS	26/32	13,500	180,000	20	40	13 1/2	12 3/4	14 1/4	7 5/8	14 9/16

FS = Foam Surround AL = Aluminium Voice Coil DD = Double Diaphragm
RS = Roll Surround PST = Polystyrene Diaphragm

Figure 8.22 Drive unit specifications, Catalogue February 1963.

The Big Decision

The most significant action of 1963, however, was Gilbert's decision that it was time to stand down as Managing Director. At the beginning of the year he had just turned 72 and at some later point he decided that he was getting too old to be in day-to-day charge of running a company. What prompted the decision is unknown, but it was unlikely to have been any single factor. The drive from Ilkley to Idle and back, especially during the winter months, was one factor. During the winter just passed he had, in fact, rented a flat for the duration not far from the factory:[14]

> *'to cut down travelling in Yorkshire fog, snow and ice'*

and although it had all the creature comforts it wasn't home. He must not have looked forward to a possible repeat of this lonely exercise the following year. 1963 was also a significant milestone. As he wrote the Conclusion to *Cabinet Handbook* in March, he noted that it was exactly 30 years since he had 'started to dabble in loudspeakers' by which he meant 30 years since Wharfedale had become his sole occupation. Reflecting that he had worked for only a couple of years less in the 'Rag Trade' before this would have added to the sense that he had been in full-time harness long enough. A final, and probably, critical factor, was the fact that running Wharfedale was no longer the same as it had been before the sell-out to Rank. Slowly but surely the influence of the parent company had encroached: non-local directors appointed by Rank overseeing operations; financial controls and systems introduced which required conformation to a central model and the requirement that Gilbert be personally accountable to the Rank Organisation directorate for Wharfedale's performance. He was also under pressure to expand the product range beyond loudspeakers, as described in the next chapter. Continuing growth could only mean more big-company bureaucracy so Gilbert had to be philosophical about the consequences of his sell-out.

When Gilbert made his decision known to Rank or, indeed, how much notice he had to give them is not known but he stepped down as MD on 10 January 1964. A short piece in the Bradford *Telegraph and Argus* (headline: 'Retiring' to build an organ) reported[15] that he would be staying on as Chairman, but that he would have more time for his hobby—music. Gilbert said he had plans to build an electronic organ and, true to form, added that if it was successful the firm might possibly begin making them. The report also noted that his successor as MD was Leonard Chapman, a director who had joined the Wharfedale Board two years previously (the only change to its composition since late 1961 when the departed Emsley and Cooke had not been replaced).

Chapman's appointment, and Gilbert's role as Chairman, rather suggests that his retirement plans had caused problems. Although Chapman had experience of Wharfedale through being a director, he had other responsibilities within the Rank Organisation and was not based at Idle. The obvious successor was Bill Escott, Gilbert's deputy. According to his younger brother, Phil, Bill was offered the job as MD more than once; whether this was the first time, although highly likely, is uncertain. In any case, he consistently declined the invitation on the grounds that he wanted to spend his time running a thriving operation, not having to be at the beck and call of Rank's head office—which is probably an indication of the effect he thought this had on Gilbert. What happened appears to be a holding position whilst a satisfactory solution was found.

At the end of January, to mark the end of Gilbert's 31 years at the helm, a series of pictures of all the employees was taken. These are shown in (8.23). Included in the picture of male 'speaker and

A Pair of Wharfedales: The Story of Gilbert Briggs and his Loudspeakers

transformer makers' are men involved in all the other occupations such as warehousing, packing and van driving. The total number of employees had now reached about 100. In the picture of the management, technical and office staff (8.23a) the following people can be identified: back row left to right; (1) Harold Reynard, (2) Colin Rhodes, (7) John McEvoy, (8) Eric Barker, (9) Joan Hodson and front row left to right; Dorothy Stevens, Ezra Broadley, Sheila Greenwood, Ken Russell, Gilbert Briggs, Leonard Chapman, Bill Escott, Dorothy Dawson, Bill Jamieson.

8.23a

8.23b

Chapter Eight: 1959–1965

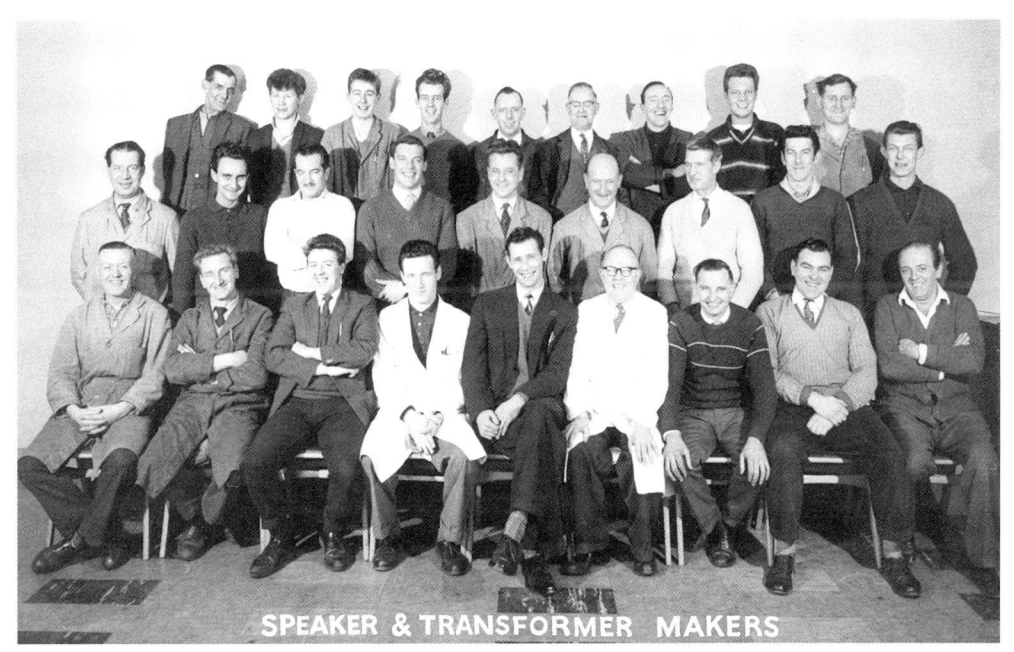

8.23c

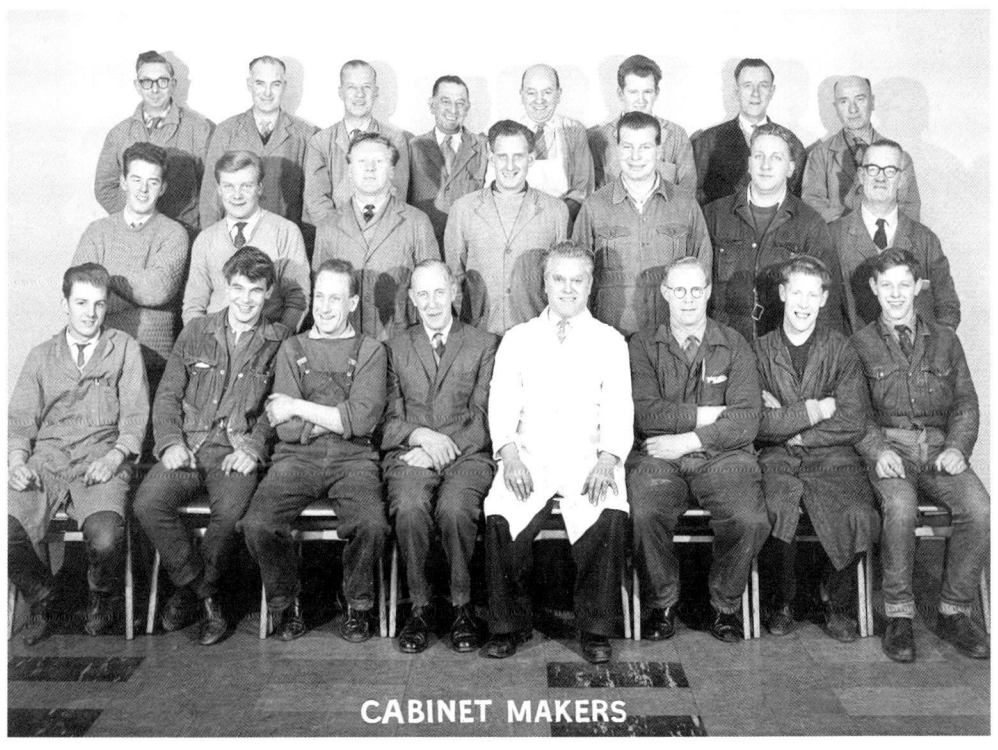

8.23d

Figure 8.23 Employees at Idle, January 1964. (Copyright—Bradford Museums and Galleries. From the C.H. Wood photographic archive negative numbers F56140–43. Photos courtesy of the Escott family.)

During 1964 there were no major additions to the product range. The 8" and 10" Bronze chassis were given roll surrounds and double diaphragms and two new crossover units were introduced: the QS/800 quarter section network with a crossover at 800 Hz and the HS/400/3 half section network with crossovers at 400 Hz and 3000 Hz. These replaced all the existing crossover units and as if to underline the end of an era the rest of the product range was also rationalised. The final list of Wharfedale offerings of Gilbert's time are reproduced in (8.24) and pictures of the chassis are reproduced in (8.25).

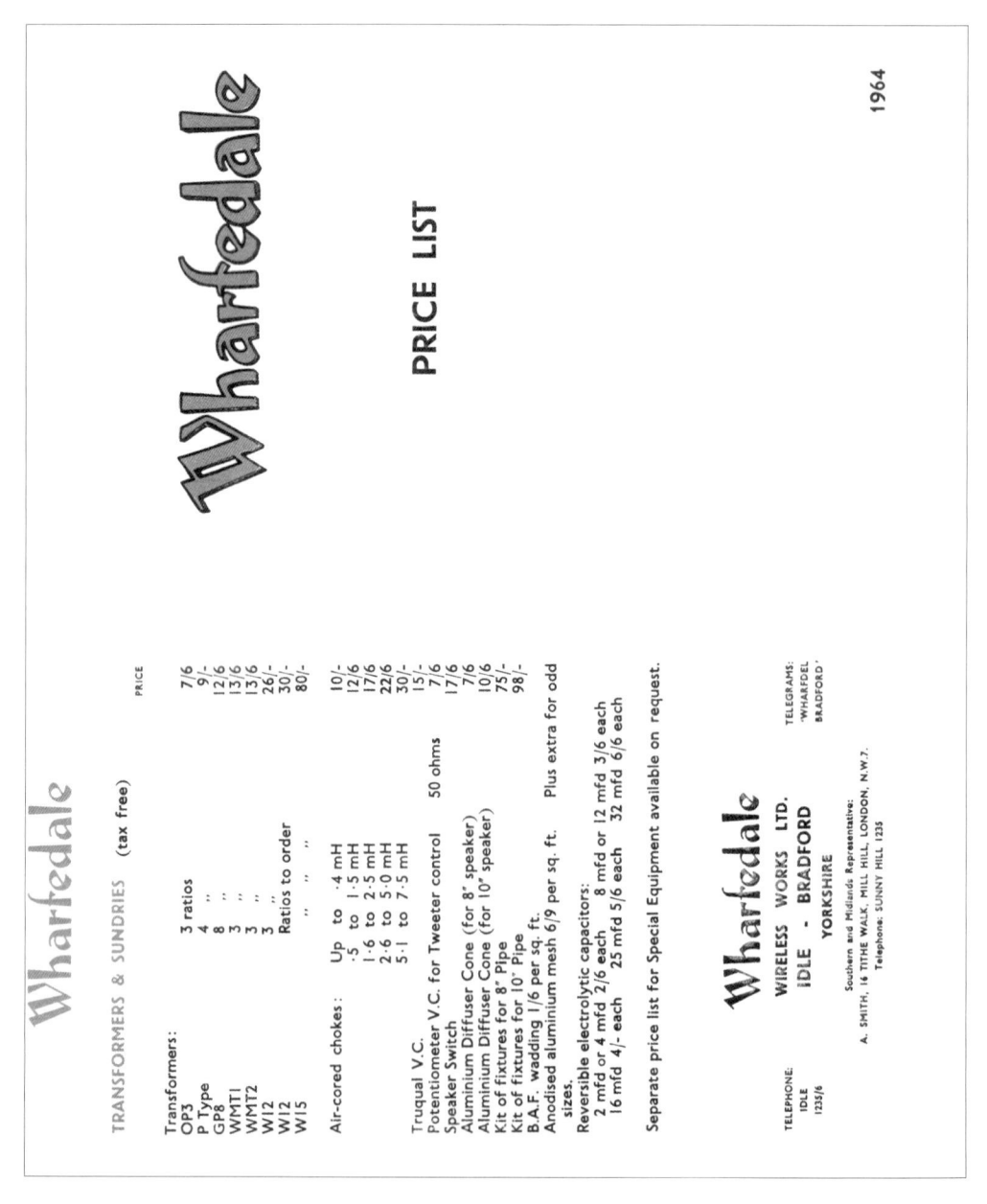

Figure 8.24 August 1964 price-list.

Wharfedale

CHASSIS	LIST PRICE	PURCHASE TAX	TOTAL
Super 3	100/-	16/8	116/8
PST/4	55/-	9/2	64/2
8" Bronze	50/-	8/4	58/4
8" Bronze/RS/CD	65/-	10/10	75/10
Super 8/RS/DD	115/-	19/2	134/2
10" Bronze/RS/DD	79/6	13/3	92/9
Golden 10/RS/DD	135/-	22/5	157/5
Super 10/RS/DD	187/6	31/2	218/8
W12/RS/PST	215/-	Nil	215/-
RS/12/DD	230/-	Nil	230/-
Super 12/RS/DD	350/-	Nil	350/-
W12/EG	210/-	Nil	210/-
W15/RS	350/-	Nil	350/-
W15/EG	350/-	Nil	350/-

EXTENSION SPEAKERS

Reflex Baffle	51/5	8/7	60/1
Bronze Baffle	95/-	15/10	110/10

CABINETS (without speakers)

	PRICE	Extra for tropical ply
PST/8 Polished	£10-10-0	£1-10-0
Unpolished	£9-10-0	£1-10-0
Whitewood	£7-10-0	£2-10-0

	Oiled Teak	Sprayed Grey	Whitewood
EG 12	£11-0-0	£10-0-0	£8-15-0
EG 15	£15-10-0	£14-0-0	£12-10-0

CABINET MODEL SPEAKER SYSTEMS

	PRICE
Bookshelf 2. polished or oiled	£16-10-0
unpolished	£15-15-0
(All Bookshelf 2 cabinets fully tropicalised)	
Slimline 2. polished or oiled	£22-10-0
unpolished	£22-0-0
Tropical 30/- extra	
W2. polished or oiled	£29-10-0
unpolished	£28-10-0
Tropical 35/- extra	
W3. polished or oiled	£39-10-0
unpolished	£38-10-0
Tropical 40/- extra	
W4. polished or oiled	£49-10-0
unpolished	£48-10-0
Tropical 50/- extra	
Airedale. polished or oiled	£65-0-0
unpolished	£64-0-0
Tropical 55/- extra	

All Wharfedale cabinet models are now available in the following finishes :-

Polished walnut or mahogany veneer.
Oiled teak veneer.
[Polished oak veneer (light, medium or dark) available to special order.]

CROSSOVERS

		PRICE
QS/800	800 c/s	£2-17-6
HS/400/3	400 & 3000 c/s	£6-5-0

Figure 8.24 (continued) August 1964 price-list.

Super 3

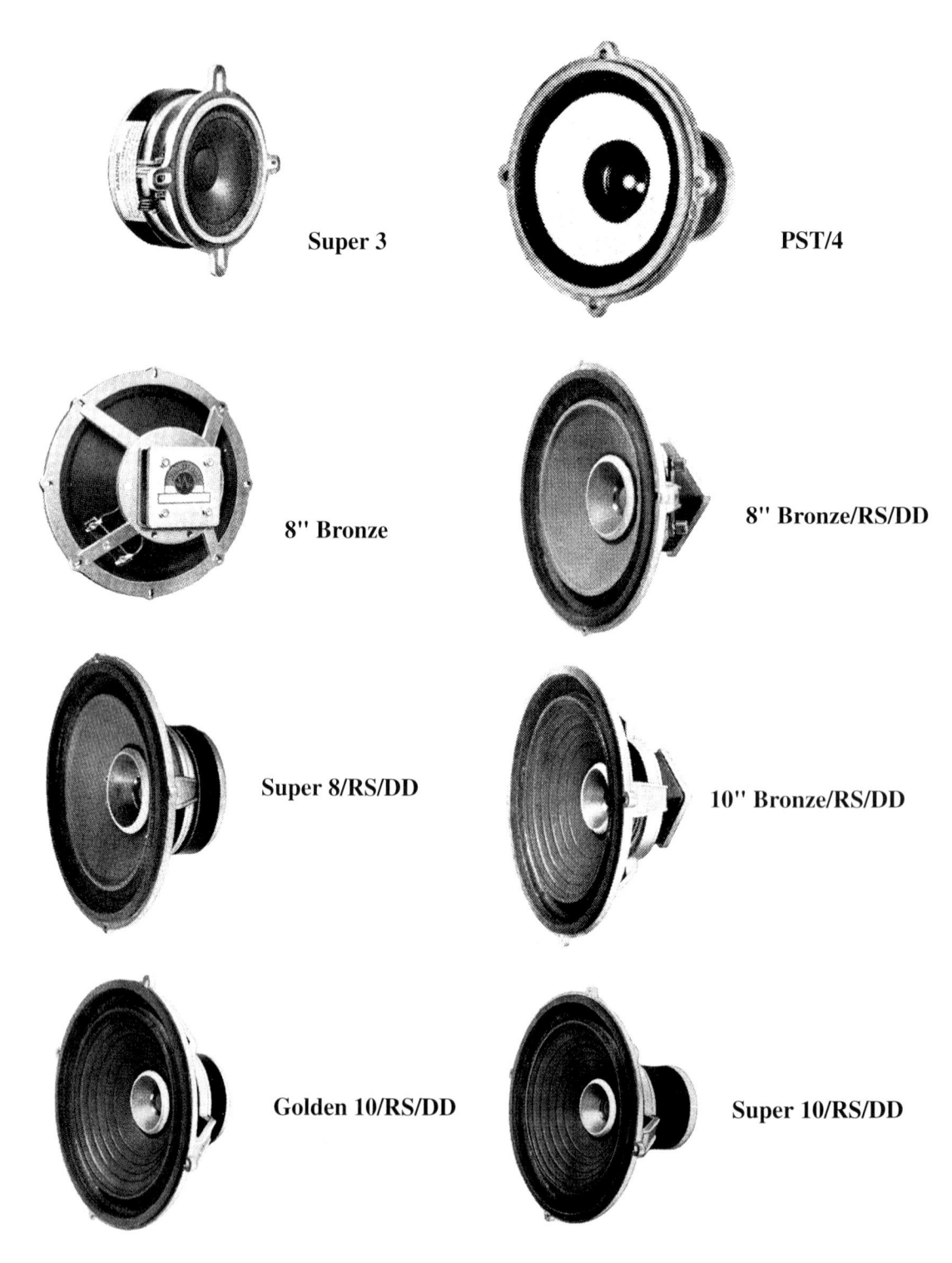

PST/4

8" Bronze

8" Bronze/RS/DD

Super 8/RS/DD

10" Bronze/RS/DD

Golden 10/RS/DD

Super 10/RS/DD

Figure 8.25 Chassis listed in 1964.

Chapter Eight: 1959–1965

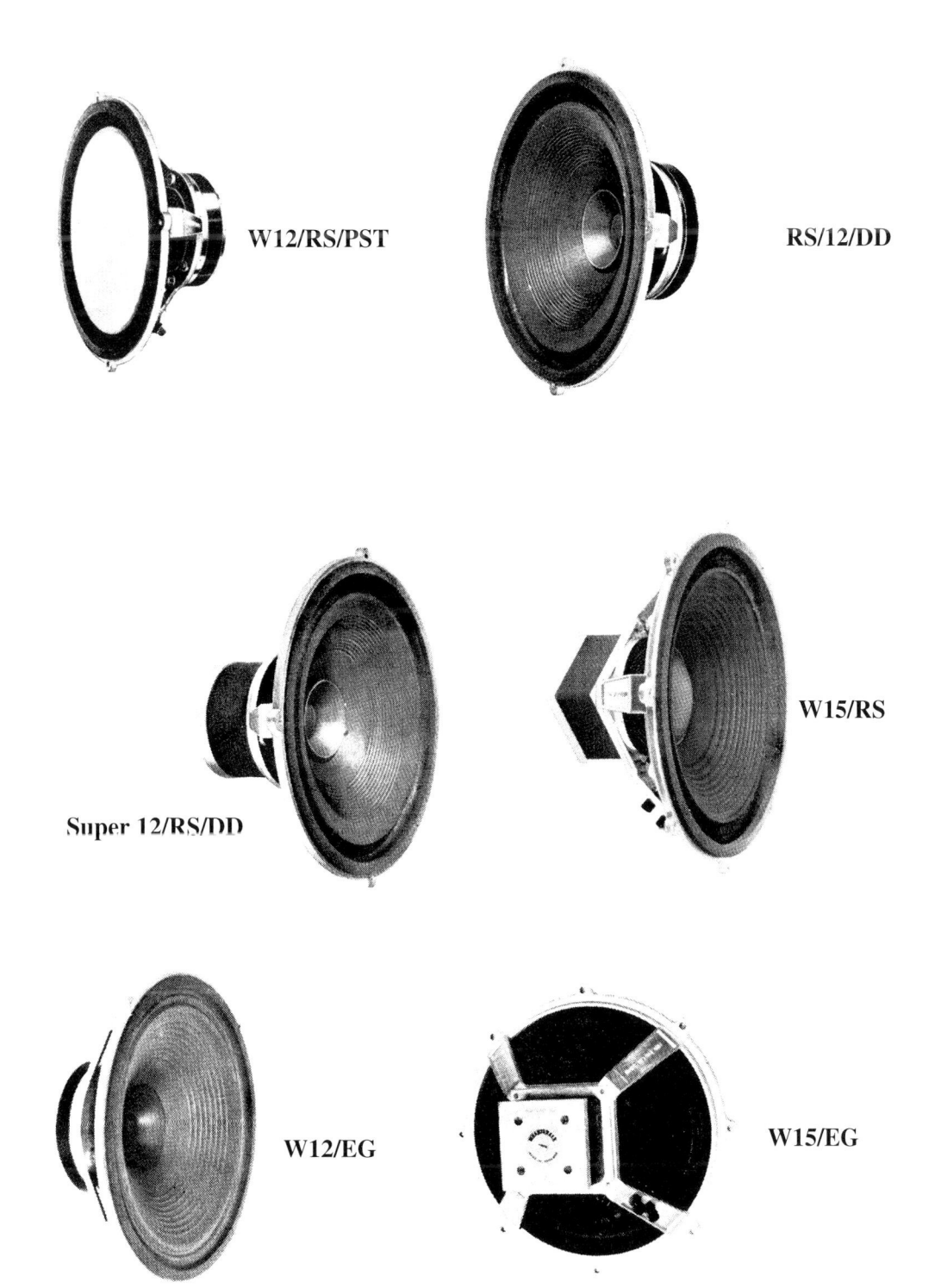

W12/RS/PST

RS/12/DD

Super 12/RS/DD

W15/RS

W12/EG

W15/EG

Figure 8.25 (continued) Chassis listed in 1964.

BIC-Wharfedale's Achromatic Systems

Since the introduction of the Wharfedale 60 by BIC-Wharfedale in 1960 the American range of cabinet systems had been increased to include, by 1963, the W40 and W60 (both two speaker systems), the W70 (three speakers) and W90 (six speakers) all based around the concept of sand-filled cabinet panels. In late 1964 the range was updated and the MkII models were referred to generically as 'Achromatic Speaker Systems'. BIC's advertising was light years away from the norm in the UK. In a 31 page brochure introducing the range, and associated display materials, for potential dealers in 1964/5 they distinguished between two rapidly diverging markets:

'One market is the group of novices who tend to buy recommended promotional systems; or are influenced mainly by budget. The other consists of people with high standards, both in music and equipment. These are your best customers. . . . But discriminating buyers are also in a position to shop critically, and today, they select speaker systems offering a particularly high degree of character and distinction, as well as technical excellence. To satisfy these customers, the dealer requires a line of speakers systems with special appeal—a complete line with an established reputation for quality in all the key price brackets . . . Wharfedale's Achromatic Systems cover the entire quality range (with the exception of exotic or unusually costly units).'

An illustration of the, hugely successful, type of double spread advertisement with a highly educational message, which BIC used for these systems, is shown in (8.26). References to G.A. Briggs as well as the Carnegie Hall concerts—even though nearly 10 years earlier—were usually included to add to the caché.

Although the BIC-Wharfedale systems were positioned towards the top end of the market, the marketing people also wanted to be able to sell higher volumes of smaller, cheaper units. Conversations with Idle personnel about this started during 1962 when Ken Russell went over to BIC at Port Washington. A subsequent visit to Idle by someone from BIC became legendary. He introduced himself to Gilbert, who was only ever referred to by *any* of his employees as 'Mr Briggs', with "Hey, Gilbert—I'm Larry!" The deferential witnesses were shocked to the core. Their boss, who had turned making outrageous statements for effect into an art form and who could slip into the American idiom on touching land in the USA, would have been secretly amused by their reaction. Two new systems emerged from this project during Gilbert's year as Chairman, as described in the next chapter.

Retirement

Gilbert's writing showed no let up during this transition period with *Audio and Acoustics* (sub-editor James Moir) published in November 1963, followed by *Aerial Handbook* (Technical Editor R.S. Roberts) in October 1964. Both were part of a project to rewrite *Sound Reproduction* which had gone out of print in 1962 (see Chapter 12). He was still not short of ideas for further books to work on when he really did retire from Wharfedale. This happened in February 1965 when Peter Dye, another Rank appointee, took over as Managing Director and the role of Chairman disappeared. After Gilbert stood down as MD Dorothy Stevens says he 'just faded away' and the change in early 1965 was merely another point along the way. His lingering association with Wharfedale still had a long time to run.

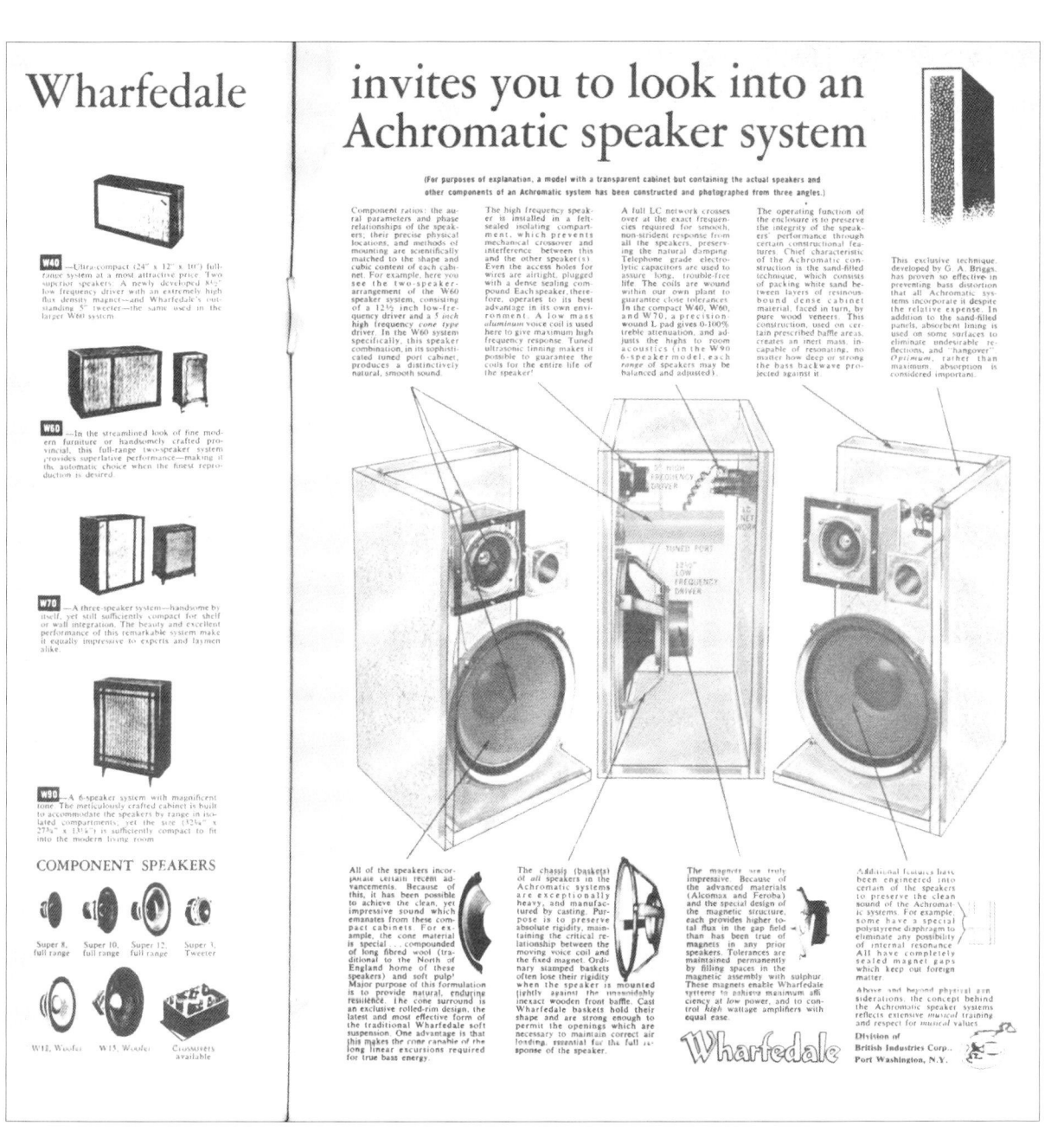

Figure 8.26 BIC-Wharfedale advertisement, 1964/5.

Part 2. Away from Wharfedale

In the summer of 1959 Gilbert's younger daughter, Valerie, was married. She had met her husband, John Pitchford, in Durham whilst she was working on the planning of new towns in the county and he was in his final year at the University. They moved to Cambridge, where John was a theology student, and it was during one trip to see them there that Gilbert asked to be taken to King's Lynn to see the school to which he had been sent following his father's death (see Chapter 2). Ninetta had her fourth child (a girl) in 1960 so when Valerie had her first (a boy) in 1961 Gilbert tally of grandchildren had reached five. John became a curate in Hereford and there Valerie bought a harmonium for £5.

At the Keyboard

The only picture* of Gilbert at the keyboard which has come to light is of him playing this instrument during one of his visits (8.27). Valerie ran a Sunday School which also had a harmonium—she persuaded her father to play this for the children whenever he came to stay. Gilbert was a most reluctant performer in public, although he enjoyed playing the piano for and with family members (playing another instrument or singing). Dorothy Stevens tells of one rare occasion when he succumbed to pressure. She had been talked into giving a dancing performance at a Wharfedale 'do', probably following a Christmas lunch, and she required a pianist to accompany her. Somehow she got Gilbert to agree to do this and then they needed to rehearse. The first rehearsal caused some embarrassment because he had to take instructions from her—a reversal of their normal roles. However, he recovered his poise at the second rehearsal by giving Dorothy an autographed copy of Gerald Moore's book *The Unashamed Accompanist*!

Figure 8.27 Gilbert playing the harmonium, about 1963. (Photo courtesy of the Briggs family.)

*This is a Polaroid print taken with a Model 120 camera. Gilbert claimed it was one of the first to be imported into the UK—he had been interested for some time, having seen Polaroid cameras in the USA. The 120 first appeared in 1961 and his was manufactured in Japan. He bought it from Wallace Heaton Ltd in London.

Chapter Eight: 1959–1965

The Pitchfords had two more children (both girls) and so by 1966 there were seven grandchildren. Gilbert loved to be with them and follow their progress—a rather interesting statistic is that four of them became professional musicians—so there was a lot of travelling between the two families during the 1950s and 60s.

On the Radio

In January 1960 Gilbert communicated with audio enthusiasts through a new medium—national UK radio. It wasn't the first time he had been on the radio; there had been at least two occasions when this had happened in the USA, both of which gave rise to reminiscences in print. The first was during the 1953 Audio Fair in New York when the Editor of *Audio*, Charles G. McProud, invited Gilbert to take part in a hi-fi programme (answering questions) in which his magazine was collaborating with one of the big broadcasting companies. Gilbert finished his description of the experience thus:[16]

'Our programme ended at 3.30 [pm] prompt. At 3.31 the telephone rang and a man called me to it, saying: "Your fans are ringing up." Feeling very important, I dashed to the 'phone. It was a fellow Bradfordian, living in New York, who had rung up to say how thoroughly he had enjoyed hearing an authentic Yorkshire accent once again. This, I have to admit, was the only reaction to my New York broadcasting premier.'

The second followed the Carnegie Hall concert-demonstration in October 1955, when he and Peter Walker were interviewed by local radio station WQXR:[17]

'With Mrs P.J. Walker looking on from the distinguished stranger's ante-room, the announcer introduced Mr Walker to the listening multitudes as my assistant! This made my day'

More Yorkshiremen (and women) would have tuned into the programme *Sound* on BBC Network Three, which shared the wavelengths of the Third Programme, known now as Radio 3. The first editions of this series, the brainchild of Mrs Marguerite Cutforth, had started a year earlier. They covered a wide variety of topics which fell within the broad scope of the title, often explored through interviews. The first presenter was John Borwick who, in January 1959, had recently left the BBC, where had had been a Studio Manager, followed by various training roles, to become a freelance journalist. Following Gilbert's farewell concert-demonstration at the Royal Festival Hall in 1959, Borwick invited him to do two programmes at the start of the 1960 series. The first on 17 January was about loudspeakers in concert halls whilst the second, a fortnight later, was about loudspeakers in general. The programmes were repeated in the following week. Borwick recalled[18] that from his point of view Gilbert was 'easy money'. He (Borwick) received a fee for both the original broadcast and the repeat. With many of his interviewees achieving a flowing conversation was difficult and he had to ask many questions during the programme. All he had to do with Gilbert was ask the initial question and the rest of the programme took care of itself! Of course, Gilbert approached these 'talks' with his usual thoroughness, collecting material from a variety of sources. A request to his concert-demonstration stalwart, the famous oboist Leon Goossens (see Chapter 11), for permission to use an extract from a BBC performance, resulted in the revealing note reproduced in Appendix 1. Later programmes included reviews of *A to Z in Audio* (8.1.61) and *Audio and*

Acoustics (29.3.64) whilst Gilbert was interviewed in programmes relating to the Audio Festival (12.4.64) and Audio Fair (10.5.64) events of 1964.

Betting

Gilbert had a fascination with gambling. This would have been an activity frowned upon during his Weslyan upbringing and he, himself, rarely indulged (although he let a broker 'gamble' with his money on the stock market!). However, small wagers over trivial matters seem to have added spice to his interactions with friends and colleagues. Ninetta recalls how he suddenly bet someone, in the middle of a conversation, that he was wearing odd shoes. He owned several pairs of near-identical black Oxfords and never knew whether he was wearing an original matching pair. He was incredulous that the bet was taken. The odds were, of course, that he was wearing 'odd' shoes and he duly won the bet. Ken Russell once found him studying something in his *Daily Telegraph*; Gilbert, out of the blue, bet him sixpence that he could not tell him the difference in temperature between the highest and lowest maxima in the world table. He said to the startled Russell, "I'll give you a clue—this was the number for yesterday …". Gilbert loved all kinds of information tabulations and obviously worked this out on a daily basis. It may have been a similar interest in the racing results that prompted him to write up an experiment on racing tips in 1961. The wry humour which characterises his books, and which was a key element of his lectures and demonstrations, is on full display as seen in (8.28).

Animal Farm

There had been no diminution in Edna's enthusiasm for animal husbandry over the years, quite the reverse in fact, and so long as it did not directly impinge on him, Gilbert was remarkably tolerant of the situation. However, over the years part of the grounds of Woodville (in what had originally been an orchard, walled off from the rear garden) had become home to a variety of animals and the house itself was under constant threat of turning into a farmhouse. The episode of the piglet on the loose, described in Chapter 6, was but a portent of the future. Bill Jamieson, who had been to the house several times, half-joked to Ken Russell, when he arrived at Wharfedale in 1962, that anyone who rang the front door bell was more likely to be met by a goat than a human occupant! Eventually, Gilbert tried to avoid inviting anyone to the house because he could never be sure that it would be 'safe' to do so. Once he was spending more time at home than at work, though, his tolerance was severely tested and matters came to a head when he was phoned one day to be told that some animals had escaped and were roaming the neighbourhood. This was by no means an isolated event—as described in Chapter 6 the goats were always on the move for fresh grazing—but Edna was not on the premises. When she returned Gilbert said in exasperation "You need a farm, I'll have to buy you one!"

In due course a farm near Blubberhouses, not too far from the 'holiday cottage' which Edna had found during the War (see Chapter 6), came up for auction. She felt that Meagill Hall Farm would be just the job and decided to bid for it. She went to the auction alone, with a warning from Gilbert not to bid more than £10,000. Although experienced in buying furniture at auction, Edna got carried away in the bidding and went well over this limit before it was knocked down to her. Of course she was expected to settle on the spot but did not even have her cheque book. She then tried to explain that her husband, the actual purchaser, might not agree to pay for the farm because she had gone over his limit.…, then the realisation of the fix she was in almost made her faint. No doubt this re-

Chapter Eight: 1959–1965

<pre>
                    APRIL 1st 1960

                 TO WHOM IT MAY CONCERN
                    by G. A. Briggs
                    ------ -------

        In the interests of my fellow man, I have been making an
    investigation into horse racing and tips. The date of this report
    aptly reflects the wisdom of the exercise.
        I sent a cheque for £50 to my turf accountants and instructed
    them to put £1 each way on the winners tipped by Hotspur of the "Daily
    Telegraph" each day until I cried Halt!   Here is the result.
                                 Lost   Won
           25 March 1960   £4. 0. 0   £2. 7. 6
           26   "   "   £10. 0. 0   £6. 5. 0
           28   "   "   £ 6. 0. 0   £8.17. 6
           29   "   "   £10.17. 9   ---
                                £30.17. 9   £17.10. 0

    At this point I said "Stop", with a net loss of £13. 7. 9. On two days
    five of Hotspur's "winners" out of six did not even get a place.

        On the 30th March I decided to pick my own winners, and I put
    £2 each way on three horses selected without rhyme or reason.   One
    of them came in third - the favourite - so I won 10s. 0d. on it, but I
    lost £8 on the other two.

        Total loss in five days was £21.17. 9.   I have asked the bookie
    to send back my £29.2.3 credit balance.

        This scientific research confirms what the reader already knew,
    that backing horses is a mug 's game, but it also shows that, as a tipster,
    I am quite as good as Hotspur.

    P.S. If any kind-hearted reader feels that he (or she!) would like to
          share the cost of this valuable financial experiment, will they
          please address contributions to G.A.B., c/o Westminster Bank
          Ltd., Bradford, a/c Bookies' Benevolent Fund, or to Menston
          Asylum.
</pre>

Figure 8.28 Gilbert's horse racing experiment, copy given originally to Raymond Cooke. (Scan courtesy of the Briggs family.)

sulted in another phone call to Gilbert, but the farm was hers and sometime during 1966 she started her farming career proper. Although Edna was nearly 11 years younger than Gilbert she was still approaching her 65th birthday when she took on this challenge, but she was a very determined lady and no stranger to hard work. Most days she would drive her increasingly battered open Landrover over to Blubberhouses (a typical Dales cross-valley route which demands respect even in decent weather), spend several hours on the farm, and drive back in time to make the evening meal.

213

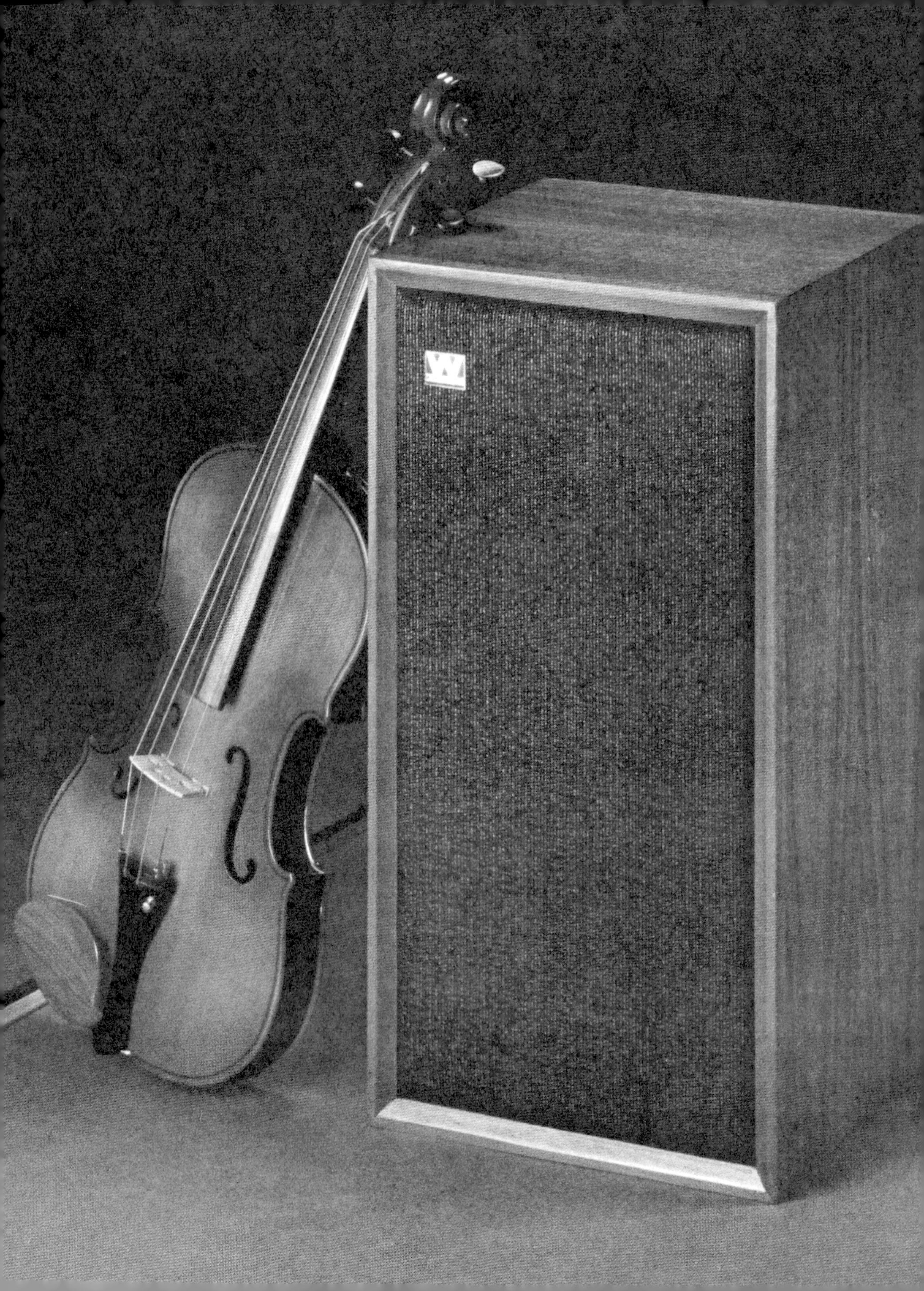

CHAPTER NINE: 1965–1978

The Evolution of Wharfedale under Rank, Bradford Road and Highfield Road, Idle

Two management decisions made not long before Gilbert stepped down as MD sowed the seeds of significant changes to the company's products and their appearance which started to take effect just as he finally retired. The first, an internal move based on the recognition that the external finish of the cabinet systems was rather dated, was to take on an appearance-design consultant. The second, driven by Rank, was to set up a small electronics team so that the product range could be broadened.

Finish Design and Electronics

No one at Wharfedale knew any professional designers so they turned to the Council for Industrial Design (now the Design Council) for help. They forwarded a list of possible consultants and, more-or-less at random, one was chosen. Robert Gutmann was invited to Idle, probably towards the end of 1963, for an exploratory consultation. Far from having to be briefed, he had already researched Wharfedale and its products and came armed with a portfolio of ideas and sketches. The practical implications for cabinet finishing were, of course, significant. Harry Bale, who would have to implement new design features, had to be won over but once he had 'bought into the future' the way was clear and Gutmann was hired.

Getting help with diversification into electronics was rather easier. Gilbert had a tame electroacoustics engineering consultant in Stanley Kelly, a fellow Yorkshireman from the Leeds area with whom he also shared a birthday. Kelly had been a sparring partner over the relative merits of competing technologies (and his articles about Wharfedale's products and other activities) for years and he had often helped Gilbert out with specialist contributions to, and informal commentary on, his books. His knowledge, experience and inventiveness were second-to-none. Whether he was consulted for advice on new product development or where to look for appropriate recruits etc. is not known, but his contribution was probably wide-ranging. The upshot was the setting up of a small electronics team during 1964 charged initially with the development of amplifiers as part of a move into PA systems, but with the objective of adding hi-fi amplifiers and tuners in due course. George W. Tillett was recruited to establish the team as a member of the Wharfedale Board of directors.

The First Linton

The collaboration with BIC on smaller systems resulted in two models which were added to the 'Achromatic' range in the USA in 1965. The existing models, described earlier, had been updated and designated W60D etc. so the new models were the W30D (19 × 10 × 9¼") and W20D

(14 × 9¾ × 8½"). Both had acoustic suspension enclosures with a new 8" driver and a new 3" polyester-domed, omni-directional pressure tweeter. The 8" unit incorporated a newly developed roll surround made from neoprene, which the marketing people had decided to call 'flexiprene', which was a key factor in its performance. These units were not available for purchase but a 12" driver with the new surround, the W12/FRS, was added to the list in the UK.

The W30D work led to an almost equivalent UK model which was called the 'Linton'. Its styling was the first manifestation of Gutmann's input (9.1) and he also styled updated versions of the 'Slimline 2' (named the 'Dalesman') and the W2 ('Dovedale') which were also introduced in 1965. Both of these made good use of the new W12/FRS bass unit. The annual catalogues for 1964 and 1965 adhered to the same template which Gilbert had used for years, so either he was responsible for them or someone else did a very good job of imitating his style in the new entries. The 1965 catalogue was the last of its type and as it was reprinted during the year one entry underlined the end of the era. Gilbert's loyal 'Southern and Midlands Representative', Albert Smith retired, with over 30 years' service, to be replaced by a Mr N. Eighteen.

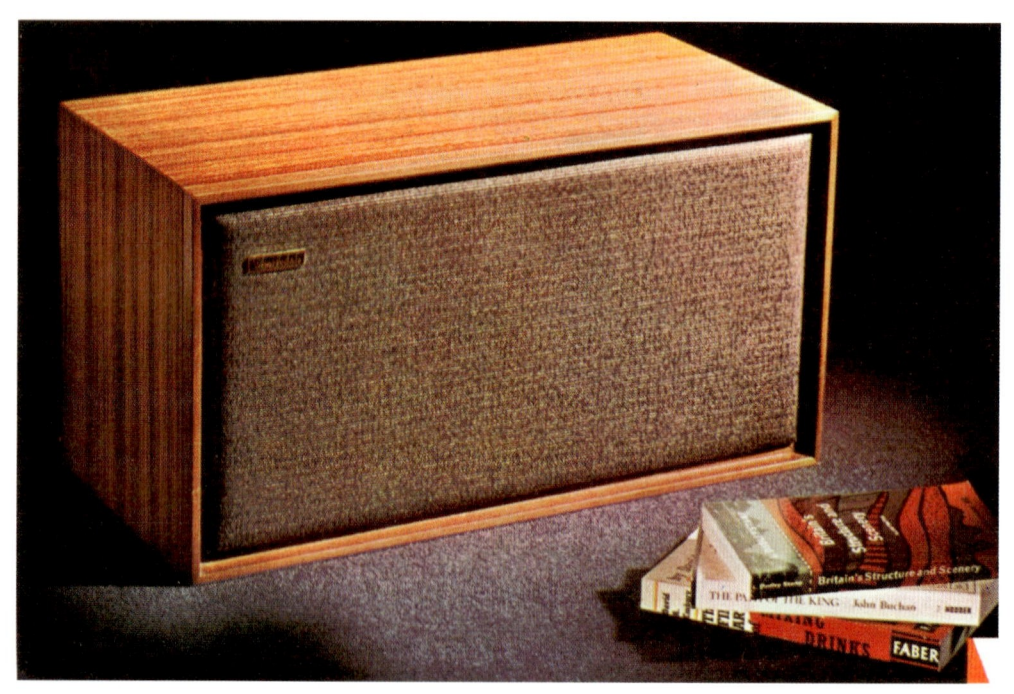

Figure 9.1 Linton, 1965 brochure.

Rank Wharfedale

By 1966 the company had been renamed Rank Wharfedale Ltd and the Gutmann influence was reinforced by a new emphasis on marketing, with more than a nod in the direction of the BIC approach across the Atlantic. Out went the 20-year-old 'Wharfedale' logo in its slightly oriental font, as in (8.23), and in came the 'W' logo, which was itself destined to last for 15 years. This first appeared on a 1966 brochure for the complete range which included the same kind of product information as in Gilbert's catalogues but in a style much closer to that current in the USA. As the front panel shows (9.2) each loudspeaker system in the range now had its own brochure. The 'Airedale' had

Chapter Nine: 1965–1978

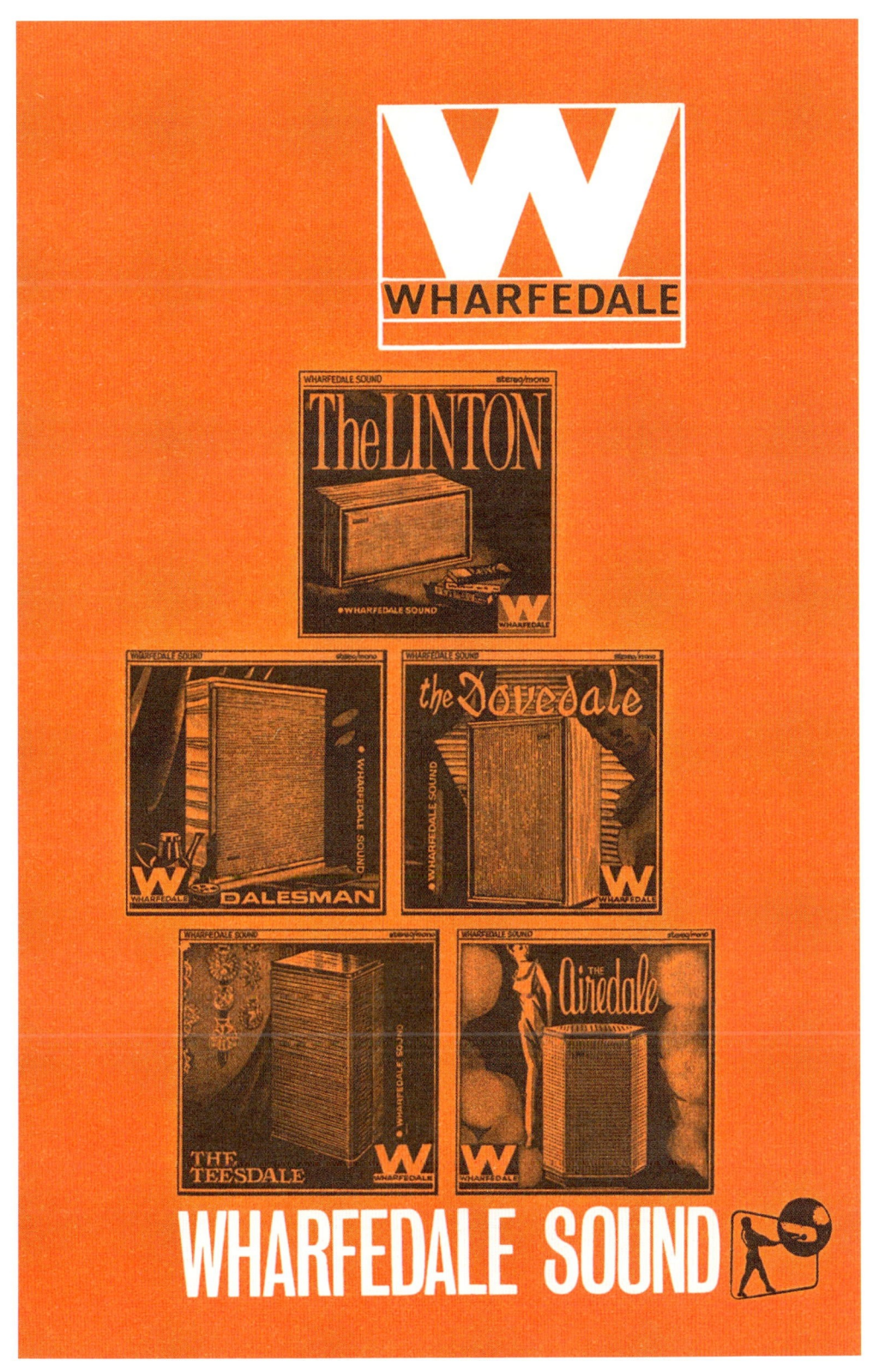

Figure 9.2 1966 catalogue.

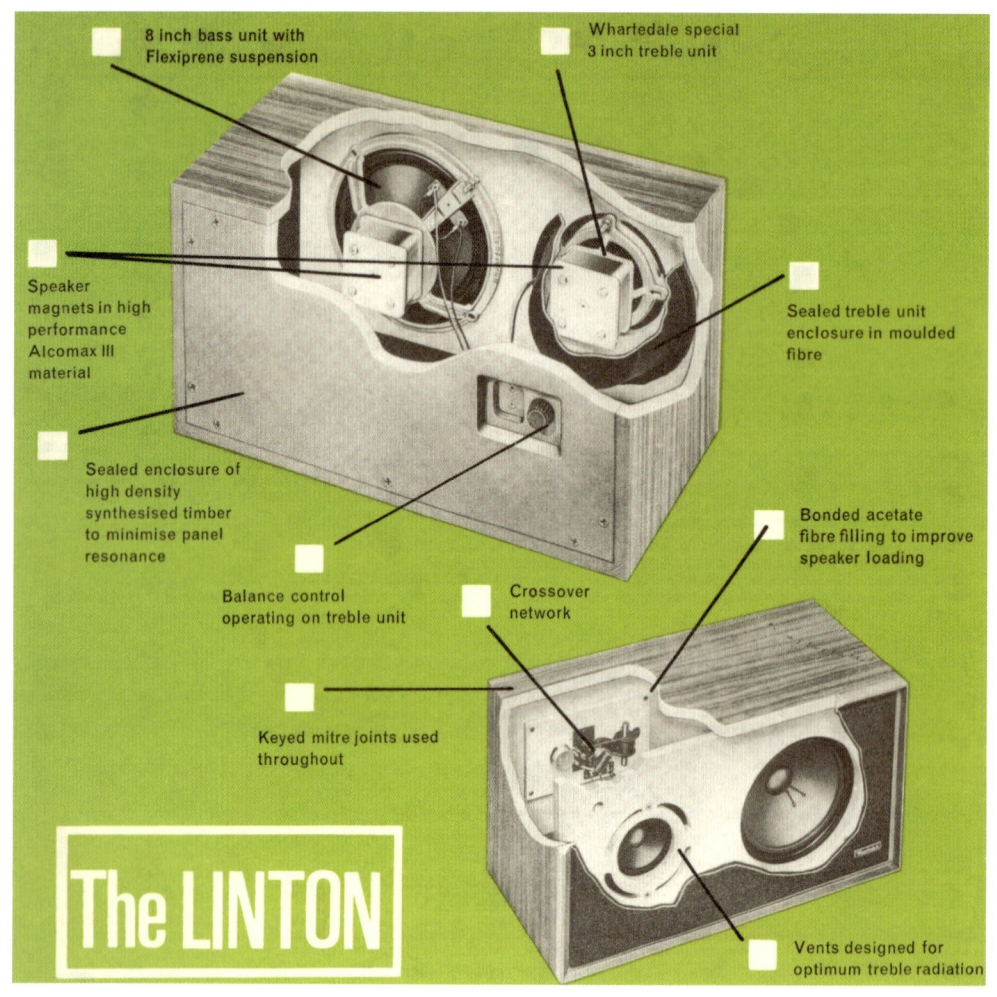

Figure 9.3 Linton brochure.

been restyled and the W3 replaced by a three-way model called the 'Teesdale' with styling similar to the smaller 'Dalesman'. The individual system brochures now contained the cut-away explanatory diagrams which had gone down so well in the USA. That for the Linton is shown in (9.3).

PA and System 20

1966 was also the year that the PA project bore fruit, with the introduction of a family of new products. There were three solid state amplifiers, two of which were for general purpose PA and rated at 20 W rms (GP20) and 50 W rms (GP50) respectively; the former was a portable unit (mains or battery powered) whilst the latter could be housed either in its cabinet or in a standard 19" rack. The third, rated at 30 W, was part of a complete portable PA system (PA30), also mains or battery powered, housed in a three foot long case with an integral line source speaker (six 5" units) as shown in (9.4). There was also a matching extension speaker (XT30). Other speakers included two wedge-shaped, wall mounted units (GP503 and 505), a cine speaker (GP510) and a family of units with a flared design: the wall mounted GP545 (single 8" driver); the line sources GP575 (with

Figure 9.4 PA30 from a flyer of about 1964.

seven 5" drivers) and GP585 (seven 8" and five 3" drivers). These line sources were of sophisticated design and intended for speech reproduction. The GP600/B, a combination of a line source with six 5" drivers and a bass reflex cabinet (essentially an update of the LS6/B), was intended for speech and music reproduction in auditoria.

It was during 1966 that John Collinson joined Wharfedale. As described in Chapter 11 he had been a key member of Gilbert's concert-demonstration team in the 1950s, whilst working for Peter Walker at Acoustical Manufacturing (Quad). After 13 years there he decided it was time for a change and put out feelers into the industry. Gilbert had tried to interest him in joining Wharfedale some years earlier, but whether he played any role in this move seems unlikely. At any rate his arrival at Idle came as a complete surprise to Ken Russell. Collinson's forte was electronics—amplifiers and tuners had occupied his time at Quad—and although his experience was a useful addition to the electronics effort he was more interested in getting into loudspeakers for the first time. About a year earlier Russell had been allowed to hire a graduate assistant who overlapped with Bill Jamieson for a while before Bill left during 1966. The step change in the human resources available, both qualitative and quantitative, by 1967 meant that some serious background research into loudspeaker behaviour could be contemplated for the first time.

Meanwhile, the electronics team had delivered 'System 20'—matching stereo-tuner, amplifier and turntable, all styled by Gutmann—which were launched during 1967 (9.5). The VHF tuner (WFM-1) came in two versions, either self-powered or powered from the 20 watts per channel amplifier (WHF20). As Wharfedale's first foray into hi-fi electronics they were well received in magazine reviews. The turntable unit (WTT-1) was sourced from Thorens (TD124 professional transcription turntable Series II, TP-14 pickup) and housed in a matching teak box. At the same time two new two-way speakers were introduce which were specifically marketed as matched pairs for stereo. The first was an improved version of the 'Linton' ('Super Linton') and the second, the smaller 'Denton', was derived from the BIC-Wharfedale W20D of 1965. There was also a radiogram incorporating System 20 units, named the 'Selby', but the cabinet manufacture was outsourced.

Figure 9.5 System 20. (Photo courtesy of IAG.)

By this time the workforce had increased to over 200 and expansion on all fronts meant that the Bradford Road site was soon going to be too small. Rank purchased a site on Highfield Road, about a mile from the Bradford Road site beyond the Five Lane Ends roundabout, which was virtually green field apart from a large dwelling called Shaw House. A project then started to build a completely new factory on this site.

DIY Kits

The only new introduction of 1968 was a kit (Unit 3). The cabinet construction sheets had probably not survived for very long after Gilbert's departure and the sale of drive units and crossovers as separate items was phased out from around 1967 (after 1969 only the RS/DD versions of the 8" Bronze, Super 8 and Super 10 chassis were still available), but the DIY enthusiasts had not gone

Chapter Nine: 1965–1978

> Your first venture into build-it-yourself hi-fi? Choose Unit 3, most popular in the Wharfedale range of kits. End product is a neat, space-saving cabinet which will stand on a bookshelf, and give amazingly balanced reproduction considering its size.
>
> ## Specification
>
> | | |
> |---|---:|
> | Power handling capacity | 15 watts |
> | Impedance | 4-8 ohms |
> | Frequency range: | |
> | Cabinet K1 | 65-17,000 Hz |
> | Cabinet K2 | 40-17,000 Hz |
> | Bass resonance of 8" unit | 35 Hz |
>
> 8" Bass/Mid Range speaker (12,000 oersteds).
>
> Crossover network (crossover frequency 1750 Hz) to ensure optimum response from each speaker.
>
> Acoustiprene Tweeter (10,500 oersteds)
>
> As in all Wharfedale kits, the necessary acoustic wadding, bolts, connecting wires, etc., are included.
>
> **External cabinet sizes**
>
> K1 Smallest cabinet size recommended 14" wide × 9¾" high × 8¾" deep (35·4 cm × 24·8 cm × 22·3 cm)
>
> K2 Largest cabinet size recommended 11½" wide × 21" high × 9½" deep (29·2 cm × 53·1 cm × 24·2 cm)

Figure 9.6 Unit 3 kit leaflet.

away. The kit rather cleverly ensured that this following could concentrate on building the cabinet whilst all the other components for a two-way speaker were provided. As the Unit 3 leaflet (9.6) shows the smallest recommended cabinet size resulted in a speaker equivalent to the Denton, whilst the largest was similar to the Super Linton. Since each kit cost £10.50, two speakers could be built for much less than the price of factory-made equivalents (Super Lintons were £41.20 a pair and Rank Wharfedale's estimate of the cost of cabinet construction materials for a pair was under £5).

221

A lengthy article in the February 1969 issue of *Hi-Fi Sound*, entitled 'Build a Bookshelf Speaker', supplemented the manufacturers assembly instructions by describing in detail how to build a Unit 3 cabinet and commented that the performance of the finished speakers was 'far superior' and 'of an entirely different breed' to factory-made 'budget' bookshelf speakers costing as much. It also pointed out that even the cabinets could be purchased in kit form from Eaton Audio Fitments, who were also suppliers of the column speaker kits at that time.

One event of note during 1968 was a flood which affected part of the Idle factory. This happened in June when torrential rain resulted in a massive accumulation of water on the adjacent International Harvesters site, higher up Bradford Road (see Fig 7.1 in Chapter 7). A retaining wall collapsed and water and dirt cascaded over the bank into the buildings below, where much of the non-cabinet production took place, and caused significant damage to both machinery and finished electronics stocks. Not long after this Dorothy Stevens, on completing 25 years with Wharfedale, left to make dance teaching her main occupation.

Rank Acquires Leak

Later in the year, events in London were shaping up which were to have long-term consequences for Wharfedale. For reasons including his own health, but more immediately that of his wife, Harold Leak decided to retire and put H.J. Leak and Co. up for sale. There were several interested parties and a deal with A.C. Farnell of Leeds was close to completion when John Davis at Rank was made aware of the situation. He had spare cash at his disposal, from a failed bid for De La Rue, and clinched the deal by offering £1,034,000 (over 10% more than the offer on the table).[1] The transfer to Rank took place in January 1969.

The formation of Rank Leak was a huge blow to the Leak employees because the decision was quickly made to consolidate all the Rank hi-fi activity in the Leeds–Bradford area. Coach loads of employees (most worked at Brunel Road in Acton, London but others were at Downham Market in Norfolk) and their wives visited Yorkshire to help them decide whether or not to move. Ultimately most opted for redundancy. Ted Ashley, the Chief Engineer, moved to Yorkshire (and into management) and had to begin assembling a new electronics engineering team almost from scratch. Since there was no room on the Bradford Road site and the Highfield Road development was in its early stages, 'temporary' premises were rented: Airedale Works at Apperley Bridge, some two to three miles north-east of the Idle sites. New product development projects were initially undertaken there and production in London either phased out or transferred when appropriate. During 1971 factory operations were fully re-established 'up North'.

For the Rank Wharfedale operation the effects of the Leak incorporation were more subtle, but did have a significant impact. The two companies had been competitors to a greater or lesser extent in different sectors of the hi-fi market. H.J. Leak and Co. had been founded as a one-man band, only two years after Gilbert founded Wharfedale Wireless Works, and specialised in building amplifiers. The growth and diversification of the firm has been described in great detail elsewhere,[1] but by 1969 Leak supplied the full range of hi-fi products including loudspeakers based on the so-called piston-action, sandwich-cone, drive unit.[2] In the electronics sector, therefore, Wharfedale were just starting to compete with a well-established player whereas in the loudspeaker sector the Wharfedale market share dwarfed that of Leak. Obviously, Rank wanted to create a win-win situation by capitalising on the strengths of the two brands, with their different technologies and design philosophies, whilst achieving some economy of scale and greater market share, from the combination. However, growing two brands which both offered integrated hi-fi systems and which had previously been in competition posed several challenges, not least of which was the marketing strategy.

Marketing at Wharfedale had moved on since Gilbert's low key approach, with influences first from the BIC-Wharfedale example in the USA and then the style guru Robert Gutmann, and Idle had its own marketing people. Their approach seems to have been to introduce high specification electronics, whose performance would be beyond doubt, and emphasise their association with the loudspeakers and their magical reputation. This changed with the need for a strategy to market both Wharfedale and Leak ranges and the function moved to a Rank site in London. For whatever reason, the strategy which emerged identified Leak as the premier brand, which today we would refer to as the 'high end', and Wharfedale as the second string. The Wharfedale engineers, at least, were livid. The early manifestations were apparently consistent: the Wharfedale System 20 range was dropped but the Leak loudspeakers were retained. It is also certain that these events hastened the rationalisation of the whole Wharfedale product range with the elimination of nearly all the non-cabinet system items (PA products, chassis, crossovers etc.).

Transformations

John Collinson's arrival at Idle during 1966 brought a different perspective to bear on moving coil loudspeakers. Whilst at Quad he had lived through, but not been directly involved in, the development of the famous full-range electrostatic speaker (first launched in 1955) and came with an understanding of the theoretical advantages of the ESL, but also the practical difficulties in realising them. He wanted to get to grips with the realities of the electromagnetic approach to transduction, believing that what was on offer in the marketplace was, from an absolutist point of view, still not that good. His questioning led to research, which he and Ken Russell carried out together, into the truth about the various forms of distortion (harmonic, inter-modulation, Doppler, delayed resonance) which were supposed to plague reproduction. Although progress was still very limited by the available measurement technology (see below) Collinson was convinced that much more work needed to be done on the mid-range units and he did this as part of a project to replace the W2-Dovedale, the 1965 update of the W2 (1959). The result was the Dovedale III, a three-way speaker with 12", 5" and 1" drivers in a cabinet with the same dimensions as the W2 (24 × 14 × 12"). The mid-range driver was acoustically isolated and the brochure boasted that the cone had an additive, 'Mistar', which helped control break-up—the result of research specifically into this phenomenon. In fact this was a plastic dope which impregnated and coated the cone material. Geoffrey Horn, reviewing the loudspeaker in August 1969, started with this personal comment:[3]

> 'It is as well to admit right at the start of this report that the various Wharfedale loudspeakers and units have fallen out of favour with me for the last several years. In my humble opinion the most recent landmark was the W3 three-speaker system which appeared in 1962 [sic] to give all the world's competition a hard time for some years—and still sound pretty good to this day. Most of its successors appear to me to have a rather old-fashioned sound and recent designs have the same mid-range colourations that appear to have crept into their individual units to give a somewhat hard and overly present flavour to the sound. Note that I said was their most recent landmark: there is a new one, the Dovedale III, now under review. Perhaps a new landmark for others too—for the W3 was the brainchild of the one and only Gilbert Briggs himself.'

Later on in the piece:

'I think the most difficult section for any loudspeaker is the part that receives least publicity, namely the mid-range from, say, 300—3000 Hz. There is no doubt that this is where the human ear is at its most sensitive and most easily upset by deficiencies of any sort. This was the worst area of the older W3 landmark, not because the response was poor or notably irregular, but because of the presence of overhangs or delayed resonances. . . . In the Dovedale III steps have been taken to minimise such troubles and taken together they have produced a most marked improvement.'

and finally:

'VERDICT: The Dovedale III is an impressive example of what can be done by further work with well-tried principles and it must be regarded as a welcome addition to the ranks of the middle-sized, middle-priced loudspeakers'.

At the same time that this work was being carried out, there were projects to introduce new models which had, in their original incarnations, been designed for the US market and assembled by BIC-Wharfedale. The 'Melton' was a two-way system (12" bass and 3" tweeter), slightly larger than the 'Linton' and derived from the W40D whilst the 'Rosedale'—a new flagship to replace the aging 'Airedale'—was a three-way version of the four-way W70D, incorporating a 15" bass unit. Unusually, the Rosedale kept the furniture-style cabinet of the US model and was marketed as 'for the connoisseur' and 'for the traditionally furnished home'. These derivatives did not have the sand-filled panels which characterised the US 'Achromatic' range. So, by the end of 1969 there had been a transformation in the Wharfedale range and the marketing was emphasising the leisure aspects of listening to well-reproduced music in the home. The five systems in the range are shown in (9.7)—Gilbert would surely have approved of the way in which the cabinets nestled against real musical instruments, even if a piano was not involved.

Big Ambitions

The following appeared in *The Gramophone* towards the end of 1969:[4]

'With their recent acquisition of H J Leak Ltd, and expansions of the Wharfedale factory by 50,000 square feet plus another 100,000 square feet being built, Rank Audio Visual seem set fair to become a real force in the audio world. The joint Managing Director, Gus Smith, told us at a press show recently that this is a planned operation to lift at least a part of the UK industry out of the 'cottage industry' phase so that it can compete with the larger Japanese and other foreign corporations. When I asked if this was not incompatible with their importation of Japanese Akai and Rotel products, he was quite clear that these lines were a useful stop-gap so long as Rank were not making comparable items.'

Although part of the Audio Visual division of the Rank Organisation, this was not delineated in Wharfedale marketing which (until 1971) only identified the products with Rank Wharfedale Ltd. For many years Wharfedale had exported about two-thirds of its output with about one-third going to North America. Whereas the remainder had initially gone mainly to the Commonwealth

Chapter Nine: 1965–1978

Figure 9.7 Loudspeaker range, as depicted in a brochure printed in December 1969. Left side, top-bottom: Rosedale, Dovedale III, Super Linton. Right side, top-bottom: Melton, Denton.

countries, by this time Europe was becoming the principle destination. Significant sales were made to Japan which was not yet posing the threat to loudspeakers which was already affecting the electronics sector. Rank's strategy to meet this threat is very clear from the above quotation: the operation had to be very large and centrally managed. The expansion at Highfield Road was intended to create huge capacity but, Ken Russell and his colleagues argued, this would have to be underpinned by a greater research and development commitment to enable Wharfedale technology to become, and stay, world class.

Alex Garner had joined the team, straight from University, in September 1969 and he was quickly followed by Tim Holl, who moved from Standard Telephones and Cables (STC) having worked on microphones. The following year a sustained effort to build up human and technical resources commenced. The most promising engineers were recruited from wherever they could be found and soon a talented team was assembled which had the opportunity to carry out original research as well as respond to the demands from marketing for 'next year's models'. These models also included 'Leak' branded speakers which, to maintain differentiation, had to be built around sandwich-cone drivers. All the Wharfedale products from 1970 to 1978 (when this story ends) are included in Appendix 4 and their individual development will not be described in detail. There was a notable lack of Wharfedale introductions in 1972/3 and at this time a new range of speaker systems to complement the Leak 'Delta' range of electronics was being developed.

In 1970 the 'Triton' was added to the range shown in (9.7), being essentially a three-way bookshelf system (approx. 22 × 9 × 9", sold in pairs) whilst the 'Aston' was a wall-mounting speaker to replace the 'Slimline 2'. This was finished in 'arctic white', the first experiment in a finish other than wood veneer. The Unit 3 kit had sold well and so further kit versions of the most popular systems were produced—Unit 4 for the 'Melton' and Unit 5 for the 'Dovedale III'. 1970 also saw the reintroduction of Wharfedale electronics units, beginning with the 100.1 'stereo multiplex receiver' (a tuner–amplifier with AM/FM reception). John Collinson had gone back to his roots to engineer this high power model (35 watts continuous per channel).

Rank Radio International

In 1971 the Wharfedale operation was moved, with Leak, into Rank Radio International, a division dominated by Bush and Murphy televisions. The new speaker models of that year were all updated versions of existing units and, to capitalise on the huge popularity of the 'Linton', an amplifier and turntable with the same name were introduced to provide a basic hi-fi system (a Linton receiver appeared in 1973). The DD1 headphones were the final complement, but sourced from Japan. The prices of the members of this family were deliberately mid-range, Leak systems carried a much higher price tag. Interestingly, the DC-9 Dolby cassette recorder, also of 1971, was only the second of its type on the UK market and there was no matching Leak product. Although it contained Japanese components it seems to have been assembled in the UK. The 1971 speaker systems featured such innovations as: front mounted bass units with long throw spiders; acoustically isolated mid-range units with transmission-line rear loading; and treble units with cellulose acetate butyrate (CAB) cones or domes, phase correction and wide-angle dispersion.

One of John Collinson's projects during 1971/2 was the development of an entirely new type of headphone. Existing products were either miniature moving coil types with a cone diaphragm (as in the DD1), microphone capsule types with a domed diaphragm or electrostatic types with a flat diaphragm. His idea was to replicate the advantage of the electrostatic type, where the diaphragm

Chapter Nine: 1965–1978

is evenly driven over most of its surface area, but to do this electromagnetically. The patented design involved the use of thin ceramic magnets incorporated into a rubber sheet to provide the magnetic field and a polyimide film diaphragm onto which was etched an electrical circuit. The resulting product, named the 'Isodynamic Headphones' was extremely lightweight and could be manufactured (and sold) for significantly less than competing electrostatic types. They reached the market during 1973, which turned out to be a bitter-sweet year for their inventor, as described below. It was marketing practice in the early 1970s to include frequency response curves in the specification details, reproduced directly from the laboratory measurements which were initialled. For historical interest, the data for the Isodynamic Headphones (signed-off by Collinson—JDC) and for the Rosedale loudspeaker (signed-off, like all the others reported, by Ken Russell–KFR) are shown in (9.8).

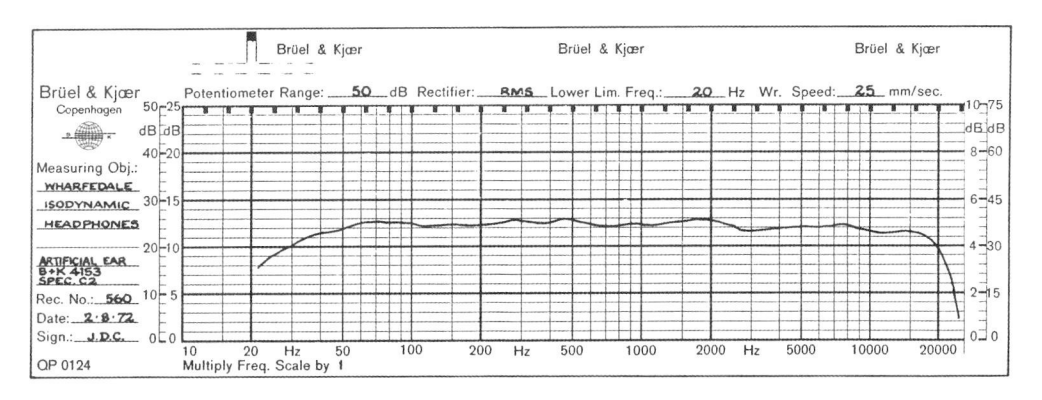

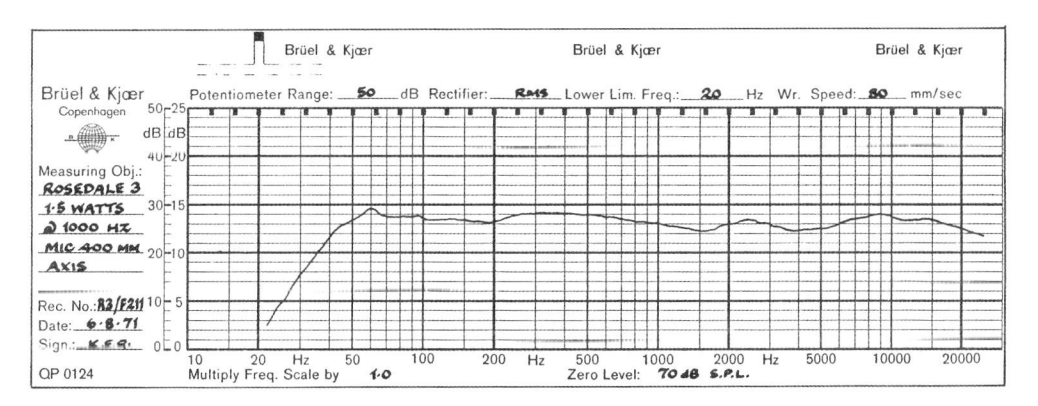

Figure 9.8 Response curves for the Isodynamic Headphones and Rosedale 3 loudspeaker.

Expansion at Highfield Road

As the Highfield Road site was developed there was a phased transfer of production from the Bradford Road site accompanied by an expansion of output. The acoustic engineering team had moved into Shaw House (albeit temporarily), the existing building on the site, along with some administration, by 1968. When the first phase of the new cabinet factory was complete, around the end of 1969, the audio engineers returned to Bradford Road, into the cabinet production area completed in 1960, with room to expand the research and development effort. By the end of 1971 the cabinet factory had stepped its output up to 1000 units per day, and the working week for everyone grew longer to keep loudspeaker unit assembly 'in phase'. Storage of finished products had been a problem even at Bradford Road and a disused warehouse, Globe Mills, in Bradford was occupied for this purpose. It became the distribution centre but after a few years its capacity was outstripped and additional space was taken over at the old Avro factory in Yeadon (33,000 square feet on a single floor). This extraordinary increase in size of the Wharfedale/Leak operation over a short space of time was not achieved without tensions. The management of Rank Radio International was based in distant Chiswick, London and their interventions were often seen as unwelcome interference. General Managers, on site, were appointees from Rank (i.e. did not come from within the local staff) and they came and went with some frequency, creating tension through changes in management style and the inevitable political manoeuvring. Bill Escott, effectively in overall charge until 1965 or later, was particularly affected and found his role being restricted in scope to commercially oriented activity. He would have been much happier working under the conditions which the Rank manager had referred to[4] as 'cottage industry', and concluded during 1973 that he had had enough of being in this kind of (highly political) organisation. John Collinson was also unhappy about the way things were turning out and he and Escott decided to go into partnership (9.9). A third partner was recruited in Alan Cunnington, the Financial Controller, and they left in late 1973 to set up Castle Acoustics Ltd—named after the castle in Skipton where they established themselves.[5]

Figure 9.9 Bill Escott (left) and John Collinson (right) with one of their secretarial assistants, early 1970s. (Photo courtesy of Ken Russell.)

Chapter Nine: 1965–1978

Figure 9.10 Denton 1 loudspeaker.

During 1974 all the activities had been finally concentrated at Highfield Road and the Bradford Road site was put up for sale. Employment was now at about its peak with around 1100 people working on the site, which in total had 170,000 square feet of occupied space. The product range for that year had changed significantly again with updates to existing speakers (Denton 2, Linton 2 and Dovedale 3), the new Glendale 3 (10", 4" and 1" drivers) and the Kingsdale 3 (15", 5" and 1" drivers) to replace the Rosedale. The trend was to higher power ratings all round. Bucking this trend, though, was a small speaker with a single twin-cone drive unit, the Denton 1, which had a

229

completely different cabinet from the variations on teak/walnut veneered box designs. As shown in (9.11) it was not only curvilinear, but made from white polyurethane. Under pressure from innovative styling trends in Japan, Wharfedale had finally taken on their own appearance designer during 1972. New electronics units included the Denton Amplifier, Linton Amplifier Mk II, Denton Tuner–Amplifier, Linton Tuner–Amplifier, and WHD-20D Cassette Deck. These marked a major change of strategy. For mid-price electronics it had been clear since around 1970 that competing with Japanese products would soon be impossible. A recent arrival, Peter Wall, was responsible for liaising with Japanese manufacturers to produce these units to Wharfedale specification. There was also a new turntable (W30) and this together with either an amplifier or tuner-amplifier was bundled with speakers of the same name to provide a choice of 'Denton' or 'Linton' hi-fi systems. The kit range was also revamped, with the new kits representing the Linton 2, Dovedale 3 and Glendale 3 systems.

Pioneering Research

Meanwhile, the Acoustics Engineering Design team had been doing pioneering work which started to feed through into loudspeakers from late 1974. Getting a handle on the complex behaviour of a vibrating loudspeaker cone had always been a dream of loudspeaker designers, but this was elusive. Techniques such as following the static patterns in lycopodium powder dusted onto the cone surface, as a function of frequency etc., and using a stroboscope to 'freeze' the motion of the cone (limited to relatively low frequencies) were used, but were woefully inadequate. The arrival of the pulsed ruby-laser in 1965 allowed the development of holography for studying small displacements in objects taking place at high speed. Holograms (images viewable in 3D) recorded on the sub-millisecond timescale could be superimposed to give a clear indication of cone surface behaviour at any frequency, which could then be related to audio distortion. Peter Fryer was responsible for harnessing this technique for Wharfedale at an early stage in its availability and in the XP (extra power) range, released in late 1974, one of the benefits was apparent. The range included the Denton 2XP, the Linton 3XP, now three-way, and the Glendale 3XP. As shown in (9.11) the 4" mid-range driver in the Linton, and larger Glendale, had a plastic cone with a pattern of small perforations to prevent the propagation of distortions (cone break-up).

Another fundamental problem area of loudspeakers, first highlighted by D.E.L. Shorter at the BBC in 1946, was that of delayed resonance—the fact that a sudden pulse, or transient, cannot be exactly replicated in the speaker's sound output because of energy dissipation in the cone, which takes place over tens of milliseconds after the pulse. In music, for instance, any resonances during the decay period of one note overlap with the next note leading to 'colouration'. Again, getting a handle on this very complex problem had defied progress—over the 20 year timescale of Gilbert's books he had little progress to report, rather, more indications of the complexity involved. The Wharfedale team came up with a measurement technique which allowed the equivalent of the usual response curve to be recorded at any chosen time interval (to well below 1 ms) following the input signal. Gilbert, who always maintained that response curves hid a multitude of sins—and hence placed much more reliance on listening tests—saw this in action in 1974, as described in the next chapter. Here, at last, was a way to relate what the ear recognised as 'colouration' to the frequency and temporal distribution of delayed resonances. Coupled with the holographic characterisation of the cone behaviour described earlier, this technique greatly reduced the empiricism previously involved in developing new systems. It was first brought to

Chapter Nine: 1965–1978

Figure 9.11 Linton 3XP with Denton 2XP.

bear in the XP designs. Ezra Broadley, the long-time Works Manager who had worked through the entire lifetime of Wharfedale, was suffering from ill-health by the early 1970s and this led him to retire in 1974. Latterly he had taken on the role of progressing new models from the prototypes emerging from the design team through pre-production and onto the shop floor—a process that took 12 months or more. This range was to be his last.

A third area of innovation was sparked off when John Collinson managed to loan a new Hewlett-Packard programmable scientific calculator in 1971. He brought it into the lab; Alex Garner recalls that he was raving about it and gave it to him saying: "Here, take it home with you, see what you think of it!" Garner was quickly hooked; compared with the laborious use of slide rules for solving complex formulae this was a fantastic advance. They persuaded Ken Russell to buy one at great cost to his budget (£250, equivalent to around £2800 in 2010) and the digital design revolution was underway.

Garner and Collinson began analysing speaker systems using the equivalent electrical circuit approach, drawing on recent papers by Small, extending earlier work by Thiele, in the Audio Engineering Society journal. This led on to work which Garner did with Peter Jackson, using electrical high pass filter theory, which proved from first principles that there was a fixed relationship between sensitivity, cabinet volume and bass cut-off frequency (which Raymond Cooke had been convinced of, but could not prove, back in the early 1950's), but also that the drive unit parameters would have to be differently optimised for either a closed box or a bass reflex design. This meant that the marketing people could be provided with options for any two of the three basic parameters—sensitivity, cabinet volume and cut-off—and the system could then be optimised for the third. By this time (1974) the computations had moved on to a remote mainframe computer which was accessed via a terminal and phone line in the lab, a rather advanced procedure at that time.

Marketing Excellence

The first development application majored on bass extension. The Dovedale SP and Airedale SP (super power) models which appeared at the end of 1975 were the result. The SP models were sensitive reflex designs with cut-offs some 10 Hz lower, and from smaller cabinets, than comparably sensitive systems on the market. Another member of the XP range, the Chevin XP (smaller than the Denton with a dual cone unit derived from the old Super 8/RS/DD) was added at about the same time.

With the introduction of the XP range, and particularly the SP range, the marketing people went to town on the technological innovations, which had been presented in papers at audio engineering conferences, and discussed the key features in surprising detail. They used the slogan 'proving that Britain still leads the world in speaker design' and claimed not to be concerned about giving away secrets to the competition because by the time they could make use of them Wharfedale would have made the next advance.

Before John Collinson left in late 1973 the other members of the Acoustics Engineering Design team with Ken Russell were Tim Holl, Alex Garner and Steve Scaife, whilst between 1974 and 1976 it had also lost Tim Holl but gained Graham Bank, Peter Jackson, Peter Maltby and Gareth Millward. Peter Wall and Peter Fryer were mainly occupied with research, rather than product development, from about 1972. Some of the team were featured in the early SP marketing material (9.12), an event which took them by surprise and which they found rather ridiculous—it was not repeated!

Chapter Nine: 1965–1978

Figure 9.12 Five members of the SP team: left to right Ken Russell (Technical Manager), Brian Pearson (lab technician), Alex Garner (senior development engineer), Graham Bank and Peter Jackson (development engineers).

Rank Hi-Fi

By the time the XP range appeared, around late 1974, Rank Radio International was making huge loses. TV production dominated the division and this was suffering greatly from Japanese competition. Similarly, the audio electronics products still made in the UK (now only the Leak models) were completely uncompetitive; it was not even possible to purchase the components, many of which also came from Japan, for the price of the imported models. However, the investment in loudspeakers at Idle, both in terms of the Highfield Road site (9.13) and R&D, had been massive and the products had ready export markets, so to protect Wharfedale and, to a lesser extent, Leak they were taken out of Rank Radio International and merged with other audio-visual activities to form Rank Hi-Fi. The strap-line in marketing and advertising for Wharfedale now became 'Britain's most famous loudspeakers'.

1976 saw the replacement of the existing kits by new ones representing the Denton 2XP, the Linton 3XP and the Glendale 3XP and the introduction of a new XP cassette recorder. In 1977 two members of a new 'E' range appeared which were designed using the computer optimisation programme to major on sensitivity. The E50 and E70 set the trend for professional disco speakers. During 1977/8 most of the existing models were further improved to give the XP2 range—Chevin, Denton, Linton, Glendale plus the new Shelton—shown in (9.14) and the SP2 range (Dovedale plus new Teesdale). Whether by coincidence or not these speakers, the last Wharfedales to be introduced during Gilbert's lifetime, were also the last to bear names derived from his beloved Yorkshire dales.

Figure 9.13 Highfield Road site, Idle, at about the time of Gilbert's death. Shaw House is centre left, Briggs House is top left.

Chapter Nine: 1965–1978

Figure 9.14 The XP2 family: left to right Chevin, Shelton, Glendale, Linton, Denton.

Wharfedale
LOUDSPEAKERS
IDLE · BRADFORD
A DIVISION OF THE RANK ORGANISATION
TELEPHONE IDLE 1235

CHAPTER TEN: 1965–1978

Wharfedale Book Department, Briggs-Wharfedale Studio, Ilkley

When Gilbert made his intention to retire known, one of the issues to be addressed was 'what happens about the books?' The titles still in print, with four emerging in the recent period 1962–1964, were selling well and they were, of course, Wharfedale publications. Apart from the modest income they generated, they were much more valuable for their role in keeping the Wharfedale brand to the fore. As noted in Chapter 8, Gilbert still had plenty of ideas for further books and was already part way through a project to rewrite *Sound Reproduction* as a series of smaller, more focused volumes. Publication of new books through Wharfedale, as before but with Gilbert taking a royalty, would have been possible but discussions soon brought up related matters, notably the secretarial support upon which both the book projects themselves, and the correspondence with 'fans', depended. Although the books were available to individuals through specialist shops and Wharfedale dealers, many were ordered direct from Idle and agents all around the world (often not the agents for the speakers, etc.) needed to be supplied, which was really a distraction for the shipping department. These considerations led to some kind of deal whereby what became known as the 'Book Department' was established in Ilkley to look after all these aspects, run by Gilbert in the new role of Consultant Director, and paid for by Rank.

The Briggs-Wharfedale Studio

Without much warning Gilbert asked Dorothy Dawson if she would like to continue as his secretary in this new capacity. It would mean travelling to Ilkley by train, but she decided to stay with him. An office—the 'Briggs-Wharfedale Studio'—was set up on the first floor of 13 Wells Road, conveniently sited close to the railway station and the Post Office. The front view of the building today is shown in (10.1). The ground floor was occupied by an electrical shop and next door, at number 15, a local painter, Gordon Barlow, had a studio above his gallery. Gilbert's office was actually at the back of the property and pictures of the occupants taken inside, shortly after they started operations, are shown in (10.2) and (10.3).

In addition to the office there was the 'book room' in another building close by. Here the books were stored and Mrs Ann Graham worked as required. She picked up the orders from Dorothy, together with the address labels she had typed, did the packing and took the parcels to the Post Office for dispatch. Not long after Gilbert's retirement, Valerie gave her father the harmonium which she had bought in Hereford (see Chapter 8). This he installed in the book room, so he was never very far away from a keyboard.

A Pair of Wharfedales: The Story of Gilbert Briggs and his Loudspeakers

Figure 10.1 13/15 Wells Road.
© Judy Smith 2010.

Marks of Retirement

Shortly after Gilbert left Idle, in March 1965, he was visited by John Davis, Chairman of the Rank Organisation and the picture shown in (10.4) was taken. It was probably intended for inclusion in a piece for an internal Rank publication, or possibly a hi-fi magazine, marking Gilbert's retirement from Wharfedale. The setting is Yeadon, now Leeds–Bradford, airport and one of the earliest occasions on which Wharfedale products were shipped by commercial air-freight from there. However, this was by no means the first time that Wharfedale speakers had left Yeadon by air.

Yeadon had been a relatively small civil aerodrome from the early 1930s but during the War it was transformed by the necessity to test fly Lancaster bombers and other aircraft built by Avro's at their neighbouring site. Gradual development and expansion of scheduled services took place from 1947 but in 1959 the Leeds–Bradford Airport Joint Committee took over the airport and major investment started. It was probably at this point that a liaison committee was set up to ensure a good working relationship between the airport and the local population. Bill Escott lived in neighbouring Rawdon and became very involved, getting to know the key people at the airport including the air-traffic controllers. There was a good market for speaker systems in the social clubs of the many American bases in Europe, particularly in Germany. Albert Smith, the sales representative, had worked in Germany before joining Gilbert in the early 1930s and he developed this market. Some-

Chapter Ten: 1965–1978

Figure 10.2 Gilbert in Briggs-Wharfedale Studio, April 1965. (Copyright—Bradford Museums and Galleries. From the C.H. Wood photographic archive negative number 62469, photo courtesy of the Briggs family.)

Figure 10.3 Dorothy Dawson in Briggs-Wharfedale Studio, April 1965. Copyright—Bradford Museums and Galleries. From the C.H. Wood photographic archive negative number 62468, photo courtesy of the Briggs family.)

239

how, through his airport connections, Bill Escott was able to make arrangements for American planes to fly into Yeadon to collect the speakers! Dorothy Stevens even recalls an aircrew turning up in her office having flown down from Iceland to collect speakers—and being so pleased with her arrangements for collection and the performance of the speakers that they came back for more. This was a time when Wharfedale could not turn out enough of their best systems and, since Gilbert would not sacrifice quality, waiting times could be many months. Smith's enthusiasm to develop this market, in particular, and Dorothy's insistence that orders had to be satisfied according to position on a waiting list, produced the occasional showdown.

In another mark of his retirement, the Yorkshire magazine *The Dalesman* ran an article about Gilbert in its 'People' section. This was the result of an interview with the journalist Bill Lang and the piece, entitled 'Wharfedale' Rules the Radio Waves, began as follows:[1]

'At that indefinable point where art becomes a science and science becomes art, stands a pioneering Yorkshireman, Mr Gilbert Briggs, who probably more than anyone else has helped to bring about the sound we can hear in our homes today—the sound known as "Hi-Fi".'

The article, which appeared in July 1966, included a picture of Gilbert, in the Wells Road office, holding the German speaker which started the whole Wharfedale story in 1932 (and reproduced in Chapter 4). Amongst several interesting titbits about Gilbert's life which appear nowhere else, the article disclosed that the 'Book Department' was sending out an average of 1000 books per month from Ilkley, so there was no lack of activity. Also, it revealed that his next book would be a change from audio—he was planning a humorous one.

Figure 10.4 Gilbert with John Davis at Yeadon airport, March 1965. (Photo courtesy of the Escott family.)

Chapter Ten: 1965–1978

Books and Letters

Gilbert's sixteenth book, *Musical Instruments and Audio*, was published in October 1965. It had been in preparation for about 18 months and for several months was running in parallel with *Aerial Handbook*, published a year earlier. A significant proportion of the book was devoted to electronic music, and electronic organs in particular, and this is doubtless the source of Gilbert's whimsical plan to build an electronic organ as a retirement project (which never happened!). A large number of oscillograms were included and the data were collected in the Wharfedale lab. during 1964, whilst he still had some influence over Ken Russell and Bill Jamieson. It was the realisation that such data collection would now be much more difficult, if not impossible, which made him think about other subject areas for new books.

Gilbert's penchant for reading humorous books and *Punch* has been noted earlier and his own books had included an increasing number of cartoons as time went on. Being a devoted solver of the *Daily Telegraph* crossword he had noticed the trend in paperbacks of collected crosswords and since he also liked puzzles and brain teasers he decided to produce a small book incorporating these into a humorous setting. The problem was that this could not be part of the Wharfedale canon so it would have to be published another way. His solution was to set up a company, G.A. Briggs Ltd, to enable him to be in control as usual. It was incorporated on 2 November 1965 with Edna and Dorothy Dawson as Gilbert's fellow directors and Harold Reynard as company secretary. *The Puzzle and Humour Book* (10.5) appeared in April 1966, a small format 96 page volume in card covers, not unlike the early *Loudspeakers* editions. Two-thirds of the puzzles etc. were original, concocted by Gilbert with help and support from Dorothy. How widely Gilbert tried to advertise this book is not known, but just as in New York with *Loudspeakers* nearly 20 years earlier, he hawked it round suitable outlets in an area he could cover. This approach was not unsuccessful but it did not shift the required fraction of the print run. Unfortunately, the venture ended up losing money (about £1000) and G.A. Briggs Ltd was wound up after a couple of years.

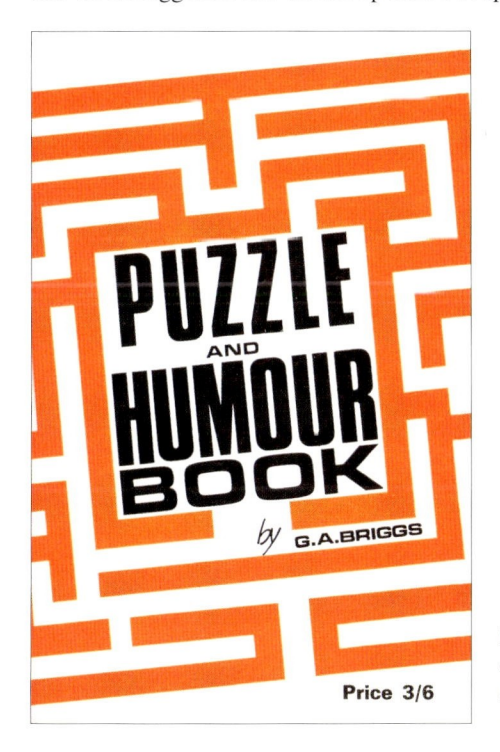

Figure 10.5 Front cover of *Puzzle and Humour Book* published in 1966.

Around the same time, for family reasons, Dorothy wanted to work part-time. She thought it was perfectly possible to do all the work in fewer days and so they agreed on Monday, Wednesday and Friday. Whilst Gilbert went to the office every day, if not away, they settled into this routine for the next 11 years. There were two final books published through Rank Wharfedale: *About Your Hearing* (May 1967) and a second edition of *Aerial Handbook* (with James Moir, January 1968). Neither of them required technical support from Wharfedale, but Ken Russell did help out with the former in other ways. Once the publishing era had ended, Gilbert's main occupation in the Wells Road office was answering the mail that continued to be generated by his books—in and out of print—and the many regular correspondents who had accumulated over the years. Many of Wharfedale's agents around the world had become personal friends, along with many others involved in all aspects of the audio business, including journalism. Gilbert's rule from the earliest days was to respond to letters within 24 hours if at all possible and he tried to maintain this regime. Writing in 1961 he said:[2]

'The most agreeable result of the books has been the receipt of friendly letters from readers in different parts of the world. I have no exact records, but the total during 13 years must be about 10,000. One of the most amusing reached me only a fortnight ago. It came from a Mr Mansfield of America and contained a few questions, ending with a request not to reveal his name and address in a Question and Answer section* of any of our future publications as he had an ex-wife who was anxious to discover his whereabouts. Naturally we agreed to maintain secrecy, but in return I suggested that he might give me his ex-wife's address if she happened to be Jayne Mansfield.'

Gilbert would arrive at the office about 10.30 am and stay until lunchtime, when he would take off for a drink and a spot of lunch at one of the nearby pubs. In the early days of the Briggs-Wharfedale Studio he would come back in the afternoon for an hour or so, but eventually he stopped doing this. On his way home in the afternoon he regularly called in on his widowed mother-in-law, Louise Mart. She had an interesting, if not exotic, background and a formidable intelligence; although 90 in 1965 she was still very sharp and Gilbert enjoyed her company. She was 101 when she died. On Fridays he would take Dorothy for her 'treat'—a leisurely lunch at a good restaurant—sometimes in the company of old colleagues from Wharfedale. The Crescent Hotel in Ilkley was a regular destination and the Devonshire Arms near Bolton Abbey the favourite for special occasions. One boost for Gilbert, as he became less inclined to travel very far once he turned 80, was that his two oldest grandchildren both attended Leeds University. Since it was an easy journey to Ilkley they were able to visit regularly and between them they were able to do this over a five-year period in the early 1970s.

Recognition by the Audio Engineering Society

Life went on rather quietly until July 1973 when Gilbert received a letter from the secretary of the British section of the Audio Engineering Society (AES), Rex Baldock, to inform him of the Society's plans to devote their first meeting of 1974 (scheduled for 8 January at the Institute of Electri-

---

*Footnote. Several books had Question and Answer sections at the end which used actual questions from correspondents to allow points of general interest to be discussed in a way which was not always possible in the formalised chapters (see Chapter 12).

cal Engineers) to an appreciation of his pioneering contributions to Audio. Although delighted by this, Gilbert did not feel up to travelling to London, especially in the winter, to attend the meeting in person. Instead it was decided that he would make a 15 minute tape recording about his early activities in loudspeakers, books and concert-demonstrations, which could be played to the audience. He knew in advance that the Chairman of the meeting would be Percy Wilson, an expert on gramophones and one-time Technical Editor of *The Gramophone*. This was an apt choice; Wilson was of Gilbert's generation (only three years younger) and had, in fact, been a pupil at Heath Grammar School in his native Halifax when Gilbert was at Crossley and Porter's School (then for orphans, later a grammar school) only a short walk away. The other contributors were to be former collaborators in one or more of Gilbert's three areas of audio activity: Raymond Cooke, Peter Walker, Leon Goossens and James Moir. The tape recording was duly made in the October by Ken Russell and John Collinson and sent to the Society. After that things started to unravel. An industrial dispute by power workers led to rolling power cuts across the country and the meeting in January had to be postponed. Eventually it was rescheduled for 29 May 1974 at the Royal Institution, but the event was still jinxed. Percy Wilson was taken ill just beforehand, Leon Goossens had a last-minute engagement clash and James Moir was delayed en route and arrived too late to contribute! These absences, however, had an unexpected benefit. After Gilbert's tape recording, Peter Walker and Raymond Cooke gave their appreciations and the spare time was taken up by inviting contributions from the floor. Since the whole proceedings were taped for Gilbert to listen to at home, the survival of this recording has preserved reminiscences from a variety of people who would otherwise have gone unheard. One of these, from Ray Dolby, is described in Chapter 11 whilst others have been referenced in earlier chapters.

Gilbert received the tape recording in early June and typically decided to write and thank all the contributors. Rex Baldock was no doubt called upon to provide names and addresses for this purpose before the letters could be sent. One of these was to Alex Garner (10.6) who had been working at Wharfedale in Ken Russell's team since 1969. Like many others he owed his early interest in hi-fi to Gilbert's books and he still treasures, along with his letter, a red leather-bound second edition of *Sound Reproduction*. Garner moved to Tannoy, as Chief Acoustics Transducer Engineer, in 1976, later becoming Technical Director.

By a strange quirk of fate Garner met Gilbert face-to-face only three months later. Gilbert was periodically invited to Idle to be shown developments and had not been back there for a couple of years. He was invited in September and made the visit on the 27th. Whilst touring the Technical Department he was introduced to a new method for characterising speakers (delayed energy measurements using a gated input signal). A picture was taken of the demonstration (10.7) and this eventually found its way into *The Gramophone*, which reported the visit in its January issue of 1975. At the end of the visit another picture was taken and this is shown in (10.8).

The public appreciation of Gilbert's pioneering contributions by the British section of the AES may have been the catalyst for international recognition by the parent AES in the USA. In March 1975 Gilbert was honoured by being made an Honorary Member ('A person of outstanding repute and eminence in the science of audio engineering or its allied arts'). He was very proud of the citation document (10.9) and it may well have been handed to him in person by Raymond Cooke (10.10). Whether or not Gilbert knew that his hero Paul Voigt had been honoured in the same way, only a year earlier, is not known but it is very likely that Cooke would have told him. Cooke himself was made a Fellow of the AES in 1975, whilst Chairman of the British section (1974–1976).

BRIGGS - WHARFEDALE STUDIO
13 WELLS ROAD ILKLEY YORKSHIRE LS29 9JB
Telephone: ILKLEY 4246
OFFICE HOURS 10 a.m. to 4 p.m.

21st June 1974

GAB/DD

A. Garner Esq.,
Technical Department,
Rank Radio International,
Bradford Road,
Idle,
Bradford BD10 8SF.

Dear Mr. Garner,

I have been listening to the recording of the AES Convention and I was interested in your spontaneous contribution.

I hope you will enjoy your work and that you will continue with Wharfedale speakers for many years.

Yours sincerely,

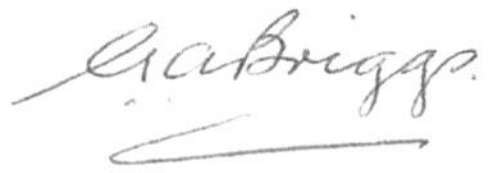

G A Briggs

G. A. BRIGGS Author of
LOUDSPEAKERS   PIANOS, PIANISTS & SONICS   A TO Z IN AUDIO   AUDIO BIOGRAPHIES   CABINET HANDBOOK
MORE ABOUT LOUDSPEAKERS   AUDIO AND ACOUSTICS   MUSICAL INSTRUMENTS AND AUDIO
ABOUT YOUR HEARING   AERIAL HANDBOOK
CONSULTANT DIRECTOR RANK WHARFEDALE LTD. IDLE, BRADFORD

Figure 10.6 Letter from Gilbert to Alex Garner, June 1974. (Scan courtesy of Alex Garner.)

Chapter Ten: 1965–1978

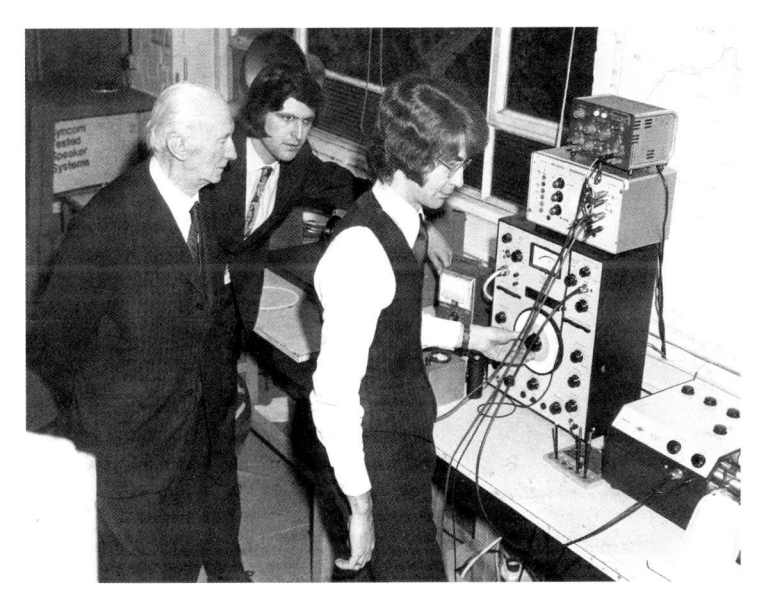

Figure 10.7 Visit to Bradford Road, Idle on 27 September 1974. Left to right: Gilbert Briggs, Dr Gareth Millward, Alex Garner. (Copyright—Bradford Museums and Galleries. From the C.H. Wood photographic archive negative number R12230/74, photo courtesy of the Briggs family.)

Figure 10.8 Visit to Bradford Road, Idle 27 September 1974. Left to right: David Bullough (General Manager), Gilbert Briggs, Ken Russell (Technical Manager), Ronald Sharp (visiting from Rank in London). (Copyright—Bradford Museums and Galleries. From the C.H. Wood photographic archive negative number R12231/74, photo courtesy of the Briggs family.)

Figure 10.9 AES citation. (Scan courtesy of the Briggs family.)

He went on to serve on the Board of Governors for several years before becoming the first ever British President of the Society in 1983. From the AES convention in Los Angeles, the following year, he sent Gilbert the postcard reproduced in (10.11).

More Interviews

Not long after this Gilbert was being tape recorded again, this time for a programme called *Oasis* put out by Radio Leeds. The musical magazine programme included all kinds of items relating to the music scene in the area covered by Radio Leeds, with live and recorded interviews one of the features. The presenter and producer was Peter Byrne, whilst Roger Beardsley, the co-presenter, was the audio professional on the team responsible for the sound recordings. One day during a planning meeting Byrne recalled the Wharfedale concert-demonstrations at the Royal Festival Hall, some 20 years earlier, which he was able to attend because he was working for the BBC in London. Beardsley, knowing that nearby Idle was the home of Wharfedales, felt that Gilbert should be recorded for the programme, if he was still alive. It did not take long to track him down. They interviewed him in the Ilkley office and because the material was longer than normal it was broad-

Chapter Ten: 1965–1978

Figure 10.10 Raymond Cooke with Gilbert Briggs at Ilkley in 1975. (Photo courtesy of the Cooke family.)

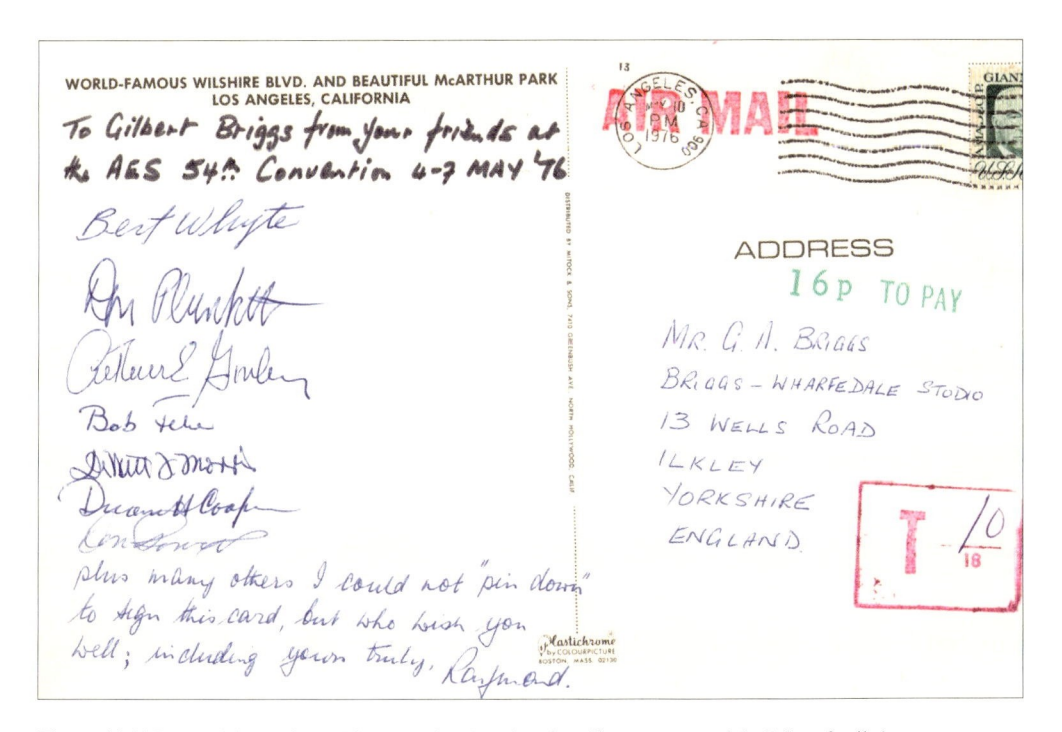

Figure 10.11 Postcard from 1976 AES convention, Los Angeles. (Scan courtesy of the Briggs family.)

cast in an extended (75 min) programme on September 21st 1975. Beardsley recalls that editing the interview was non-trivial because from time to time Gilbert's teeth would click. The irony of having to surgically remove these annoying transients would not have been lost on the old man!

Peter Bryne was not the only person thinking back to the Royal Festival Hall concert-demonstrations at this time. Adrian Hope, a journalist working for the magazine *Hi-Fi for Pleasure*, had always wanted to meet Gilbert and hear from the man himself about the famous concerts. Following the Northern Audio Fair in 1975 (always held in Harrogate) he arranged to visit him in nearby Ilkley and an article based on their discussion of the concerts was published in the December issue of the magazine. The final paragraph gives this salutation[3]:

> *'Briggs would like to see recordings again played alongside live sound at the Festival Hall, but the Wharfedale team at Bradford feel that any such concert should involve Briggs himself. But Briggs, now astonishingly in his eighty-fifth year, says he cannot possibly be involved. So, for the time being at least, there is an impasse. There is, of course, nothing to stop any of us booking the Festival Hall and doing for ourselves what Briggs did twenty years ago. The problem is that none of us has the guts to put money where our mouths are, as Briggs did.'*

Briggs House

Shortly after this article appeared and two weeks before his 85th birthday Gilbert was reunited with some old colleagues when he opened a new building on the Highfield Road site of what was now Rank Hi-Fi. 'Briggs House' (10.12) was a three story block occupied by the acoustic engineering and administration departments together with a new canteen. Ezra Broadley had retired because of ill-health in 1974 (he died in 1976) and this meant that Arnold Hatton, recently returned to the engineering department from working with BIC-Wharfedale in the USA, was the longest serving employee. Harry Bale, who had joined after the War and been responsible for cabinet finishing for many years, was now the safety officer. He and Hatton were photographed with Gilbert during the event (10.12).

Captured on Film

A curious episode involving a filmed interview took place around this time.[4] Stefan Sargent, an Australian film-maker based in London, was asked to do some filming of the Wharfedale Works; the footage was to be included in a Rank Organisation film for internal use. He was keen to do this because Gilbert had been a schoolboy hero and the original *Loudspeakers* an influential text. His school friend Robert Parker[5] had actually built a brick reflex corner enclosure in the family living room, from the Wharfedale design given in the book. Whilst at Idle, Sargent discovered that Gilbert was still alive and living in Ilkley so he arranged to capture his hero on film. His reminiscence ends as follows:[6]

> *'The next morning we drive north to visit Gilbert, who is delighted to tell his story on camera. Gilbert looks directly at the lens and says, "The worst mistake in my life was selling my company to The Rank Organisation." His interview was never shown at the conference. Funny about that.'*

There is no doubt that towards the end of his life Gilbert was disillusioned by what had happened to his company under Rank. Whilst he was still working he felt both he and Wharfedale had been

Chapter Ten: 1965–1978

Figure 10.12 Opening of Briggs House, 17 December 1975. left to right: Ken Russell (Technical Manager), Geoff Craggs (general manager), Gilbert Briggs, Alex Kokinis (personnel manager). (Photo courtesy *Telegraph* & *Argus*, Bradford, www.telegraphandargus.co.uk.)

Figure 10.13 Opening of Briggs House, December 1975. Left to right: Harry Bale, Arnold Hatton, Gilbert Briggs. Reproduced from *Electrical Trader* of 16 January 1976.

very fairly treated and he enjoyed a good relationship with John Davis, who would later go out of his way to call in on Gilbert at Wells Road when he visited Idle. But by this stage his life revolved around interactions with people, either through his continuing correspondence or through talking with and about old friends and colleagues. It was probably the consequences of the changes under Rank, described in the previous chapter, for just these people that really affected him.

In July 1976 Gilbert was interviewed again, this time for Radio London. John ('Johnny') Longden and David Clifton travelled to Ilkley and made the recording in the Wells Road office. This was subsequently broadcast in the programme *Sounds Good*, presented by David Simmons.

The Final Days

In the second week of January 1978, shortly after he had turned 87, Gilbert worked with Dorothy Dawson on the Monday as usual. He seemed fine. On her next day in the office, Wednesday the 11th, he arrived at his normal time but she thought he looked terrible. He admitted to not feeling too well, so she called for a taxi to take him home. Within an hour of his departure, Dorothy received a phone call from Edna telling her that Gilbert had died. His old enemy, bronchial pneumonia, had caught up with him and induced heart failure. On the Friday the *Ilkley Gazette* announced that the funeral would be at the Wells Road Methodist Church on the following Monday, 16 January. At the service Ken Russell gave the tribute to a packed congregation and then Gilbert was interred in the Ilkley cemetery, next to his son Peter, in a corner plot very close to the river Wharfe (10.14).

Obituaries on both sides of the Atlantic referred to Gilbert as the 'father of hi-fi' and, indeed, some commentators thought he had invented the term 'high fidelity'.[7] There was no mistake, however, in the belief that, through his audio books, his audacious concert-demonstrations and his drive over 30 years to free loudspeakers of the stigma of being the weakest link in the sound reproduction chain, he had done more than anybody to make high fidelity sound a popular reality.

The only letter of condolence which Edna kept, undoubtedly because she knew how much it would have meant to Gilbert, was sent on 14 January by Paul Voigt from Canada. Remarkably, Rex Baldock, the Audio Engineering Society secretary, was not only aware of Gilbert's death, but within hours he was phoning people in the audio world to pass on the sad news. He managed to reach Voigt late that same night (UK time). In his letter Voigt recalled with pleasure Gilbert's visits to him during Canadian trips, when he was living in Toronto, and the occasion when Edna and Valerie were there too. He had just read an article in the current issue of *Audio* by Bert Whyte (audio pioneer, one-time producer and engineer for Everest Records and Associate Editor of *Audio*) which he thought apposite. Whyte's article was a tribute to the conductor Leopold Stokowski, who had died the previous September, and with whom he had worked closely on recordings for Everest. Voigt quoted this passage:[8]

'In his study, surrounded by still more stacks of scores, was Stoky's hi-fi system. When I set this up for him in 1959, he told me he needed a good system, but not something in the very top category. He said he wanted to approximate to the kind of system used by the average hi-fi enthusiast. This was in the early days of stereo, of course, so we settled on a fairly large pair of Wharfedale speakers, a mid powered H. H. Scott receiver and a Garrard changer with a Pickering cartridge. He liked the system, but a few years later he wanted more power and better bass, so we installed a McIntosh pre-amp and amplifier.'

Figure 10.14 Gilbert's gravestone, Ilkley cemetery. © D. Briggs 2009.

He then pointed out that the Wharfedales did not have to be replaced to achieve the desired objective and that what Stokowski heard at home through these speakers would undoubtedly have affected his approach to making later stereo recordings (several of which are now regarded as classics).

Perhaps the last word should go to the people to whom Gilbert was both colleague and friend:

Ken Russell[9]

'Gilbert Briggs possessed all those very finest qualities which are unfortunately often regarded as old-fashioned, but which are the hallmarks of human greatness: absolute integrity, uprightness, straight speaking and straight dealing, ingenuousness—he set standards which few of us could easily reach.'

Raymond Cooke[10]

'He also had a great integrity and this, I think, led so many people to open their doors and their hearts to him and provided the foundation for lifelong friendships. Nobody who ever met Gilbert Briggs, or knew him for very long, could ever forget this quite remarkable man.'

Peter Walker[11]

'In one word, he was a lovable man.'

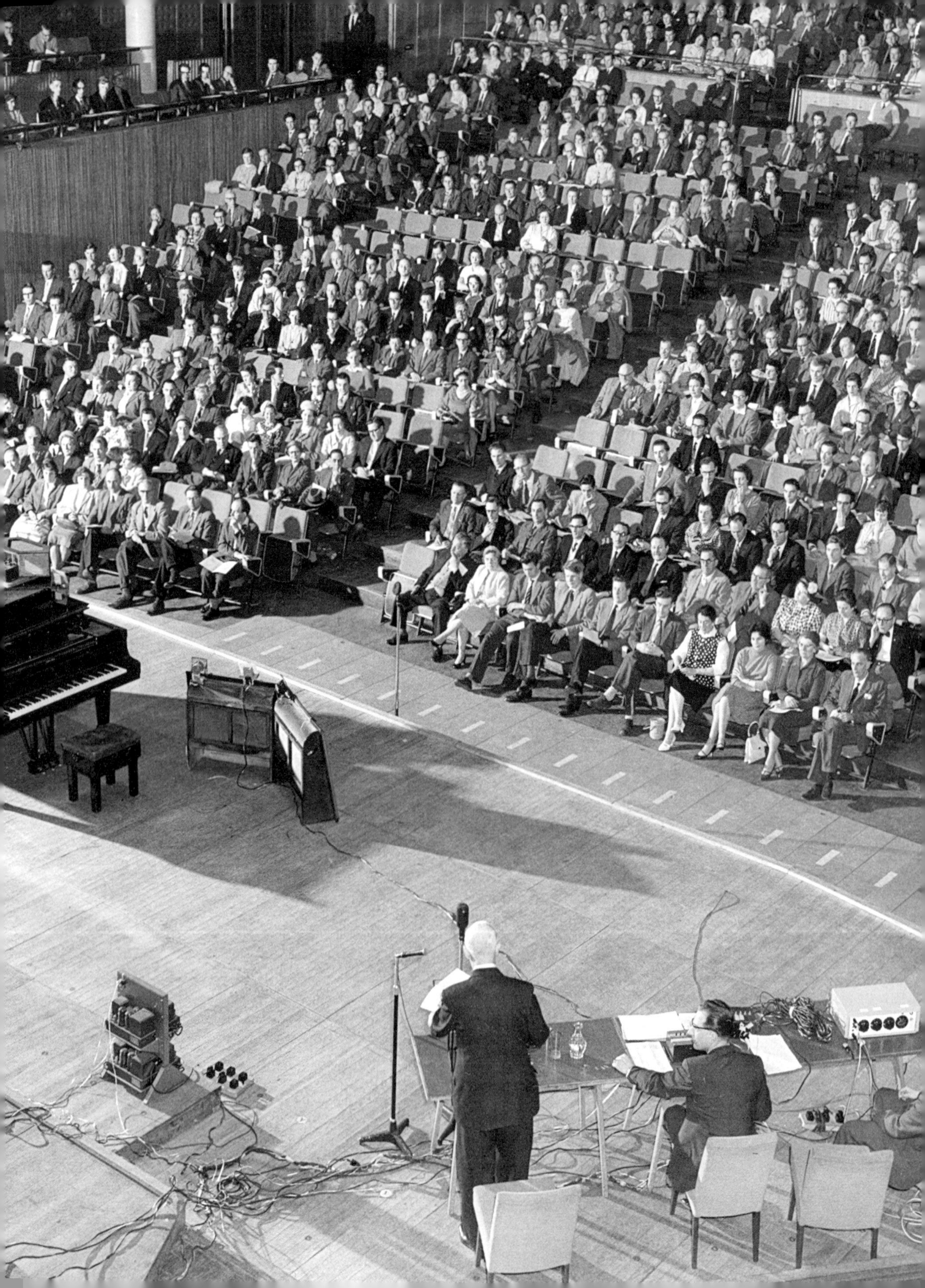

CHAPTER ELEVEN

The Concert Hall Lecture-Demonstrations

Beginnings

The demonstration of loudspeakers is a vital part of selling and Gilbert had been fully engaged in this process from the earliest days of Wharfedale Wireless Works in 1932. Before the Second World War the emphasis was very much on improving the quality of reproduction from radio sets since the reproduction of the original sound in the broadcast was already far better than that achievable by the loudspeaker. Arnold Hatton recalled[1] being Gilbert's assistant at pre-war demonstrations at which his job was to switch the output to different speakers when given his cue.

After the War, in the UK, the dramatic improvement in reproduction from disc recordings marked a major change in emphasis, ushering in the age of 'high fidelity'. The audio industry response was to organise, through the British Sound Recording Association (BSRA), annual exhibitions to showcase to the public the advances in records, pickups, amplifiers, loudspeakers etc. The first of these was held at St Ermin's Hotel, London on the weekend of 29–30 May 1948. This was a relatively small affair with 15 firms exhibiting and several hundred attendees. The second was at the Waldorf Hotel in May 1950, with 21 firms and by the third in May 1951 there were 23 firms and about 2000 attendees. At this meeting communal demonstrations were started, at which about 250 people seated in the Waldorf ballroom could listen to 15 minute presentations by the exhibitors. Novelty of approach and wit in presentation were key to making a lasting impression and this was an opportunity for Gilbert to reconnect with the extrovert side of his personality which loved an audience.

By this time, through the early books, his name was becoming familiar to those interested in serious musical reproduction and lecture-demonstrations to relevant societies started to evolve. In October 1953 Gilbert went to the USA to attend his first Audio Fair, held at the New Yorker Hotel in New York, and his Canadian agent, J.B. (Barney) Smyth, used the opportunity to raise the Wharfedale profile in his territory. He arranged for Gilbert to give a talk on Sound Reproduction to the Society of Music Enthusiasts at Toronto University on 28 October. Gilbert recalled:[2]

'The affair was held in a lecture room at the University and the equipment was rigged up and made ready for use only about half an hour before opening time. A ribbon pickup, one 15 watt amplifier and a three-speaker corner assembly were used. The audience numbered about three hundred, and I was myself astonished how well such a modest array of apparatus performed in a fairly large hall, especially as I had just spent four days at the Audio Fair in New York where 30/50 watt amplifiers working in hotel bedrooms shook every corridor.'

[Exhibitors were allocated bedrooms for demonstration purposes.]

Figure 11.1 University of Toronto lecture-demonstration. Gilbert Briggs extreme left, Barney Smyth extreme right. (Photo courtesy of the Briggs family.)

Thinking about the experience, he soon realised that good recordings, i.e. those which contain only a limited and acceptable amount of ambience or 'room colour', sound much better in a large hall than in a small room. A good hall adds little of its own colouration and wall reflections are minimised, which particularly affects the low frequencies (long wavelengths). Conversely, given the right recording, any defects in the reproduction chain in general, and loudspeakers in particular, will be much more apparent when demonstrated in a large hall. As a result he decided to put this idea to the test in St. George's Hall, Bradford as soon as possible. Furthermore, he would try the ultimate experiment of directly comparing live versus recorded music.

'Live versus Recorded' in St. George's Hall

There are as many stories about how Gilbert came up with this idea as people who have told them and with succeeding recollections in print the originator himself acknowledged different influences. An appeal to the notion that reproduced sound might be indistinguishable from the original is as old as the sound recording industry and its advertising. Small-scale direct comparisons of a live performance using a human voice or a solo instrument followed by reproduction via a microphone was used to demonstrate some aspect of the reproduction process by several pioneers at audio shows and sales demonstrations. The earliest one that Gilbert recalled was of Paul Voigt demonstrating his 'Tractrix' horn speaker with poetry readings, going back to about 1933. He also learned from D.E.L. Shorter, during his 1938 interactions with the BBC Research Station, of Shorter's practice of testing loudspeakers by comparing reproduced speech with the actual voice, sometimes in the open air to avoid room effects.[3] Probably the most telling example of this type of comparison came from his association with Cecil Watts who, in 1930, was pioneering direct disc recording. Watts was also a talented multi-instrumentalist and included in the exceptional recordings he produced, on nitrocellulose lacquer-coated aluminium discs, were some of his own playing. When Gilbert first visited Watts at his home, Darby House in Chiswick, (which was also his workplace) in 1949 he listened to recordings of duets played by Cecil (piano) and his wife, Agnes, ('cello) alternating with the real thing from an adjoining room. Gilbert was quite unable 'to tell t'other from which'.[4]

St. George's Hall was booked for 25 March 1954 for 'a public lecture recital' which was locally advertised and tickets were free. The hall capacity was around 2000 and nearly 1400 people turned up. The records used covered a wide range from sound effects (breaking glass and tugboat sounds) through solo voice, harpsichord and piano, trio and dance band to full orchestra. These were carefully chosen to 'match' the acoustic characteristics of the hall and to demonstrate the best of the relatively new 33.3 rpm microgroove (long play) recordings which were replacing the standard groove 78's. The equipment was similar to that used in Toronto, including a Ferranti ribbon pickup, a single 15 watt Quad II amplifier and a single Wharfedale three-speaker corner unit. Gilbert introduced the items and explained what they were intended to demonstrate, with his characteristic witticisms. Bill Escott was responsible for playing the records and adjusting the amplifier controls (in full view of the audience, Fig. 11.2 and 3) whilst Arnold Hatton sat at the back of the stalls and rapidly indicated, by signs, any necessary adjustments in volume or top and bass. This was necessary because any rehearsal settings would be affected by the presence of the audience. The live versus recorded items were of solo piano and for these Gilbert had persuaded his old friend and teacher Edgar Knight to perform. This was quite a feat for two reasons. Knight had to mimic the style of the performer on the chosen recording which, as Gilbert remarked, would probably have been flatly refused by any other professional (but for which Edgar had a particular talent). Also, by now in his mid-50's, Knight had become extremely nervous when performing before large audiences; he much preferred teaching and performing in more intimate spaces. The two live versus

Figure 11.2 St. George's Hall, Bradford 25 March 1954. Edgar Knight at the piano, Gilbert Briggs at the table, left, Bill Escott operating turntable, right. (Photo courtesy of the Escott family.)

A Pair of Wharfedales: The Story of Gilbert Briggs and his Loudspeakers

Figure 11.3 Audience, St. George's Hall, Bradford, 25 March 1954. (Photo courtesy of the Escott family.)

recorded items were: Years of Pilgrimage by Liszt (recording by Wilhelm Kempff, Decca LXT 2572) and Preludes by Rachmaninoff (recording by Geza Anda, Columbia SCB 117).[5]

The local *Telegraph and Argus* published a report the following day:

'CRASH!!
A good proportion of the audience at St. George's Hall last night were women, and no doubt many of them were housewives who had broken many a piece of crockery. But never, surely, did they imagine that breaking china or glass could make such a terrible crash as they heard reproduced during Mr G.A. Briggs's lecture demonstration on sound reproduction.
He had brought with him a number or records, popular as well as classical and novel—and a special loudspeaker. This latter was designed to eradicate resonance and to give a smoother tone to the record. The record of glass breaking was a perfect symphony of tinkling crashing glass. Another record was of a tug-boat on the Thames. One could almost smell the sea and the oil which persistently attaches itself to such a craft.
The lecture was highly informative and amusing. One good thing from the layman's point of view was that Mr Briggs never became too technical. Fine orchestral records helped to make the evening an enjoyable one.
So perfect was the tone and reproduction of some of the records that, when Mr Edgar Knight played a piano piece after the piece had been heard on a record, it was hard to distinguish the one from the other.'

Chapter Eleven: Concerts

The experiment was deemed to have succeeded and much had been learned in the process, not least the value of testing loudspeakers in such a space, which Wharfedale did regularly thereafter. In fact, referring to the photograph shown in (11.2) some years later Gilbert commented:[6]

'The small and rather weird looking loudspeaker on the left of the platform should be ignored. It sounded dreadful and was not used during the concert. Incidentally, this acoustic orphan gave us our first lesson in the value of concert hall conditions in assessing loudspeaker performance.'

By 'acoustic orphan' he presumably meant that it had no parentage, i.e. it was highly experimental. However, the nature of the cut-out in front of the speaker cone suggests that that the design was partly inspired by the R-J cabinet introduced later that year (see Chapter 7).

On to the Royal Festival Hall

Gilbert felt that there was clearly scope for improvement on the St. George's Hall event and the conviction that 'the bigger the hall the better' led naturally to a new challenge. The relatively new Royal Festival Hall, opened in 1951 as part of the Festival of Britain, had a seating capacity of 3000 and Gilbert was already familiar with its clear acoustics. It seemed the obvious choice, but it was virtually booked-up for the next 12 months. The first evening available during the remainder of 1954 was 1 November, so Gilbert made the booking more-or-less on the spot. The cost was £225 (about £4500 in 2010), with a deposit of £75. As with his illustrated lectures on composers to the Wesley Guild nearly 40 years earlier, he could not resist the challenge to cap the previous performance, but similarly he had consigned himself to months of stress during the preparation.

By the end of April Gilbert had secured the support of EMI and Decca and he wrote the letter to the BBC shown in Appendix 1. His experience the previous year of trying to use a BBC recording in his Audio Fair demonstration in New York (see Chapter 7), ultimately unsuccessful, lies behind his careful choice of words. This particular request was also unsuccessful, although Gilbert did manage to use a BBC recording to illustrate echo effects. The problems created by copyright and performing rights restrictions were daunting, with the Corporation adopting a very cautious stance, but, perhaps because they were unable to deliver on music recordings, the BBC did help Gilbert with questions about their experience of making recordings in the RFH. This resulted in their engineers being involved in rehearsals and ironing out several problems. Altogether, Gilbert wrote 17 letters to them over the next five months!

The 'In Brief' section of the May issue of *Wireless World* carried a few lines on the St. George's Hall demonstration, but this was obviously not submitted by Gilbert because it erroneously stated that the piano recording followed the live performance; the reverse was true as described by the *Telegraph and Argus*. The actual announcement of the RFH concert in the magazine, which Gilbert did submit, came out in the editorial pages of the August issue (Fig 11.4) along with a full-page advertisement in the advertising section.

Gilbert wanted to improve on the live versus recorded items, if at all possible, by having the same performer for both, and to broaden the musical range. There was clearly more scope for this if the performances could be specially recorded, rather than using commercial recordings. The first problem was to identify suitable performers and persuade them to take part, the

> **"Live" versus "Recorded" Sound**
>
> ENCOURAGED by the response to his efforts in the provinces, G. A. Briggs has booked the Royal Festival Hall in London for 8 p.m. on November 1st for a lecture-demonstration on sound reproduction.
>
> The audience will be invited to compare high-quality tape recordings with "live" performances of the same items by such distinguished executants as Denis Matthews (piano), Stanislav Heller (harpsichord) and Ralph Downes (organ).
>
> The reproducing equipment will be of the same calibre as that used for high-quality sound reproduction in the home, and many may think it a bold experiment to attempt to demonstrate it in so large an auditorium. Early scepticism has already receded as the result of two successful rehearsals, and there can be little doubt that an enjoyable and instructive experience awaits those who can "make it a date." Tickets (3s 6d reserved, including tax) will be obtainable on and after August 16th, from the Festival Hall booking office, from dealers in audio equipment in the London area, or from Wharfedale Wireless Works, Idle, Bradford.
>
> WIRELESS WORLD, AUGUST 1954

Figure 11.4 *Wireless World* announcement of the first RFH concert-demonstration.

second was to work out how best to make the recordings. Meanwhile the overarching technical challenge was how to achieve the required sound volumes within the vastness of the RFH.

The Amplification Question

It had already been noted at St. George's Hall that for the choral, orchestral and organ records reproduction 15 watts was insufficient, so significantly more power would be required. Rough calculations based on parameters for comfortable listening in a typical room and the scale-up required to give the same pressure changes in the ear in the RFH, with around 400 times the volume, suggested powers of the order of 60 watts would be needed, with the lowest harmonic distortion. Gilbert knew Harold Leak, whose company produced high power, low-distortion amplifiers for businesses in addition to their hi-fi products, reasonably well. In fact, he had flown to New York only six months earlier with Leak ('via Scotland, Iceland, Newfoundland and Boston',[7] which he did not enjoy) and they stayed in the New Yorker Hotel together during the Audio Fair. As noted in Chapter 7, they may even have put on a joint demonstration—British Industries Corporation had been the USA agent for both Wharfedale and Leak since late 1949, when they took on Leak amplifiers. His first approach was to Leak for a collaborator in the event, since good amplification was going to be a key ingredient, and Leak initially went along with the idea. However, either just before or just after the second rehearsal in early July he pulled out; the reasons why are not known but it is not difficult to imagine problems arising. Leak had always been immersed in electronics and was an out-and-out businessman. Hi-fi equipment was his business and he had no real interest in music as such.

Gilbert, whose raison d'etre was his passion for music, was not interested in the promotion as a profit making venture, so Leak probably saw little in it for his company. The early rehearsals would have amply demonstrated that the event could be a source of major distractions.

Some considerable time had therefore passed before it became clear that an alternative collaborator had to be found. In the two earlier demonstrations a single 15 watt Quad II amplifier from Acoustical Manufacturing of Huntingdon had performed well so this was the obvious next choice. Gilbert had only met Peter J. Walker, its MD, a few times at industry events. Their first meeting, at the BSRA Exhibition of May 1951, was not too propitious. Both had, unknowingly, turned up with a copy of the first *Audiophile* recording made by Ewing D. Nunn in the USA (in December 1950) and supplied by their common Canadian agent, Barney Smyth. This 78 rpm microgroove disc included the track 'Pop Goes the Weasel' which was state-of-the-art for demonstrating transient and high frequency response and both thought they had an 'ace' in their demonstration pack. Walker was amazed, and disappointed, to hear Gilbert play this before he had his opportunity![8]

Like most of the audio industry, Peter Walker, had been amazed when he read Gilbert's announcement of the RFH lecture-demonstration. It seemed to be a venture of such boldness as to be bordering on madness. The idea that domestic hi-fi equipment could be used in such a vast hall was just counter-intuitive. So, he was surprised to receive a phone call from Gilbert one Monday morning wanting to know if he would be a collaborator. The job was to take responsibility for the pickup and amplifier requirements for the event, with a short deadline for delivering appropriate amplifier power ('by Friday', as Walker recalled[8]). At first he was understandably reluctant, but Gilbert talked him into it. Unlike Harold Leak he was very interested in music and, like Gilbert, he was a musician (by then playing flute in the Huntingdon Orchestra, but he had played sax in dance bands when younger).

The task of delivering the amplifier unit was handed to John D. Collinson who had only recently joined 'Quad' (in November 1953) from the Admiralty. The 60 watts requirement was met by connecting up four Quad II units in parallel and mounting them in a portable rack, with connections to control units which could be operated from a distance.

The Build-up

By now the forthcoming event was causing a real stir within the sound recording and hi-fi world. As Collinson observed on becoming a part of the audio industry:[9]

'I was happily surprised to find that it was largely composed of amateurs (in the best sense of the word). It seemed that that the art of high quality sound reproduction was not in any response to public demand but resulted from the ambition of individuals to get the best possible with the means available.'

This being the case, it is not surprising that Gilbert's initiative and his consequent overtures to the record companies and the BBC for assistance with various aspects resulted in a great deal of support. For all concerned this promised to be a hugely satisfying project, if it could be pulled off. Gilbert had a very high opinion of recordings which Cecil Watts had made of Thurstan Dart playing a Goff harpsichord. He thought these delicate pieces would be the ideal opening for the live versus recorded items. Unfortunately, Dart was unavailable on the chosen date but Gilbert stuck with the recordings and Stanislav Heller was booked to play the harpsichord. Piano pieces were a must and

Denis Matthews, whose playing Gilbert admired (and who had contributed to the *Pianos* book), was engaged. Finally, he wanted to unleash the power of the RFH organ which had been unveiled earlier in the year. Ralph Downes, the designer of the organ, was booked to play it. Downes was the recently appointed Professor of Organ at the Royal College of Music and organist at the London Oratory.

Decca suggested a number of their recent 33.3 rpm microgroove recordings would be potentially suitable for reproduction in the RFH and eventually several were chosen. EMI also made available disc recordings and undertook to make studio recordings of the pieces which Denis Matthews would perform live and recordings of Ralph Downes in the RFH itself, using their own BTR/2 tape recorders operating at 30 in/sec. Engineers from both companies and the BBC attended the early rehearsals and they were critical to the diagnosis of a number of problems, which were eventually overcome. One of the issues concerning the live versus recorded items was the exact matching of pitch. With the disc recordings this was easily remedied; the variable speed transcription turntable used (in this case a Garrard 301) could be slightly altered to achieve the desired pitch. For the tape recordings there was no such easy solution. John Collinson produced a frequency control unit (11.5) with 30 watts output for insertion in the mains lead to the tape drive and this became one of the most important items of the concert equipment. In fact, the design derived from a unit he had worked on for Goodmans Industries in connection with fatigue testing of the de Havilland Comet 1 jetliner—a result of the tragic crashes of early 1954.

A particularly bright idea was to show the audience the amount of power being fed into the loudspeakers as the music was played. Ernest Price, Gilbert's long-time consultant, designed a 'watt indicator' which was placed on top of one of the loudspeakers. A series of neon lamps was used to indicate progressively higher powers as shown in (11.6, 11.7). Price, himself, looked after these during the RFH performance.

Figure 11.5 Frequency control unit designed by John Collinson. Reproduced from *Stereo Handbook*.

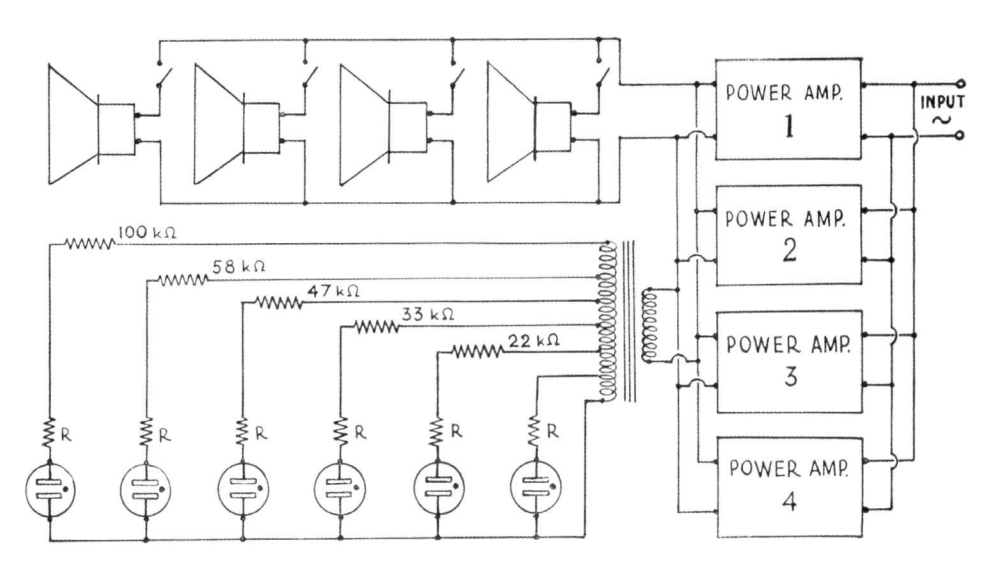

Figure 11.6 Watt indicator circuit. Reproduced from *Loudspeakers*, 5th Edition

Figure 11.7 Generic watt indicator unit with interchangeable scale. Reproduced from *Loudspeakers*, 5th Edition.

The effort which went into the preparations was remarkable, with five afternoon rehearsals required just in the RFH itself (11.8). The time spent there building up to the actual concert did, however, allow Gilbert to get used to the idea as he recalled:[10]

'I remember having a severe attack of the jitters in 1954 when walking over Waterloo Bridge just before the first rehearsal in the RFH, and I saw this huge building dominating the skyline. The mere thought of appearing there was to me an appalling prospect.'

In addition Gilbert designed the printed programme—the format remained virtually unchanged over the entire series of concert hall demonstrations except for the number of pages used—and worked on his 'script' which, like his books, was interspersed with humour. The musical jokes, to be delivered in a spontaneous manner, were as carefully rehearsed as the rest of the programme!

Figure 11.8 Peter Walker sitting facing, John Collinson and Gilbert Briggs, standing, in Box 25 at an RFH rehearsal, 1954. (Photo courtesy of the Escott family.)

The RFH Programmes cost 1 s. (The layout of the more compact Programme from the subsequent Bradford event is illustrated later in Fig. 11.10.) The RFH clearly had low expectations for bookings because when they had printed the tickets they only kept 250 for sale through their booking office, the rest were sent to Wharfedale for mail order. When the cheaply priced 3/6 tickets went on sale, a few weeks before the performance, a queue formed around the Hall and desperate requests were made for more to be returned. The enquiries for tickets direct from Wharfedale, however, ensured that the event was a sell-out within only four days, with hundreds of would-be attendees disappointed. So far, so good.

The Big Event

For some reason Gilbert decided that the concert should have a Chairman and he invited J. Raymond Tobin, the prolific author of music tutors and editor of *The Music Teacher* magazine, whom he had got to know through *Pianos, Pianists and Sonics* (see Chapter 12), to perform the role. On the night—the performance commenced at 8 pm—all went more-or-less according to plan, although the timing started to slip and following the warning at the end of the Programme that 'Last minute alterations to the programme may be made according to circumstances', Gilbert deleted three items and cut short one or two others. Family and friends gave him some stick for this afterwards. (Details of all the major concert programmes are collected in Appendix 2.)

As shown on the programme, the first item was entitled 'Barrell Symphony'. This was actually an EMI demonstration recording of various percussion instruments, each recorded with low and wide frequency range. It was produced by W.S. Barrell who had been in charge of recording for

Columbia and then EMI since 1925 (at the time of the concert he was a director of EMI) and it may have been intended to simulate the dramatic improvements in high frequency recording between the original mechanical and later electrical disc cutting techniques. The next item, 'Glass Breaking', was a demonstration disc that Wharfedale often used. It had been sent to Gilbert originally by its originator, Robert W. Bradford, a pioneer of direct disc recording best known for his work on cutter heads and the use of negative feedback in the associated circuitry. 'Memories of you' was another early Audiophile recording by Ewing Nunn. After these three items the audience was rather uncertain about the proceedings but the next item, 'Thames Tugboat Noises', with its local significance (and deliberately added distortion for added effect) finally broke the ice and the audience started to enjoy themselves (11.9).

Little did the audience realise the drama unfolding in Box 25 where the turntable and amplifier controls were situated, operated by John Collinson and Peter Walker respectively. As Gilbert recalled with a shudder six years later:[11]

'During the playing of the Tugboat record I noticed that it stopped and restarted and chugged in an unusual manner, but I shrugged this off as a strangely developed fault in the record. At the end of the concert—not during the interval—they told me what had actually happened. The ball bearings in the tone arm mounting had come out and rolled all over the floor of Box 25. It was fortunate that this happened during the playing of a "noise" record instead of a musical one, but it was even more fortunate that we had a spare turntable and tone arm assembly all ready for use in the Box. Failing this we should have been compelled to stop the proceedings.'

Figure 11.9 RFH audience reaction to a Gilbert Briggs witticism. Reproduced from *High Fidelity*.

For many of the items only a single, very efficient, three-speaker system (corner unit type) was required, but for the orchestra and organ pieces, three of these were employed. For the final item only, an extract from a recording of Vaughan Williams' Sea Symphony, the full 60 watts at 0.1% harmonic distortion was slightly under-powered. Gilbert made the apparently spontaneous suggestion to the audience that Ralph Downes double the organ part. Just at the point in the extract when more power was required the live organ came in, allowing the amplifiers to be somewhat overloaded, since the slight increase in distortion was drowned out by the organ.

For Walker and Collinson, who were responsible for ensuring that the disc recordings were reproduced exactly as determined during the final rehearsal, it was a tense experience (and not only because of the ball bearing scare). Long after, both recalled the moment when the first live versus recorded item was due, the harpsichord comparison (item 5). According to Collinson:[12]

'The disc was to be played first and when Mr Briggs gave the cue there came from the loudspeaker a very thin faint noise. P.J. Walker was quite convinced that he had misread his notes and set the gain control incorrectly; I was sure that something had died in the amplifiers; some of the audience just looked puzzled. A very long two minutes passed and when the real harpsichord took over we were all relieved to hear it give a most creditable imitation of a very thin faint noise.'

Walker joked that this was evidence that Gilbert was 'a real showman'. The audience would indeed have found it difficult to distinguish between the recorded and live performances because it was almost impossible to hear either![8] His objective of gaining the audience's rapt attention at this point was certainly met.

Reaction

In the course of his *Wireless World* review, the editor Fred Deveraux wrote:[13]

'The most courageous of Mr Briggs' experiments—the immediate comparison of live performances ... with disc and tape recordings—proved to be the highlight of the evening. The delicacy and precision of the harpsichord playing, with every gradation of tone crystal clear in the recording made by C.E. Watts, were exactly matched in the impeccable playing of Stanislav Heller. The background noises in the Hall, which fell to a level creditable for an audience of three in a country cottage rather than 3000 in the heart of London, was an even more eloquent comment than the applause which followed.'

The report had started:

'When G.A. Briggs announced his intention of taking the Royal Festival Hall for a lecture-demonstration on sound reproduction there was much shaking of heads.'

and it ended memorably:

'There was still much shaking of heads as the crowds left the Festival Hall but it was noticeable that whereas six months ago the polarisation was horizontal it had now changed to vertical.'

The reaction of the audience was very positive; Gilbert had invited feedback and scores of letters were received. These gave opinions about the various items, especially the live versus recorded ones, requested more information about technical aspects of the programme and its staging and took the opportunity to ask general questions about sound reproduction, especially in the home. Gilbert had lavished 16 free tickets on various national daily and Sunday papers, hoping for some wide publicity after the event. He was sorely disappointed. A rather brief mention in the London edition of the *Manchester Guardian* was the only result. However, *The Gramophone*, was impressed and its editor, Percy Wilson, invited Gilbert and Peter Walker to report on the experience. This they did, published in February and March 1955. Gilbert discussed the problems encountered—'the difficulties were so great' that had there been the opportunity for a rehearsal before booking the RHF they would probably have decided not to proceed (but having paid the deposit, no Yorkshireman was going to quit!) He then gave more details of the watt indicators, load matching techniques and the power level comparisons between reproduction of the recordings in the RFH and in his own home.[14] Walker used an analysis of the letters to make observations about the wide variation in response of individuals to the same reproduced sound in relation to actual and perceived volume levels, ambience, balance and filtering etc.[15] The following month a further article by Gilbert and Percy Wilson tried to summarise what had been learned from the experiment and looked forward to the follow-up events already scheduled, with Wilson being somewhat indiscreet about the next RFH concert, whilst opening up the whole question of judgement of aesthetic quality in listening tests.[16]

By the time these articles appeared the 'Sound Reproduction—A Non-Technical Lecture-Demonstration by G.A. Briggs' bandwagon was really rolling. A second Bradford performance was scheduled for 1 April, The Royal Festival Hall had been rebooked for 21 May and Carnegie Hall (New York) booked for 9 October. Of these three events, the first was to bring the benefits of the RFH experience back home, the second was to give all the disappointed RFH applicants, and others, a second chance and the third was a case of 'onwards and upwards': the challenge of taking the show to the USA.

For the Bradford programme the number of 'effects' recordings was reduced to two and only four other recordings were the same as used at the RFH. Great care was taken to ensure that the chosen recordings were acoustically suitable for St. George's Hall. The way this was done became the standard approach for all the subsequent events. The record companies sent Gilbert copies of new recordings, presumably not entirely out of a sense of altruism, which he—sometimes helped by his daughter Valerie—tried out in his Music Room at home. A stack of candidates was then taken to the concert hall and played through to a small audience consisting of Wharfedale people and invited friends and colleagues. Each member was a given a clipboard and asked to comment on the recordings, giving some kind of figure of merit. Gilbert analysed these results and produced a probable final choice of the records to be used. Copies were sent to Quad at Huntingdon so that Peter Walker and John Collinson could have their say. An important consideration was the ease with which John could pick out the particular point on the record at which Gilbert wanted to start the extract. As Gilbert put it:[11]

'I always maintain that he is the only man I know who can pick out a certain note on a gramophone record by the naked eye and in the picture we see him doing just that. Like a true artist he spurns mechanical aids.' (see 11.10)

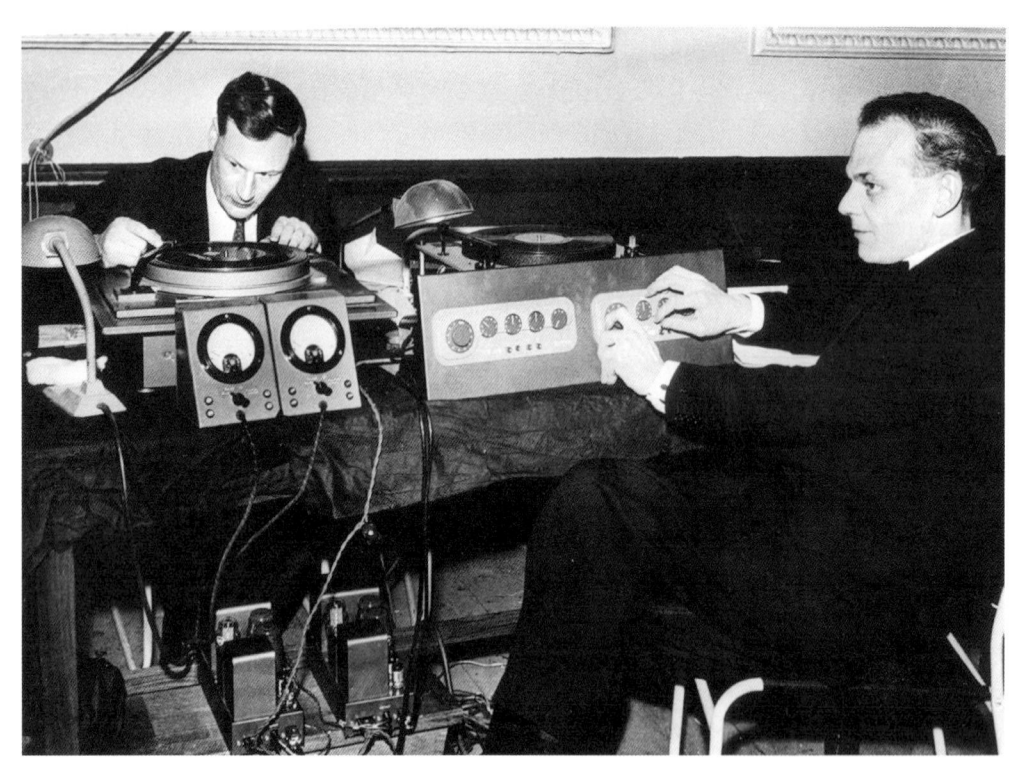

A Pair of Wharfedales: The Story of Gilbert Briggs and his Loudspeakers

Figure 11.10 John Collinson with turntable and Peter Walker with amplifier controls, St. George's Hall, Bradford, 1955. (Photo courtesy of IAG.)

Again there were three live versus recorded items, this time involving oboe, piano and organ. The world famous oboist, Leon Goossens, was to become a stalwart of these events over the years, his popularity as much for the humorous stories told when introducing pieces as for the uniquely exquisite oboe sound he produced. Gilbert regarded him as 'a raconteur of the first order'. Edgar Knight again performed in the style of Geza Anda and it is probably through Edgar that Leon Goossens became involved. The pair had planned programmes and performed together before the War and had in all likelihood first met at the Royal College of Music which they attended as prodigies (Knight was born in Bradford in 1899, Goossens in Liverpool two years earlier). The organist was Mr G. Hankin. This time the output from the 60 watt Quad amplifiers could be shared by up to four three-speaker corner units and the pickup and variable speed turntable were 'Connoisseur' units made by A.R. Sugden and Co. Ltd of Brighouse. Arnold Sugden, another of the audio pioneers, made an early two-channel (stereophonic) tape recording of the oboe/piano pieces by Edgar Knight and Leon Goossens in St. George's Hall, for the live versus recorded item.

In his programme Introduction Gilbert was at pains to reiterate that the event was non-profit-making. The tickets had been priced at 3/6 and 2/- which, after losing 1/5 and 9 d respectively to entertainment tax, left enough to cover expenses only if the event was a sell-out. The printed programme cost 6 d and two pages are reproduced in (11.11) to illustrate the format which remained unchanged for subsequent events. The inside front page was always an introduction written by Gilbert. Record details include the playing speed (78 rpm etc.) and the type of groove width (Std = standard, Mic = microgroove).

Chapter Eleven: Concerts

ST. GEORGE'S HALL, BRADFORD
(Manager : Bernard Beard)

SOUND REPRODUCTION

A NON-TECHNICAL LECTURE-DEMONSTRATION
by
G. A. BRIGGS

FRIDAY 1st APRIL 1955
Commencing at 7.30 p.m. prompt
Concluding at 9.40 p.m.

LEON GOOSSENS	Oboe
EDGAR KNIGHT	Piano
G. HANKIN	Organ

Steinway Pianoforte   Wharfedale Loudspeakers
Quad II Amplifiers—60 watts
Connoisseur Pickup (Mark 2) and Variable speed Turntable

PROGRAMME . . . . PRICE SIXPENCE
(*Subject to alteration without notice*)

Promoted in the interests of the Science and Art of Sound Reproduction by
WHARFEDALE WIRELESS WORKS LTD · IDLE · BRADFORD · YORKSHIRE

Figure 11.11 Front page of programme.

PROGRAMME

THE QUEEN

OPENING REMARKS by G. A. Briggs

1
HARPSICHORD C. E. Watts
 Sonata in G *Handel* 78 Std.
 Thurston Dart

2
ECHO EFFECTS B.B.C.
 Recorded in The Hamilton Mausoleum, Lanarkshire 78 Std.

3
CAROL Audiophile 1-A
 Lay Down Your Staffs *Wasner* 78 Mic.
 Sung by The Lutheran Choir of Milwaukee

4
TUGBOAT NOISES Mercury Sound Recordings Ltd.
 Recorded on the Thames by G. A. Elliot 78 Std.

5
BOLERO from 8 Danses Modernes Nixa LPY 122
 M. Philippe-Gerard Ensemble 33⅓ Mic.

6
ORGAN E.M.I. Festival Hall
 Toccata in D Minor *Bach* Decca LW 5095 33⅓ Mic.
 Choral Prelude *Bach* Decca LXT 2915 33⅓ Mic.
 E.M.I. Festival Hall recording by Ralph Downes
 Decca recording by Jeanne Demessieux
 Live performance by G. HANKIN

7
ORCHESTRA H.M.V. ALP 1178
 C Major Symphony *Schubert* 33⅓ Mic.
 Hallé Orchestra conducted by Sir John Barbirolli

8
PIANO Columbia SCB 117
 Prelude in G Minor *Rachmaninoff* 45 Mic.
 Prelude in G Major ,,
 Recorded by Geza Anda
 Live performance by EDGAR KNIGHT

INTERVAL — 15 Minutes

Figure 11.11 (continued) First half of programme.

Rehearsals for the second RFH event were already taking place when the Bradford programme was being considered. At one of these, on 27 January 1955 an interesting experiment in volume level setting for recording reproduction was carried out. A few days earlier, on 19 January, the BBC made the first broadcast of Vaughan Williams' choral-orchestral work 'This Day' from the Royal Festival Hall. EMI had recorded the performance and the microphone levels had been noted by the BBC engineers on a copy of the music. The recording was replayed at the rehearsal and the volume could be calibrated against the same microphones. In fact, when Peter Walker set the amplifier controls to what he thought the volume ought to be this turned out to be within 2 dB of the original, live, performance levels. Although appropriate reproduction volume levels had been established in the previous live versus recorded performances, these were for soloists, so this experiment provided a similar degree of confidence for orchestral pieces. Gilbert obtained permission from all the necessary bodies (BBC, EMI, Musicians Union and Mechanical Copyright-Protection Society!) to play an excerpt from the recording and was therefore able to describe the experiment at the Bradford event.

In the printed programme Introduction and during the performance Gilbert acknowledged the fact that the success of the event from a technical point of view was largely down to Peter Walker. After the RFH demonstration Gilbert had:[17]

'*...suggested to Mr Walker that he might derive some benefit by mentioning in his advertising that Quad II amplifiers had been used. He replied that we had booked the hall and footed the bill and he did not think it right that his firm should directly try and make capital out of it. A nice point from which he never departed.*'

Figure 11.12 Platform layout, St. George's Hall, Bradford 1955. (Photo courtesy of the Escott family.)

Figure 11.13 St. George's Hall 1955: Leon Goossens oboe, Edgar Knight and page-turner at piano, Gilbert Briggs at the table, Ernest Price alongside. Note the 'watt indicator' in front of the small speaker cabinet on the left. (Photo courtesy of the Escott family.)

The Quad equipment was, of course, noted in the programmes. However, Walker was persuaded to make some concluding remarks at Bradford, the start of a process which gradually raised his profile in the events.

Dorothy Stevens looked after all the ticketing for these events and on concert day her role was to ensure that all the participants' needs were met. When Leon Goossens arrived at St. George's Hall that day, and she met him for the first time, she asked for his dinner suit so that she could look after it. Nobody had told him that everyone would be wearing a dinner suit because they assumed he would be attired as normal for a concert, whilst he thought this was more of a lecture and so had just come in a lounge suit. Dorothy sent him off to Moss Bros., not far from the Hall, to hire the necessary clothes and then told Gilbert. At first he grumbled—'why should he foot the bill'—but, not for the first time, soon acknowledged she had done the right thing. Things might have looked odd in (11.13) had she not used her initiative!

Royal Festival Hall, May 1955

The second RFH event had been timed to coincide with the BSRA Exhibition of that year (the weekend of 21/22 May) and it took place on the Saturday afternoon at 2.30 pm. Finishing on time at 4.30, so that members of the audience could visit the exhibition in the Waldorf Hotel afterwards, was important and Gilbert could not afford the timing problems of the first event. He dispensed with a Chairman, put more of his introductory remarks into the programme and also added a section of Notes to this which commented on the items, all intended to give the maximum time over to the music itself. This time he was able to book Thurstan Dart to play the harpsichord. Ralph Downes again played the organ and Phyllis Selleck was the pianist. It was Peter Walker's suggestion that the boundary of the 'live versus recorded' envelope be pushed by trying out a small choir. EMI again undertook to record all the artists, either in the Hall itself or in their studios and H. John Dyer from the company instigated the rehearsing and recording of a choir drawn from the Goldsmiths' Choral Union, with their conductor Frederick Haggis (11.14).

Despite the sell-out of the previous RFH event a significant loss was made, as described in more detail later in the chapter, so in an attempt to prevent this happening again the ticket price was increased to 5/6. The printed programme price was unchanged at one shilling. Wharfedale Wireless Works wrote to all of the people who had been disappointed in their ticket applications for November 1954 offering first refusal on the new tickets. This and very brief announcements in the *Wireless World* and *Gramophone* magazines led to another sell-out without recourse to any advertising. Pictures from the event are shown in (11.15, 11.16).

Figure 11.14 RFH 1955 interval: Gilbert Briggs with Frederick Haggis, conductor of the Goldsmiths' Choral Union, centre and Ralph Downes, standing. (Photo courtesy of the Briggs family.)

Figure 11.15 RFH 1955. Phyllis Seleck a the piano, Ralph Downes sitting behind piano to right of choir, Gilbert Briggs standing at table, Arnold Hatton sitting alongside, EMI engineer operating BTR/2 tape recorder on right—possibly a Mr Dillnutt who performed this task in the previous concert. (Photo courtesy of the Briggs family.)

Figure 11.16 RFH 1955: Gilbert Briggs standing, Arnold Hatton sitting alongside, EMI engineer on his left. (Photo courtesy of the Briggs family.)

The audience feedback was very positive and this time a notice appeared in *The Times* entitled 'Live and Recorded Music Compared: Deceiving the Ear'. This began:[18]

'A curious experiment was made in the Festival Hall on Saturday when a Yorkshire engineer, Mr G.A. Briggs, who manufactures loudspeakers, set up a battery of four of them to demonstrate two features of their capabilities.'

From the ensuing description, Wharfedale loudspeakers gained useful good publicity. The piece went on:

'The other demonstration was to get a living artist on to the platform and let him play Box and Cox with his own recording and see if we could tell the difference.'

The writer felt that Gilbert could have tilted the scales more in his favour by changing from live to recorded more than once, but finished:

'The difference in immediacy and also a slight difference in depth (in the three dimensional sense) can only be accounted for, now that fidelity of reproduction has reached so high a pitch of indistinguishability, by the fact that we have two ears, the microphone but one.'

The event was also reported in the USA *High Fidelity* magazine. The piece included the information that because of (unexplained) problems encountered in the Royal Festival Hall, two of the corner units were specially reinforced by dispensing with the sand-filled panels and lining the wooden panels with one-inch thick tiles. Early in the session Gilbert demonstrated the effect of this on the resonance, pounding emphatically on the enclosures with a large wooden mallet. He was quoted[19] as quipping: 'I hope my name will go down in history as the first man to test loudspeakers with a hammer!'

Carnegie Hall, October 1955

Attention now focused on the next event in Carnegie Hall and this soon turned into a baptism of fire. The organisational experience gained in the four UK concerts was rather nullified by the distance and different ways of doing things. Someone needed to be there for as long is at would take to make all the necessary early arrangements. Albert Smith, who after 20 years selling Wharfedales in the South and Midlands of England was perhaps a strange choice, was rewarded with the task. Gilbert may have regarded his patience as the key virtue, in which case he was prescient! Smith arrived in New York in July during a terrific heat wave and 16 days later he had secured the most essential items—tapes of the performances chosen for the live versus recorded items. The performers were E. Power Biggs (organ), Leonid Hambro (piano) and four members of the Philadelphia Woodwind Quintet: John De Lancie (oboe), Sol Schoenbach (bassoon), Anthony Gigliotti (clarinet) and Mason Jones (french horn). The recordings were made by Columbia in Carnegie Hall itself (11.17) under the supervision of Howard H. Scott. Of the 18 items in the final programme, five were live versus recorded. Gilbert had attended several concerts in Carnegie Hall during previous visits to New York and therefore had some knowledge of the acoustics on which to base the choice of potentially suitable records. Normally, the final selection would only be made *in situ*, but since the programme had to be printed in advance this was not possible.

A Pair of Wharfedales: The Story of Gilbert Briggs and his Loudspeakers

Figure 11.17 Albert Smith rehearsing with the members of the Philadelphia Woodwind Quintet, Carnegie Hall 1955. Reproduced from *Loudspeakers* 5th Edition.

The concert date had been chosen to coincide with Gilbert's trip to New York for that year's Audio Fair (held for the last time in the New Yorker Hotel), with the performance at 3 pm on 9 October 1955. Gilbert set out from Southampton, aboard the *Queen Elizabeth*, on 22 September and arrived five days later. Peter Walker, collaborating once more, chose to fly and arrived the same day, having travelled with Arnold Hatton, whose role, as usual, was to look after the speakers and switching gear. That gave about 10 days to complete the preparations since the Audio Fair was over the weekend following the concert. There were significant differences in the acoustics compared with the RFH, as expected. In Carnegie Hall they found severe absorption at very high frequencies, but extremely good bass response. Thus the favoured harpsichord recordings were not as impressive, losing brilliance, but organ works sounded much better with all the pedal notes fully audible. Tonal variations around the auditorium were rather noticeable. It took three rehearsals to sort out the issues, and much was learned about the positioning of the loudspeakers and performers on the theatre-like platform—very different from the RFH (11.18). Barney Smyth, the Canadian agent, was taking on John Collinson's role at the turntable, and he flew down from Montreal for two of these rehearsals. The grip of the unions meant that getting anything done which required electrical or even manual assistance was time consuming and very expensive, leading to much frustration. On the other hand, the attitude of the musicians and the audio engineers from Columbia was exactly the same as back home—enthusiastic and keen to enter into the spirit of a grand experiment. In fact, the musicians only charged nominal fees and their total cost was well below that of the stage crew.

The event had been highly publicised by BIC, the co-sponsors. Again the ticket prices were modest, ranging from the then equivalent of 7/6 to 17/6, but there seemed to be a reluctance on the part of the paying public to accept the concert at face value—that it was not a commercial venture but a promotion 'in the interests of the science and art of sound reproduction' as (always) stated on the programme cover. The advertising was also a major expense but, in

Chapter Eleven: Concerts

the end, 2400 of the 2900 seats were filled. The concert went well; the musicians performed superbly and produced some of the best live versus recorded results so far. Technically some of these were very difficult with up to 20 switches between live and recorded during a single piece; Columbia engineers Messers Plaut and Meyers were responsible for operating the tape recorders (11.19, 11.20).

Figure 11.18 Stage layout Carnegie Hall 1955. A pair of R-J cabinets is placed next to the piano with four three-speaker corner units between them and the organ. Watt indicators are sitting on top of the R-J cabinets and the second corner unit from the left. Picture included in the 1956 Wharfedale catalogue.

Figure 11.19 Gilbert Briggs introducing the Philadelphia woodwind quartet, Carnegie Hall 1955. (Photo courtesy of IAG.)

275

Figure 11.20 Columbia engineers operating the equipment during the 1955 Carnegie Hall concert. (Photo courtesy of IAG.)

American Reaction

Another difference from the UK events was the reaction of the popular press. In New York the free tickets were taken up and resulted in wide reporting and generous notices. Gilbert, quick as ever to spot a marketing opportunity, published extracts from these in a full-page advertisement in *Wireless World*. This was headed by the picture of the Carnegie Hall stage (11.18) below the headline 'Supreme Test of Speaker Performance'; the rest of the page is reproduced in (11.21)

Harper's Magazine had a feature article by Edward Tatnall Canby entitled 'Mr Briggs and the Concert Hall' which began:[20]

'The main trouble with the recent Carnegie Hall demonstration of Sound Reproduction by the British engineer G.A. Briggs was that it simply wasn't loud enough'. So, at least, most of the hi-fi audience thought . . . Yet he had rather special reasons. His purpose it seems was to reproduce his recordings at the same loudness as the actual music might have upon the stage—this was to be "concert hall reproduction" right in the concert hall. The fact that it didn't work too well was perhaps his most interesting point.'

Canby then gave a considered analysis of the proceedings and the acoustical effects they had all witnessed, before concluding:

'In these curious ways the genial Mr Briggs, who is one of the most thoughtful of British engineers (Wharfedale Wireless Works Ltd), proved very nicely for us what some of us had already been told, that a large auditorium—a concert hall—marvellously exposes to the ear every technical fault and every acoustical anomaly in recorded sound (especially as compared with live music) whereas in fortunate contrast the average small living room gracefully disguises almost all of these same faults for our greater pleasure.'

'A heartening conclusion and the members of Mr Briggs' audience will think twice, I suspect, before they demand "concert hall reproduction" from their home phonographs.'

In fact, Gilbert's main concern from the outset was the American enthusiasm for amplification as evidenced by the fact that domestic hi-fi equipment tended to include amplifiers with significantly higher power ratings than was normal in the UK! On the return transatlantic voyage he drafted an article for *High Fidelity*, which appears not to have been published. In this Gilbert reviewed the experience, described the financial aspects in some detail (see later) and discussed various aspects of the equipment, the Hall and the performances.[21] In the concluding section he answered the oft-asked question 'Why do it at your age?':

'Although the entire recording and reproducing industry should benefit from such demonstrations if no mishaps occur, it would be silly to claim that I am an idealist working for others. I regret to say that I have no such altruistic motives. I suppose the idea of playing records in famous concert halls, and comparing them with actual performances of first class artists, appeals to me so strongly that I cannot resist the urge to make the experiment, knowing that if we are lucky and nothing goes wrong on the day the prestige value will probably justify the cost in the dim and distant future.'

Inspired by the London Mozart Players

From the first Bradford lecture-demonstration in March 1954 to surviving Carnegie Hall had been 18 months of almost continuous 'experimentation'. Gilbert was approaching normal retirement age and starting to think more about looking after his health, if he wanted to carry on with the job he loved. At this point there were no plans for more concerts, but the situation soon changed. Towards the end of the year Gilbert met his daughter Valerie in London and went to a Youth Concert at the RFH where the London Mozart Players under Harry Blech, with Denis Matthews at the piano, were performing. Their superb playing took Gilbert off on a new flight of fancy; such an ensemble (about 40 musicians) would be a new challenge for live versus recorded demonstrations! The seed was sown and the hall was again booked for a concert to coincide with the BSRA exhibition (2.30 pm on Saturday 12 May 1956). In addition to the LMP and Denis Matthews, Leon Goossens was booked for a second appearance. The feedback from previous concerts had included comment (even complaint) that solo violin had not featured; Gilbert remedied this in style by booking Campoli.

This was the time when stereo recording on tape was making its presence felt, certainly in the USA where Ampex professional two-channel tape recorders had been available since the late 1940's. In April 1955 Sir Malcolm Sargent introduced the first stereo tape recording demonstration in Europe to the press and public, including Gilbert, at EMI's Abbey Road Studios. (The record companies were still working on the technology, and the all-important agreements on a standard approach, for getting stereo onto disc.) It seemed an appropriate point at which to introduce commercial stereo recordings into the lecture-demonstration and EMI obliged, providing two tapes marketed under the Columbia and HMV labels. Once again EMI also made the recordings for the 'live versus recorded' items either in the RFH or in their studios. There were no less than six of these in the crowded programme of 17 items, leaving no room for organ items with Ralph Downes.

Rehearsals for the event (11.22, 11.23) created a novel situation. As conductor of the ensemble he

> An audience of 2,400 assembled to hear the lecture demonstration given by G. A. Briggs, when two R-J cabinets and four Wharfedale corner speaker systems were demonstrated. The following extracts from the New York press give an indication of American reaction to this Experiment in Sound:—
>
> **THE NEW YORK TIMES**   OCTOBER 10, 1955
>
> The English sound engineer Gilbert A. Briggs, who sold out a London concert hall for his recent demonstration of recorded sound equipment duplicated this feat at Carnegie Hall yesterday afternoon. He proved hi-fi enthusiasm to be at an equally feverish pitch on both sides of the Atlantic. A feature of the demonstration was the playing of records in conjunction with live performance of the same works by the same artists.
>
> One listener found the pipe-organ demonstration most impressive. Mr. Briggs' battery of sound equipment reproduced it all, from the highest squeal of the " mixture " stops to the boom of the pedal diapason. An observer looking away from the stage could not detect the moment at which the record stopped and E. Power Biggs, at the console took over. The mellow tone of the Philadelphia woodwinds, too, was projected with almost startling fidelity to the sound of the players in person.
>
> **WORLD TELEGRAM AND SUN**   OCTOBER 10, 1955
> **HI-FI BATTLES REAL THING IN CARNEGIE**
>
> A championship bout between live and recorded music was fought out in amiable and friendly style in Carnegie Hall yesterday afternoon. Actually, it was a stupendous demonstration of how far Hi-Fi has travelled in achieving the illusion of live music.
>
> My hunch is that both side won recruits yesterday—many Hi-Fi enthusiasts must have noted down Carnegie as a good place to visit between hook-up sessions at home. And even more concert goers, still minus a Hi-Fi system of their own, must have vowed to repair the omission at the first opportunity. Mr. Briggs couldn't have been more impartial in expressing the claims and qualifications of each contender, although his heart spoke for the record from the start. He is a white-haired man with rosy cheeks and a sense of humor to match.
>
> **STARS AND STRIPES**   OCTOBER 12, 1955
> **HI-FI ENGINEER FOOLS FANS WITH " CANNED " ARTISTS.**
>
> A Briton demonstrated high fidelity at the Carnegie Hall here and some recordings were so true to life the audience could not tell where the canned music stopped and real performance began.
>
> English sound engineer Gilbert A. Briggs showed all kinds of recording and reproducing equipment and demonstrated its high fidelity by playing tapes and records and switching to live performances by the same artists.
>
> Most convincing was a performance of organ playing where a listener—his face averted from the stage—was not able to tell when the recording was switched to the " live " organist.
>
> **WORLD TELEGRAM, NEW YORK**   OCTOBER 15, 1955
> **CROWD SPELLBOUND BY HI-FI DISCOURSE**
>
> A fascinating aspect of Gilbert Briggs' Hi-Fi demonstration in Carnegie Hall last Sunday afternoon was the audience. This was one of the most alert and earnest crowds I ever observed in Carnegie Hall, and I believe I have seen every possible variety. They followed Mr. Briggs' discourse with grave intentness, watched the technical activities on stage with studious care, laughed easily at highly technical quips.
>
> Mr. Briggs was in good form and so was the crowd. The result was a marvel of mutual understanding and respect ordinarily reserved for great musical events.
>
> And speaking of music, I must say that on both sides of the demonstration there prevailed a reverent attitude toward musicians and their manifold works. This was a moving and reassuring experience for me. These were not just faddists indulging mania for the latest gadget, or the latest twist of the latest gadget. They were, most of them, a body of music lovers sedulously seeking ways of bringing music into the home without distortion and without bringing the concert hall along, too.
>
> **Wharfedale** WIRELESS WORKS LTD. · IDLE · BRADFORD · YORKSHIRE
> TELEPHONE: IDLE 1235/6 (2 Lines)   TELEGRAMS: WHARFDEL, IDLE, BRADFORD

Figure 11.21 Part of *Wireless World* advertisement, January 1956.

had formed six years earlier, Blech was naturally used to directing operations. However, the LMP had been hired by Gilbert and he was even more used to having his own way. The closing programme item was to be an excerpt from a recording of the Hallelujah Chorus and Gilbert suggested that the LMP and Ralph Downes could accompany this piece to give a truly grand finale. When it came to the first rehearsal of this item Blech said they hadn't brought the band parts because he didn't think it was a good idea. Gilbert had to remind him, firmly, who was paying the bill. The parts were soon forthcoming! At another rehearsal Blech decided to break for lunch in the middle of work on one piece. Gilbert, who could not abide being diverted from something he was concentrating on, just refused to countenance lunch until he was satisfied with the result. Fortunately, these minor hiccups did not affect the outcome and, in particular, the final item was rendered enthusiastically by all concerned (11.24).

Chapter Eleven: Concerts

Figure 11.22 RFH 1956 rehearsal, L-R: Gilbert Briggs, Harry Blech, Denis Matthews, Wilfred L. Proctor (Wharfedale's South Africa agent). (Photo courtesy of the Briggs family.)

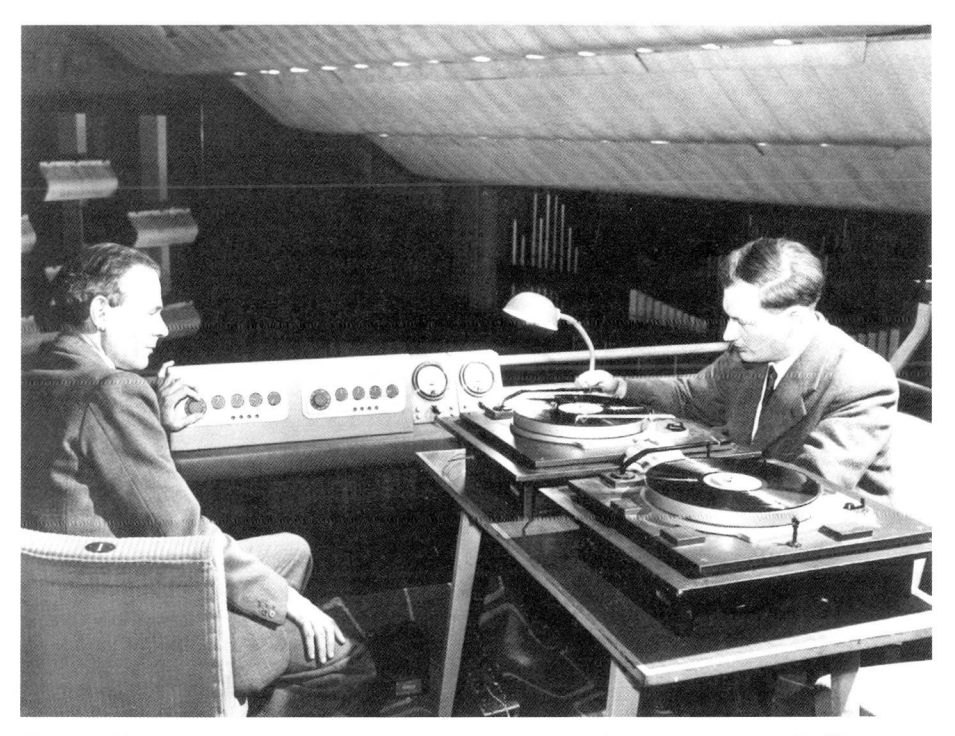

Figure 11.23 Peter Walker (left) and John Collinson, rehearsal RFH 1956. (Photo courtesy of IAG.)

Royal Festival Hall, May 1956

There was some advertising of the event, now described as 'A Concert of Live and Recorded Music' and all but 100 seats were sold. For the first time, the collaboration of Peter Walker was acknowledged on the programme cover and he gave a short address at the end of the interval. Everything went according to plan. Stanley Kelly, the well-known audio engineer/inventor, consultant and prolific writer of articles on audio and related technologies, attended with a miniature sound level meter and wrote a critique for the new magazine *Hi-Fi News*. He started by saying he hoped Gilbert's opening comment that this concert would be his 'swan song' could be taken in the same vein as the periodic 'final tour' announcements of artists close to 'retirement'. He then commented on each of the items, giving marks out of 10 and comparing his assessment with that of the audience (averaged from the feedback sent to Gilbert). He concluded:[22]

'The afternoon proved one thing if nothing else—that given good quality equipment, expertly handled, and with choice recordings, the aesthetic difference between the original and reproduced sound is as near as makes no difference.'

Not unexpectedly, the stereo experiment had a mixed response. For those in the audience who were ideally positioned relative to the speakers, the experience was a good one and a significant enhancement over listening to mono recordings. For the rest the stereo experience was 'no better than' or 'worse' than mono.

Figure 11.24 RFH 1956: Harry Blech conducting the London Mozart Players with Ralph Downes at the organ. Sitting at the table, from the left: Gilbert Briggs, Arnold Hatton and Raymond Cooke. (Photo courtesy of Dorothy Stevens.)

Carnegie Hall, October 1956

The following October (the 3rd) a second concert was given in Carnegie Hall. The circumstances are something of a mystery. The first concert had been a struggle to organise, the hall was by no means full and the financial loss was very significant (for both Wharfedale and BIC). In his write-up immediately after the event, Gilbert was not anticipating a re-booking. Peter Walker and his Quad amplifiers were not involved; instead Harold Leak was named on the programme as Gilbert's collaborator and Leak amplification was employed. Since BIC were US agent for both Wharfedale loudspeakers and Leak amplifiers it must be the case that this event was driven by them. In the programme Introduction Gilbert says that 'all the practical arrangements have been in the hands of Leonard Carduner and Arthur Gasman of BIC'. The 'live versus recorded' items did not have the G.A.B. stamp on them, apart from the organ pieces—again played by E. Power Biggs—which had been a success of the previous concert. These recordings, with their double ambience because they were recorded in the Hall, sounded much better than the live version. Several items had an avant garde character, original pieces for percussion ensemble composed and conducted by Morton Gould—one of them including a tap-dancer (11.25). The other performers were the pianists Teicher and Ferrante, playing duets. The choice of records would appear to be Gilbert's and the finale was again the Hallelujah Chorus with live organ accompaniment. Given the RFH experience, no attempt was made to introduce stereo recordings. The wording of the announcement put out by BIC did not hold back:[23]

> 'Audience and critics alike acclaimed the first sound demonstration by G.A. Briggs at Carnegie Hall last year. Now, in answer to a strong public demand for a second demonstration, a fascinating new programme has been prepared featuring artists and records entirely different from those in the first programme. Even if you know little about technical matters this is an opportunity to enjoy perhaps the most significant, most interesting high fidelity event of the year.'

Figure 11.25 Carnegie Hall 1956, Danny Daniels tap-dancer. Seated at table behind conductor, clockwise: Arnold Hatton, Gilbert Briggs, Valerie Briggs, Harold Leak. (Reproduced by permission of IAG.)

On this occasion Gilbert took Edna and Valerie with him and the latter took part in the concert, handing the records and tapes to Harold Leak before each item. The event was not as successful as the previous one, as Gilbert admitted, and even more money was lost by the two sponsors, Wharfedale and BIC. Fewer seats were sold, although the Hall was still reasonably full, more had been spent on advertising and this time Columbia charged for making the recordings.

The Shows Must Go On

This disappointment did not, however, dampen enthusiasm for playing other venues in the UK and indeed invitations to give the concerts began to arrive on Gilbert's desk. In 1958, Gilbert said:[24]

'Between March 1954 and July 1957, we gave 10 concert hall demonstrations in various cities, including London and New York, with direct comparisons of live and recorded music. The total attendance figure was in excess of 16,000, so we can at least claim to have put the problem of lifelike reproduction to the test.'

and in 1960, following a description of the first event in Bradford:[25]

'During the following six years we gave no less than twenty concerts at which live and recorded items were compared ...'

and again in 1963:[26]

'We gave 21 [live versus recorded] *demonstrations in all between March 1954 (Bradford) and March 1962 (Blackpool). The last incidentally was the worst which seems to prove that we do not always learn by experience ... '*

Clearly, these last two statements are inconsistent. Of the first ten concerts, seven have been described so far. On 14 March 1957 a demonstration was given in the Great Hall of Leeds University at the invitation of the Engineering Society. Most of the team were by now pretty experienced: Peter Walker and John Collinson looking after pickups and amplifiers with Edgar Knight and Leon Goossens giving live performances. In addition Kenneth Page, with accompanist Kenneth Clare, performed violin solos which were doubled with tape recordings made by Hollick and Taylor of Birmingham. Allen E. Stagg of the International Broadcasting Company (London) made some special two-channel tape recordings for a stereo demonstration item. Only two weeks later (28 March) almost exactly the same concert was given in the Large Theatre (today's Lyttleton Theatre) of the Birmingham and Midland Institute, central Birmingham, at the invitation of the Midland Gramophone Society. The experience with the stereo item in the previous concert could not have been satisfactory—the removal of this item was the only programme change.

Philharmonic Hall, Liverpool, July 1957

The next, and tenth, event was in Philharmonic Hall, Liverpool on 2 July 1957. The decision was taken to make another attempt at stereo demonstration, and the responsibility for this was shouldered by Raymond Cooke, who thereby gained his first recognition as contributor to an event (11.26). From Gilbert's programme Introduction:

Chapter Eleven: Concerts

'The widespread use of three-channel sound systems in cinemas (five-channel in Cinerama) has made most people familiar with the extra realism to be had from Stereo compared with normal single-channel working, provided reasonable volume level is maintained.[27] *For domestic use more than two channels are impracticable, and listeners are as a rule placed in an arc between a pair of directional loudspeakers. With a big audience this arrangement is not feasible, even if half the audience are ladies and they sit on the knees of the gentlemen. So we are today using omni-directional speakers, with the full approval and support of HMV, in an effort to produce some of the benefits of two-channel realism in all parts of the hall. (I would like to draw your special attention to the Saint-Saens 'Cello Concerto).*

Mr R.E. Cooke is in charge of the loudspeaker switching and the Stereo set-up, the vital amplifier and pickup controls being once again in the capable hands of Messrs. P.J. Walker and John Collinson.'

There were four live versus recorded items. Two of these involving Leon Goossens and Kenneth Page (violin) respectively, both accompanied by Kenneth Clare, used taped recordings of the performers again made by Hollick and Taylor in Birmingham. In the other two the pianist Gerald Gover and the organist Dr Caleb Jarvis played alongside disc recordings by other artists. The protagonists felt that this was perhaps their most satisfying event—the acoustics

Figure 11.26 Gilbert Briggs and Raymond Cooke demonstrating, with a lighted taper, the force of the air coming from the reflex port of an AF10 speaker, Philharmonic Hall, Liverpool, 2 July 1957. (Photo courtesy of the Cooke family.)

283

of the Philharmonic Hall seemed particularly well-suited to their unusual requirements. The *Liverpool Post*, in its review the following day, said:

'the concert last evening proved to be good entertainment even to those of us who are not au fait with the technics of sound reproduction'. The live versus recordings items were very well received and, in particular, the oboe items with Leon Goossens and Kenneth Clare ('. . . there was one completely perfect or completely deceptive doubling. . . . Anyone who listened to this with eyes closed, as we did, would have been hard put to detect the change from "live" to recorded sound. . . . .'). As to the presenter: Mr Briggs carried the whole demonstration off with great good humour and the minimum of technical jargon.'

Home and Away

There are references to an event in the Tivoli Hall, Lisbon put on by Valentim de Carvalho Ltd, Wharfedale's agent in Portugal.[28] Gilbert assisted with this and attended, making the introductory remarks. No programme survives, but a flyer for the event ('A concert-demonstration of high fidelity') does and this is simply dated 17 November (with 1956 added in handwriting). In print, Gilbert mentions the event as being in 1958 but this must be an error, because there was a follow-up event in Trindade-Porto on 7 December 1957. There is a surviving programme for this concert which mentions Gilbert's earlier successes in London, New York *and Lisbon*. According to the programme Gilbert performed a similar function in Porto. At these events Gilbert's normal role was taken on by a well-known Portuguese musicologist and composer and they both included a 'live versus recorded' item using solo piano and taped recordings made at the venue. The fascinating story behind these concerts is told in more detail in Chapter 7.

Rather less is known about the remaining events, with only three surviving programmes from the period 1958–1962, and uncertainty even about the number.

Early in 1958, there was a concert at the Royal Aircraft Establishment at Farnborough, Hampshire, given to the Farnborough Scientific Society.[29] This was no doubt arranged for the benefit of the legion of his fellow audio enthusiasts who worked there by, or through, Richard W. Lowden, a stalwart of the BSRA, as secretary and president for many years, and therefore well known to the 'demonstrators'. It is known that Leon Goossens took part[30] and it is fairly certain that Edgar Knight also participated.

On 30 April 1958 an event was staged at the Tempest Anderson Hall in York, promoted by Cussins and Light Ltd, a TV and radio rentals business based in York, and Wharfedale Wireless Works. All the proceeds, with no deductions for any expenses, were presented to the Leonard Cheshire Home at nearby Alne Hall. The live performers were once again Leon Goossens and Edgar Knight plus the soprano Barbara Elsey, accompanied by Stanley E. Toms. Tape recordings of Goossens playing with Knight were made by Hollick and Taylor in Birmingham, and of Elsey by IBC in London. This time Edgar Knight had to imitate Victor Schioler (on record) in a solo piano item. This concert produced a delightful picture of Gilbert with Leon Goossens which, following a great deal of restoration, is shown in (11.27).

Figure 11.27 Gilbert Briggs, right, with Leon Goossens. Tempest Anderson Hall, York, April 1958. (Photo courtesy of the Briggs family.)

Farewell to the Royal Festival Hall, May 1959

In the early part of 1958 stereo discs finally hit the UK market. This long-awaited event was accompanied by confusion and controversy (see Chapters 7 and 12) and trying the cut through this for the music lover was an obvious reason for another concert-demonstration. Accordingly, the Royal Festival Hall was booked for the fourth time for a concert on 9 May 1959. Gilbert made an announcement in *Wireless World* in March and intimated that part of the experiment would be to compare mono and stereo versions of the same recorded performance, using the best pickups. However, at the first rehearsal in February 1959, it soon became obvious that this was unworkable. The difference between the mono and stereo versions actually came down to comparisons between pickups and the vagaries of the chosen recordings; one version was often obviously better than the other, but randomly so! So this idea was shelved and a number of excellent stereo recordings were simply chosen to demonstrate the effect. There were 'live versus recorded' items involving the old hands Leon Goossens and Denis Matthews (recorded in mono by EMI), Ralph Downes (playing against himself on a commercial stereo Pye recording) and the new introduction Harold Blackburn (bass), accompanied by Gerald Gover, (recorded in mono by IBC). A picture of the 'demonstrators' taken before the concert started is shown in (11.28) and of the audience eagerly awaiting the performance in (11.29)

With the experience of the rather unsatisfactory stereo experiment at the previous RFH event three years earlier in mind, Raymond Cooke had designed an omni-directional multi-speaker assembly to spread the sound in all directions at frequencies above 400 Hz[31] (11.30). A six-speaker version of the eight-speaker unit, which Cooke used at home, is shown in (11.31)

A Pair of Wharfedales: The Story of Gilbert Briggs and his Loudspeakers

Figure 11.28 The conductor's room, RFH 1959, L-R: Gilbert Briggs, Audrey Lambert (a friend of Valerie), Valerie Briggs, John Pitchford (Valerie's fiancée), John Collinson, Raymond Cooke, Peter Walker. (Photo courtesy of the Briggs family.)

Figure 11.29 RFH 1959, Denis Matthews at the piano taking applause. At the table R-L; Gilbert Briggs, Raymond Cooke, Stewart Horne(?). (Photo courtesy of the Escott family.)

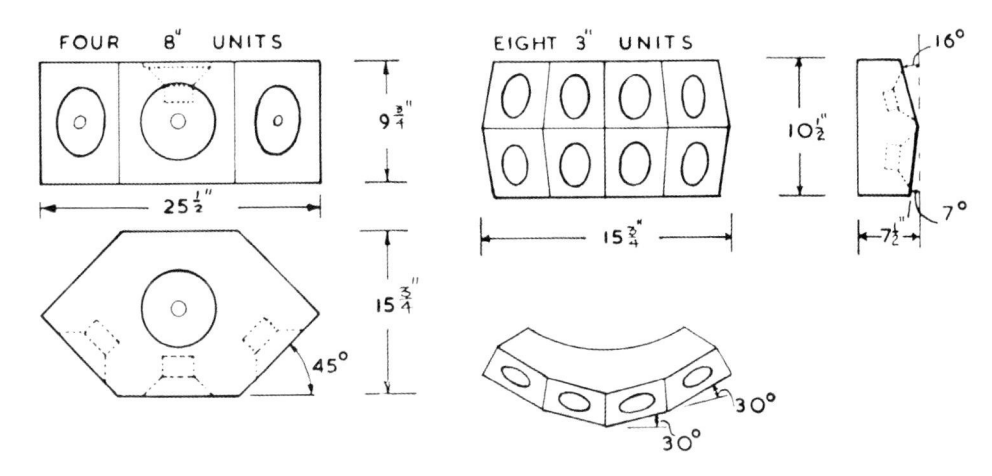

Figure 11.30 Omni-directional speaker design. The eight-speaker unit sat on top of the larger four-speaker cabinet. Reproduced from *Stereo Handbook*.

Figure 11.31 6–3" speaker omni-directional tweeter unit designed by Raymond Cooke and photographed on his living room window sill. (Photo courtesy of the Escott family.)

Two units were placed as widely apart as possible, one at each end of the platform, and brought into use for the stereo recordings. Whilst this largely overcame the original problem of the stereo experience being critically dependent on seating position, another only really came to light during the event itself. During loud passages of music from the stereo discs the audience in many parts of the hall heard 50 Hz (mains) hum and the bass speakers therefore sounded woolly and resonant. Apart from this, which brought forth a few rude letters, the full house enjoyed the experience. Interest in the subject was demonstrated by the fact that 700 applications for tickets had to be returned and also by the appearance of a notice in *The Times* which concluded:[32]

'The promoters of the demonstration were modest in their claims for "stereo", recognising no doubt that it is still beset with growing pains. Nevertheless they offered some of the most satisfactory "stereo" reproduction yet heard and there was no possible doubt as to the extra dimension introduced into the tone as a result—particular valuable in the avoidance of opaque tuttis.'

Commenting in *Wireless World*, Fred Devereux had suggested that a real test of the merits of stereo would be to compare mono and stereo recordings with live performances. The ultimate challenge which could not be resisted, and a good excuse to work in another major concert hall!

Colston Hall, Bristol, October 1959

Colston Hall in Bristol was booked purely on the basis of its acoustic reputation, without any prior tests, for 9 October 1959. For some reason Peter Walker and John Collinson were not involved, although the usual Quad amplifiers and stereo pre-amp were employed. Peter Walker's role was taken on by Ralph L West, lecturer in radio at Northern Polytechnic, London, and technical writer. Amongst other things, West reviewed loudspeakers for *Hi-Fi News*, and had been a contributor to Gilbert's books.

For the 'live versus recorded' items the performers were again Leon Goossens and Harold Blackburn, both accompanied by Gerald Gover, Edward Fry (organ) and a piano trio consisting of Gerald Gover, Kenneth Popplewell (violin) and Terence Weil ('cello). At the EMI studios mono and stereo tape recordings were made of the bass solos and the piano trio, whilst the oboe pieces were recorded in mono. Bristol and West Recording Services recorded the organ solos in Colston Hall, using microphones borrowed from the BBC. The stage layout is shown in (11.32) with Gilbert sitting at the desk and Raymond Cooke behind the pianist—the 'S' sign above his head indicates a stereo recording was being played (replaced by an 'M' for mono). The four open (sand-filled) baffles at the front were used for the bass voice and the four large corner speakers at the rear for reproducing the piano trio. The omni-directional systems developed for the previous event in the RFH are perched on top of the outermost corner speakers; these were used additionally for the stereo items. A view from the stage is shown in (11.33).

It was essential that the problem of 'hum' experienced in the RFH be eliminated and to this end Raymond Cooke undertook some extensive research. Six reasons for the increased hum (as well as rumble and noise) when playing stereo discs were identified and measures taken to reduce or eliminate the potential causes. The work was published in an article by Raymond Cooke in *Wireless World* (October 1959), and no problems were encountered in Colston Hall. The experiment was a great success and the audience, during the concert, voted overwhelmingly in favour of the stereo versions in the mono/stereo comparisons which, in Gilbert's words,[33] 'had a depth and warmth not felt with the mono version, using the same loudspeakers'.

The Final Events

During 1959 Gilbert was also occupied for some time with a potential concert in Moscow, seemingly inspired by attending a recital by the Russian pianist Emil Gilels during his first visit to the UK. He hoped to do some live versus recorded items with him. Accordingly, he sent a proposal to

Chapter Eleven: Concerts

Figure 11.32 Colston Hall, October 1959. Note the omni-directional, multi-speaker systems sitting on top of the outermost corner cabinets. Reproduced from *Stereo Handbook*.

Figure 11.33 Colston Hall, October 1959. (Photo courtesy of the Cooke family.)

the Ministry of Culture in Moscow, which had been much in evidence at the time because of a visit by the Bolshoi Ballet. Three months later he received a reply simply informing him that he should have written to the Ministry of State Broadcasting (or some such). Gilbert's response was that if it took so long to be told this, what hope was there of putting on a concert there! This had been sufficiently serious for Gilbert to have talked to Peter Walker about it; he and Raymond Cooke were greatly relieved when the project came to nothing.

On 7 July 1960 a second concert was organised in the Great Hall of Leeds University. Three photographs taken of the stage layout and preparations for the event survive in the C.H. Wood photographic archive but no other details have come to light. A lecture-demonstration was also given at Ampleforth College, the Benedictine boarding school on the edge of the North York moors, around the same time and there is some evidence that this also included 'live versus recorded' items. It came about because one of the Housemasters, Father Bernard Boyen, wanted to equip his House with a first class hi-fi system. Having been advised to buy Wharfedale speakers he visited the Idle factory and talked to Gilbert; one outcome was an invitation to put on a demonstration at the College.

Gilbert was now nearing 70 and he probably intended that this be his last concert. Raymond Cooke presided over a concert in the Abbey Lecture Hall, Dublin on 16 February 1961 which included live artists Dr Richard Hayward (traditional ballad singer and verse speaker) and pianist Julian Dawson. This was jointly sponsored by Wharfedale and the Dublin Gramophone Society, being in aid of the Concert Hall Fund. Photographs also survive of Cooke at a demonstration in Belfast, which included a pianist so this could also count in the tally of 'live versus recorded' concerts. The event at Blackpool in March 1962, which Gilbert referred to as 'the last and worst', was probably also planned by Cooke to coincide with the annual LP Fair. However, he left Wharfedale in August 1961 so Gilbert would have had to re-enter the fray. Leon Goossens and Edgar Knight again performed. Ken Russell, who had taken over as Technical Manager only a few weeks earlier, recalls that before the performance one of the keys of Goossens' oboe fractured leading to a desperate, but successful, attempt to make a temporary repair using a soldering iron!

The grand total of all mentioned events is 21, which does agree with one of Gilbert's own tallies!

The Reckoning

It is fortunate that Gilbert recorded some details of the costs associated with these events. He did so for several reasons: to dispel any notion that, despite the statement that they were promoted 'in the interests of the science and art of sound reproduction', he was in it to make a profit; to illustrate the futility of costly advertising for events which would only attract those really interested in the subject; to indicate to others who he hoped would follow his lead what to expect financially.

The most detailed breakdown given was for the Carnegie Hall concert of 1955:[21]

Net proceeds from ticket sales: $3268
Hire of Hall, organ and piano: $1200
Printing tickets and programmes: $400
Artists: $570
Stage crew: $775
Advertising and promotion: $2200
Albert Smith's expenses (July)/Arnold Hatton's (October): $1800
Loss: $3677, of which borne by Wharfedale: $1839 (roughly £600)

The second concert there the following year had greater publicity and the recording sessions had to be paid for, resulting in the cost to Wharfedale of about £1500.[34]

The largest loss on the RFH concerts (£450) was in 1956, due to the cost of hiring the LMP. For the other three the total loss was of a similar magnitude, averaging about £150 and this figure was also typical of the concerts in the provincial cities.[35] A conservative estimate of the total cost to Wharfedale of all the events might be £4500 (today the equivalent of about £90,000). To put this into context: in 1957 Wharfedale's most expensive loudspeaker driver was the Super 12/FS/AL at about £17 whilst the most expensive system was the omni-directional three-speaker corner unit, as used in the concerts, at about £73.

There can be no doubt that these events added considerably to the reputation of Wharfedale, the company and its products, and to Gilbert himself. Twenty years later journalists were still marvelling at his boldness in taking on the initial challenge of the Royal Festival Hall and the financial commitment involved, not to mention sustaining the enterprise over the following five years or more. For all those fortunate to attend one of the events it was a unique experience.

The live versus recorded items, which became the centrepiece of the concerts, may have caused some to think that Gilbert invented them. Ever keen to put things into historical perspective, he more than once attempted to list earlier efforts and in 1956, having noted the lack of any historical records, he modestly concluded:[36]

'And so it follows that the only novelty about the events described in this chapter [his own] *is the scale on which they were done.'*

In 1974, in a spontaneous contribution to the AES meeting held in appreciation of Gilbert and his work (described in Chapter 10), Dr Ray Dolby, founder of Dolby Laboratories, made the following observation.[37] In 1951 he was a high school student and working part-time for Ampex, then making professional tape recording equipment, when they decided to see if a live versus recorded test would work in a concert hall, using the best professional equipment available. This consisted of one of their own specially constructed three-track 30 inches/sec tape recorders with Altech amplifiers and loudspeakers. Recordings were made involving members of the string section and certain soloists of the San Francisco Symphony Orchestra and the live versus recorded concert was put on in the Memorial Hall. (At that time Ampex was probably very similar in size to Wharfedale in the early 1950's with about 40 employees.) The effort and expense was considerable and there were serious doubts at the outset as to the feasibility of the proposition, but the results exceeded expectations. Given this experience, Dolby was astounded that only three years later Gilbert achieved similar results using only audio equipment available to the general public. The first time that Gilbert heard about this event was very probably when he listened to the tape recording of the AES meeting, sent to him afterwards, and which I first listened to in 2008!

PIANOS, PIANISTS AND SONICS

G. A. BRIGGS

F KEIR DAWSON

CHAPTER TWELVE

The Audio Books by G.A. Briggs

All of the book covers are illustrated in colour at the end of the chapter, except that for *Pianos, Pianists and Sonics* which is shown opposite (in the original greyscale)

In all, between 1948 and 1968, Gilbert published 22 books, all but one—*Puzzle and Humour Book*—being on some aspect of audio. These 21 titles, which included a number of multiple editions, were all published by Wharfedale Wireless Works or Rank Wharfedale. Total sales reached about 260,000 copies. Given that he was 57 when he started this 'hobby' and that between 1953 and 1963 he also put on 20 or more major concert-demonstrations, this is a remarkable achievement. Gilbert described his books as 'semi-technical', and initially this was meant to convey that they were not text books—strictly for the layman and equations very few and far between. Even so, his style, which evolved quite rapidly, broke all the rules even for this genre. He included quotations culled from Bartlett's *Familiar Quotations* (usually couplets), anecdotes and personal opinions (written in the first person), couldn't resist puns and dry (and wry) witticisms, and eventually added cartoons. He gathered statistics and unusual information from a wide variety of sources and called on others for their views and specialist contributions, which were included, with personal introductions, in a highly conversational style. Whatever the specific subject of the title, the books kept up a running commentary on sound reproduction—which morphed into 'hi-fi'—and audio throughout a particularly important period of its development. But, most unusual of all, he included a great deal of new data from his own experiments, conducted either to illustrate some specific point where none was otherwise available, or to investigate something he thought needed to be better understood. This gave a further overtone of 'work in progress' and a strong sense of continuity overall.

As described in Chapter 6, his first book on loudspeakers came about because one Wharfedale dealer complained that there was an unsatisfied demand for such a book for the amateur. Once Gilbert decided to write it, his sales representative, Albert Smith, quickly gained orders just on the basis of subject and likely price, and by the time it emerged he had advance orders for some 2000 copies. The initial order for 1000 copies from BIC, gained within a few days of the first print run, showed that the situation was much the same in the USA. 'G.A. Briggs' acquired a following almost immediately and he never looked back.

This chapter is intended only as an introduction to the books by way of their chronology and aspects of their content, production and reception which relate to the overall theme of the book. Today, some of them are close to being classified as rare books, but they are all worth seeking out.

Loudspeakers: The Why & How of Good Reproduction by G.A. Briggs

First Edition May 1948, Second Edition December 1948, Third Edition March 1949 (reprinted nine times, finally September 1953), Fourth Edition January 1955 (reprinted five times up to 1958). About 90 pages. Price 5s.

In the Introduction, Gilbert noted the marked increase in interest in better quality of reproduction of radio and records and for which he gave the following probable reasons: the blackout during the war had encouraged many people to listen to, and start to enjoy, good music (mainly on the radio) for the first time; thousands of men in the forces had advanced radio experience; troops in Italy and other countries had listened to operas and orchestras and returned home with a new interest in serious music; people who liked lighter music were noting the marked difference between live and recorded performance, especially at either end of the frequency spectrum.

Therefore, his 'little book' was intended for such people who were interested in the loudspeaker and how it worked, and how results might be improved, particularly readers with no knowledge or experience of the subjects concerned and those interested in home construction (DIY). In around 90 pages Gilbert covered all the basics of design and construction of speakers, cabinets and other components, performance and behavioural characteristics, measurements for comparison and listening tests, transformers and impedance matching etc. Each subject was covered in one or two pages in an easy style. A small six-page 'supplement' at the end gave examples of interesting current designs. The book was simply stapled with cardboard covers and sold for 5 shillings (25 p, roughly £7 in 2010). Gilbert was rather taken with one American reviewer's description of it as a 'pithy pamphlet'!

The print run of 5000 copies sold out in only five months. Gilbert made a few small changes—his first trip to the USA, which took place during the pagination and printing, made him more appreciative of the use of 15" units for bass reproduction—and designated the next printing a second edition. He noted in the new Foreword that as a result of many letters received (see Chapter 6) he was going to produce a supplement to the book which expanded its coverage. The 3000 copies of this second edition had gone within two months, of which 1000 went to BIC, and the third edition was printed in March 1949. A short piece in the US magazine *Radio* and *Television News* of September 1949 included this extract:

> *'Written by an English music lover turned sound expert, this little book has been selling like hot cakes for months. You'll find it the handiest, dandiest thing ever written on the subject of sound reproduction.'*

By this time Gilbert had realised that his following wanted reproduction of records to be treated in the same way, so in the Foreword to the third edition he announced the contents of what would be his second book—the envisaged supplement plus the material on records and their reproduction—on Sound Reproduction. The third edition was reprinted nine times between 1949 and 1955, bringing total sales to 41,000, with nearly half in the USA and Canada. In the fourth edition the supplement section on current designs was updated, especially to include the new developments in electrostatic designs. By this time the book had been translated into French and Dutch, both editions published by P.H. Brans Ltd of Antwerp, Belgium. When the fourth edition finally went out of print in 1958, some 57,000 copies of the English language editions *Loudspeakers* had been sold.

Chapter Twelve: Books

Sound Reproduction by G.A. Briggs
First edition July 1949, 143 pages, price 7 s 6 d
Second Edition May 1950, 248 pages, price 10 s 6 d
Third Edition March 1953, 368 pages, price 17 s 6 d

The first edition was, as promised, in two parts. The first part expanded several sections of *Loudspeakers* and added new ones. In particular, Gilbert undertook a large number of experiments on different enclosures, in combination with variously sized drive units, to investigate behavioural characteristics. The second part was all about records, recording and playing. As described in Chapter 6, this took Gilbert into new territory since his experience of sound reproduction had been dominated by radio. He approached, and received a great deal of help and advice from, the record producers Decca and EMI, the BBC who made about 80% of their recordings for broadcasts on disc, and the M.S.S. Recording Co.. This last company was particularly significant, being originally set up by Cecil Watts and his wife Agnes.[1] When Gilbert went to visit the BBC he met Mr M.J.L. Pulling, the Superintendent Engineer, who, in the course of the conversation showed him a photomicrograph of record grooves taken by Watts (12.1). Gilbert had never heard of this remarkable man before, but rapidly commissioned him to undertake a series of experiments to produce photomicrographs, for the book, to illustrate various points relating to recordings, styli and the wear of both. It turned out that Watts, the pioneer of direct disc recording, already knew about Gilbert. As he recalled:[2]

Figure 12.1 Cecil E. Watts. with his instrument for photomicroscopy of gramophone records and styli. Reproduced from *Cecil E. Watts: Pioneer of Direct Disc Recording.*

A Pair of Wharfedales: The Story of Gilbert Briggs and his Loudspeakers

'Speaker design had not stood still during these pre-war years. The Voigt horn had become the 'ultimate' but for second best we used a 12" unit with an energised field coil, mounted on a 5 ft baffle. Rumours were heard about a man in Yorkshire who could make the same noise with a 10" unit fitted with a permanent magnet . . . Rumour for once had underestimated the truth. Side by side there was little to choose between the 'Golden Wharfedale' and the 'big boys' over the range we could get. The information 'leaked' and we used them right through the war. It wasn't until years later that I looked into the guileless face of the maker. I cannot recall a single failure with that speaker.'

A typical example of Watt's work is reproduced in (12.2). There can be no doubt that these pictures in particular, representing cutting edge research, and Gilbert's results on enclosures created

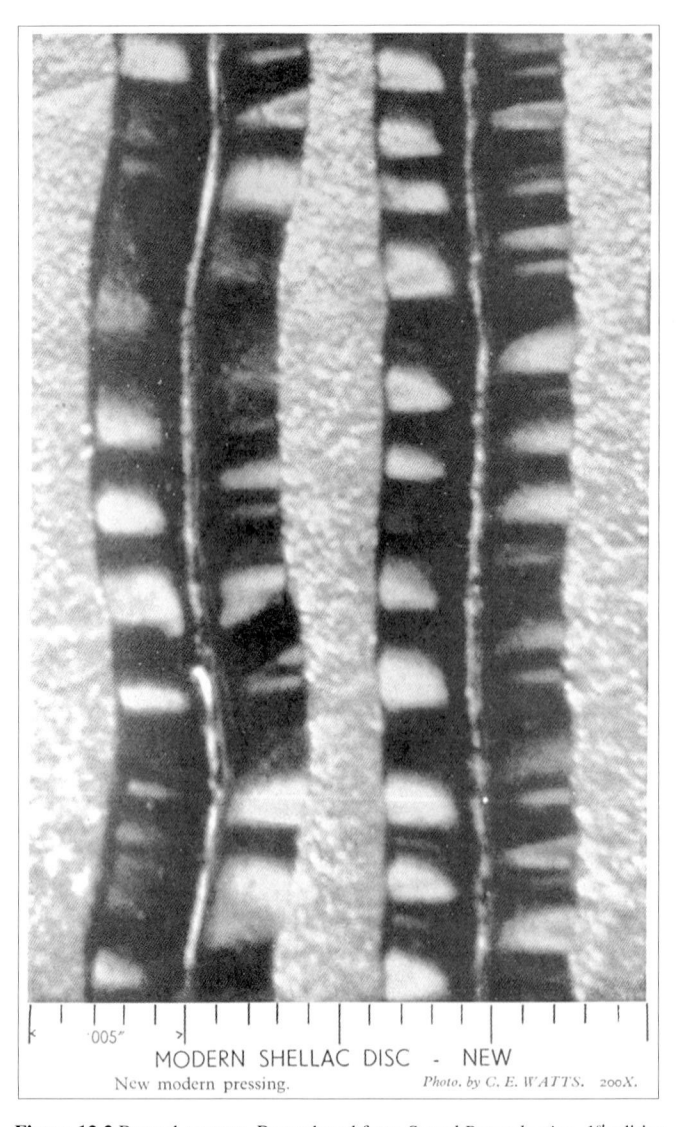

Figure 12.2 Record grooves, Reproduced from *Sound Reproduction*, 1st edition.

Chapter Twelve: Books

a great deal of interest. They also added an important new dimension to his publishing activity and his status as a communicator in the audio field. At 143 pages, the book was large enough to be conventionally bound and it was priced at 7 s 6 d. It was an immediate hit and the first printing of 10,000 copies sold out in eight months. This was something of a surprise; Gilbert was in a dilemma because, once again, he had received a sack-full of letters requesting more information and suggestions for further coverage, which he wanted to respond to in a new edition, but a second printing was already required. He decided on an expanded new edition. This was 100 pages longer with several entirely new chapters, including a 'Questions and Answers' which used actual reader's questions to address interesting points, and an expansion of all the sections which included experimental observations. He managed all this in time to launch the second edition in May 1950. In the new Foreword he remarked:

'I am well aware that these little books of mine break all the rules and probably engender a feeling of dismay in the well-ordered mind of the professional technical writer . . . yet I have ample evidence that the personal touch which has occasionally crept into these pages has been welcomed by the majority of readers. I have therefore succumbed to the suggestion of my youngest daughter that my photograph should be inserted, thus adding still more emphasis to the personal side . . . There can be no reason to suggest that books dealing with sound and music should be devoid of humour, so we can only decide to continue with the mixture as before.'

12,000 copies of the second edition were printed and these disappeared over the next two and a half years. As described in Chapter 6, Gilbert's results and conclusions on enclosures in this edition led to his meeting with Raymond Cooke and their collaboration on vented enclosure (bass reflex) design. This work, in which Gilbert made much use of oscilloscope techniques developed for his Pianos book (see below), was included in an even larger third edition of 368 pages published in March 1953. The vented enclosures chapter was in two parts with Cooke writing about his approach, followed by Gilbert's section; both had worked on the same cabinets made at Wharfedale and sent to Cooke in London.

After Cooke's section Gilbert wrote:[3]

'This brings us to the end of Mr Cooke's section and I should like to pay tribute to the diligence and energy which he has devoted to the work in his spare time. Nobody without experience of similar research can have the least idea of the scores of hours of concentration which are required to produce reliable data on such a scale. Although Mr Cooke's first approach to me implied a criticism of some of my findings, I welcomed his collaboration, as my sole object is to arrive at the basic truth of the problems.'

and in the Conclusion to the edition:[4]

'. . . working with Mr Cooke on Vented Enclosures has been rather like an old amateur boxing a young professional, but he mercifully pulled his punches and I have never been down for a count of more than 3 or 4. The net result is that the interpretation of the behaviour of reflex cabinets has been brought into line with scientific facts, although the accuracy of the tests described in the previous edition has not been disputed.'

Gilbert had called upon several others to contribute specialist sections, such as Stanley Kelly, Ralph West and Norman Crowhurst. His earlier editions had been checked for technical accuracy, and included many comments from, his friend Frank Beaumont at Ambassador Radio. This time Beaumont was 'away in South Africa' so the task was undertaken by Mr C.R. Cosens. As always, all the drawings were done by Gilbert's old friend F. Keir Dawson. This edition, with three reprints, stayed in print until 1962, by which time total sales had reached 47,500. A French translation was published by Societé des Editions Radio, Paris.

Pianos, Pianists and Sonics by G.A. Briggs
June 1951, 190 pages, price 10 s 6 d

Ever since his extraordinary series of trials of pianos, carried out mainly during the 1930s (altogether 40 passed through his hands), Gilbert had wanted to write a book about pianos. As noted in Chapter 6, he made enquiries of publishers after the war and the difficulties he encountered were directly responsible for the manner in which *Loudspeakers* was published in 1948. After the second edition of *Sound Reproduction* appeared, with two successful titles under his belt, he felt confident enough to tackle the subject of pianos. Reactions to his idea were mixed, but mainly unfavourable. Answering the response 'What do I know about pianos?' in his Introduction, he noted honestly that he now knew how much he didn't know and that his thirst for more information on the subject was a principle reason for writing the book. However, he did receive full support from those in the piano industry because he said a key feature was an investigation into the nature of piano tone—which was greatly superior to any electronic imitations he had heard. As usual the content was idiosyncratic with chapters on the history, types and construction of the instrument, tuning, care for and aspects of playing and listening to pianos, interviews with seven famous pianists and the results of a great deal of novel investigations, carried out with the help of Ernest Price, into touch and tone, vibration and harmonic analysis.

This scientific work was done on his 7 ft 6 in Steinway grand at home, in Ilkely, and occupied many hours. The equipment is shown in (12.3). The use of an oscilloscope, fitted with a camera for recording waveforms of sounds picked up on a microphone, or to record time-dependant behaviour on moving film, with a motorised camera (and no time base), was the novel feature. A large number of the resulting photographs, or oscillograms, were included in the text. It was painstaking, and expensive, work. Gilbert calculated that between 300 and 400 had been recorded, to achieve the desired illustrations, and that he and Price did well to produce about 15 in a two hour session. Working for much longer than that often resulted in him developing a migraine. One example is reproduced in (12.4).

Gilbert was unsure about how many to print so he did the following calculation.[5] He thought 10% of the two million pianos in the country were being played and that 10% of the players (or learners) would be interested in his book, nothing on the subject having been published for several years. A print run of 20,000 would give him a break-even selling price of 10 s 6 d, which he settled on. Profit was clearly not uppermost in his mind! Ernest Price thought the conventional wisdom of 2000 copies for a musical book was much safer; Albert Smith, who had to try and sell them, thought 5000 the limit. Gilbert went for 20,000! He wasn't helped by a review in *The Pianomaker* (by Miles Henslow, who went on to found *Hi-Fi News* and the *Hi-Fi Yearbook*) which included this extract:[6]

Chapter Twelve: Books

> FIG. 2/5. *Equipment used for inspection and measurement of piano sounds and vibrations. (Ilkley, 1950).*
>
> A — Mechanical Striker.
> B — Microphone.
> C — Audio Frequency Analyser.
> D — Sound-level Recorder.
> E — Vibration Adapter.
> F — Vibration Pickup on spring-loaded support for under-soundboard readings.
> G — Camera with fittings for mounting on oscilloscope.
> H — Double-beam Oscilloscope.
>   Piano—Steinway Model C.

Figure 12.3 Equipment used for recording oscillograms. Reproduced from *Pianos, Pianists and Sonics*.

'Why he wrote this book I don't know. I hope he does. And, having read through it, I can only hope that if he writes another it will be on a subject with which he is obviously more at home.'

It was the most devastating review Gilbert had ever read. Compared with the previous books, sales were slow, although quite healthy by the standards of books with a musical theme. However, Gilbert received many letters of praise and encouragement from readers*, including one from J. Raymond Tobin, the editor of *The Music Teacher*. To encourage sales, Gilbert took the unusual step of setting up a competition, with total prize money of £100, open to purchasers of the book. This was advertised in *Wireless World*, and presumably also in *The Music Teacher*, as shown in (12.5).

*An indication of the extent of Gilbert's fan mail is that when *Amplifiers* was published in March 1952, he included, in the first flyer, extracts from some of the stated 2000 plus letters already received about the previous books.

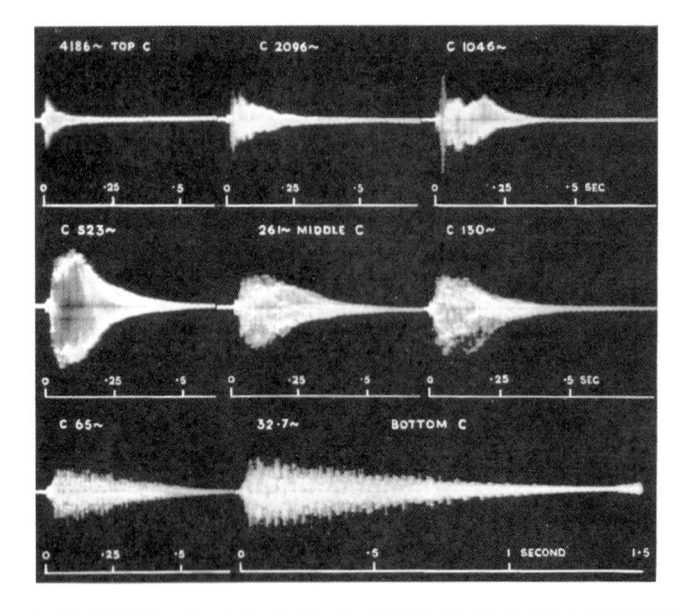

Figure 12.4 Oscillograms of piano C notes. Reproduced from *Pianos, Pianists and Sonics*.

Figure 12.5 *Wireless World* advert for *Pianos, Pianists and Sonics* competition, December 1951.

In March 1953, when the third edition of *Sound Reproduction* appeared, 4000 copies had been sold. The remaining 16,000 had, fortunately, not been bound and to save on storage, 5000 were pulped. There must have been another similar cull later, but the book kept up steady sales. It was given a new dust cover in the mid-1960s and when it was finally declared out of print around 1970, 9644 had been sold. There was an ironic finale to the story. In 1977, Frank W. Holland, the founder of the British Piano Museum Charitable Trust, who really liked the book, wanted to be able to sell copies in the museum. With Gilbert's approval, he approached Unwin's, the publishers, to see if they would reprint around 500 copies. They agreed to consider the proposal, but whether a reprint materialised seems doubtful.

Amplifiers: The Why and How of Good Amplification by G.A. Briggs and H.H. Garner

March 1952, 216 pages, price 15 s 6 d

Wharfedale started producing speakers for school use in 1949 and discussions with the Essex Education Committee (EEC) led to a new product in late 1950 named the Essex Grey. Since no Wharfedale product had ever been named in such a way before it seems likely that the EEC intended to place a significant order for a product tailored to their specification. The technical officer to the EEC was Major H.H. Garner, whose previous experience had been in radio with the army, so Gilbert got to know him. When the second edition of *Sound Reproduction* was published in May 1950 Gilbert sent Garner a complementary copy, and he (Garner) noted, with regret, that the field of amplification had not been covered and that a book on this large subject might usefully fill the gap. Gilbert agreed and suggested that Garner, who was close to retirement, was well qualified to write it. In the end it became a joint enterprise with Garner writing much of the book and Gilbert trying to ensure that the coverage was satisfactory as well as being an obvious bridge, and in a similar style, to his previous books. More use was made of the oscilloscope techniques introduced in the Pianos book, with Ernest Price helping as usual. Frank Beaumont resumed his role as technical sub-editor, making significant contributions, whilst Keir Dawson drew a huge number of figures.

10,000 copies were printed and 7,000 had been sold by the end of the year. A brief review in *The Gramophone* in July was fairly bland, remarking on the 'commendable freedom from errors' but a more extensive one in *Wireless World*, in September, by Fred Devereux, was rather damning, the main criticism being summarised as 'the "Why" is less to be relied upon than the "How"'. It seems there may have been others of similar ilk because when the need for a reprint quickly arose, the typeset was dismantled, due to the fact that the book had been 'rather roughly handled by the critics'. The orders continued to pour in over the next two years and Gilbert reckoned another 10,000 could easily have been sold.[7]

High Fidelity: The Why and How for Amateurs by G.A. Briggs

May 1956, 192 pages, price 12 s 6 d

The title page, but not the book cover, acknowledged that Gilbert had been 'assisted by R.E. Cooke B.Sc. (Eng.) as Technical Editor'. This was the first book to be published after the third edition of *Sound Reproduction*, to which Raymond Cooke had made an important contribution. During 1954 and 1955 Gilbert had been particularly taken up with the early 'live versus recorded' concert-demonstrations, which had given him new insights into sound reproduction quality, and during this period the term 'high fidelity' had fully entered the vernacular. Exactly what this meant and how it was achieved in a domestic setting by 'amateurs' seemed, to Gilbert, to be a worthy subject for demystification. It may well have been a discussion with Cooke about a follow-up collaboration on this project that led on to him being offered the job of Technical Manager at Wharfedale. In any event he was roped in as Technical Editor more-or-less on arrival and the book was completed about seven months later.

The book delved into all aspects of the subject: what 'quality' and 'fidelity' actually meant in terms of sound reproduction, how all the components of the chain behaved in this respect and what justified a high price. With *Amplifiers* out of print, the most essential sections of that book were repeated in a chapter in this one. Tape recorders were included and there was more on records and their care, courtesy of Cecil Watts who by this time had gone on to invent the 'dust bug' for cleaning the record grooves during playing. Gilbert included a chapter on his concert hall experiences and contrasted this with room acoustics in the home, finishing with another 'Questions and Answers' chapter.

W.A. Chislett, reviewing the book in *The Gramophone* (June 1956), had a few grumbles about the relative length of some chapters, especially the one on concert halls, but concluded:

> 'It is an extremely good and useful book, sound in judgement, based on wide and catholic experience, comprehensive in scope and outlook and generous, as well as just, to the author's business competitors. And I am grateful that Mr Briggs did not extend his economy campaign to the extent of expunging the terse, seemingly irrelevant but really illuminating, and often amusing, asides with which he, as usual, liberally besprinkles his main text.'

The print run of 10,000 copies had sold out within a year, but again Gilbert refrained from reprinting on the grounds that the contents were already going out of date!

Loudspeakers by G.A. Briggs
October 1958, 336 pages, price 19 s 6 d

Although listed as the fifth edition of the first book on an inside page, this was essentially a completely new work (the 'why and how . . . ' subtitle was dropped). Once again Raymond Cooke was the Technical Editor. During 1957 Gilbert, with Cooke's collaboration, had published a series of invited articles under the general title of 'All about Audio and Hi-Fi' in the American magazine *Radio and TV News*; eight of the 31 chapters in the book were based on these articles. The first two chapters were very unusual in that Gilbert decided to reminisce about the early days of Wharfedale Wireless Works ('Looking Back') and the book publishing saga to date. He had thought for some time that if others who had helped shape their industry did the same, maybe in a series of magazine articles, it would be both interesting and historically useful, so this might set the ball rolling. He was also rather fed up with his books being reviewed against text book criteria, despite his stated aims being different, which led to unwarranted negativity—and completely at odds with the reaction of the paying customers. These chapters were 'calculated to exasperate any technical reviewer' he noted in the Introduction! The rest of the book was a comprehensive treatment of the subject, with little of the original 'little book' surviving. Gilbert's old friend Fred Tetley, now a director of Swift–Levick, contributed a chapter on loudspeaker magnets and their development history, whilst James Moir did the same for cinema speakers.

As usual the reviews were mixed, but if Gilbert needed any more proof that the reviewers were fickle he got it. Their reaction to his reminiscences was very positive!

The first printing of 10,000 sold out in two years and the book was reprinted eight times between 1961 and 1972. By the end of 1964 sales had passed 13,000 and by 1971 the total was over 19,000. A French translation was published by Societé des Editions Radio, Paris.

The Process

By now Gilbert and Raymond Cooke had evolved an efficient way of working together, which Cooke described in 1974.[8] Gilbert did not sleep very well and he would wake up early, make himself a cup of tea, and start writing. His grandchildren were amused by this—he did it sitting up in bed, Churchillian-style. He would arrive at work about 10.30 and hand over his morning's output to Cooke, who would take this home and, being a 'night owl', would make his comments and suggested alterations/additions on Gilbert's handwritten foolscap pages. (His methods had not changed since 1914, see Chapter 3). These would be exchanged for Gilbert's new material next morning. The combined efforts would be refined by Gilbert, who was obsessed with the need to be succinct, as the book progressed and Edith Isles, his secretary, would often transcribe the final copy. This was never typed; either's handwriting was good enough for the printer. When the galley proofs were ready there would be another round of alterations, and these might be typed. Armed with the final set of proofs, Gilbert would decamp to a hotel by the sea, usually at Morecambe, for several days of solitude during which he would organise the page layout (pasting up the text and inserting the figures). This required total concentration, but he would break off and take the sea air, walking or sitting on the beach in a deckchair, if the weather was good enough.

Stereo Handbook by G.A. Briggs
December 1959, 146 pages, price 10 s 6 d

Disc stereo had arrived in the UK in 1958 as described in Chapter 7. The idea behind this book was to provide some clarity in a confused world, where relatively cheap stereo equipment could provide quite startling results in demonstrations which had everything to do with transients and directional effects but nothing to do with high fidelity music reproduction—many had been lured by this siren in North America to abandon decent mono systems for much lower fidelity stereo. On the other hand, gaining from stereo, whilst maintaining high fidelity, was neither cheap nor easy. Gilbert could see his good work in *High Fidelity* being undone unless stereo reproduction was given the same treatment. Once again with Cooke as his Technical Editor, he had started writing around November 1958 but made little progress until, about six months later, the way to tackle the subject suddenly became clear. In a leaflet Gilbert wrote:

> 'As stereo is still a controversial subject, the views of several experts in the field of audio have been included with contributions by N.H. Crowhurst on the American angle, M.G. Foster on BBC and stereo, S. Kelly on pickups, C.E. Watts on record wear, and R.L. West on amplifiers. At the same time, the Author and Technical Editor have not been backward in expressing their own opinions.'

Figure 12.6 Cabinets in Raymond Cooke's living room. Reproduced from *Stereo Handbook*.

There were also contributions from several Wharfedale agents in other countries and a question and answer session with other luminaries, whose comments Gilbert arranged as if emanating from a round table discussion, on the current state-of-play in the UK. Listening tests on stereo pairs in various configurations were described, both in the Wharfedale lab and in Cooke's lounge at home—which gave rise to the picture reproduced in (12.6). Gilbert clearly understood his public; the first printing of 10,000 copies sold out in six months. There was one reprint and sales totalled 14,500, by which time Gilbert thought it had achieved its purpose. U.M. De Muiderkring of Bussum, The Netherlands, published a Dutch Edition.

A to Z in Audio by G.A. Briggs
November 1960, 224 pages, price 15 s 6 d.

In February 1960, not long after *Stereo Handbook* was published, Gilbert was out walking with one of his old friends, Mr E.E. Ladhams, when the latter suddenly announced that Gilbert should write a new book, a sort of glossary which would answer most questions which puzzle the amateur—but it would have to be in the normal style: interesting, readable and spiced with humour. Gilbert mentioned the idea to Raymond Cooke and he came back 24 hours later with a list of about 500 possible entries. They whittled this down to about 400 and although drawing on earlier Wharfedale material, much of it was entirely original. Gilbert liked to use the illustration count as a marker for originality and in this case 110 out of 160 were specially prepared, 34 were from earlier publications and 16 came from elsewhere. Although Cooke again

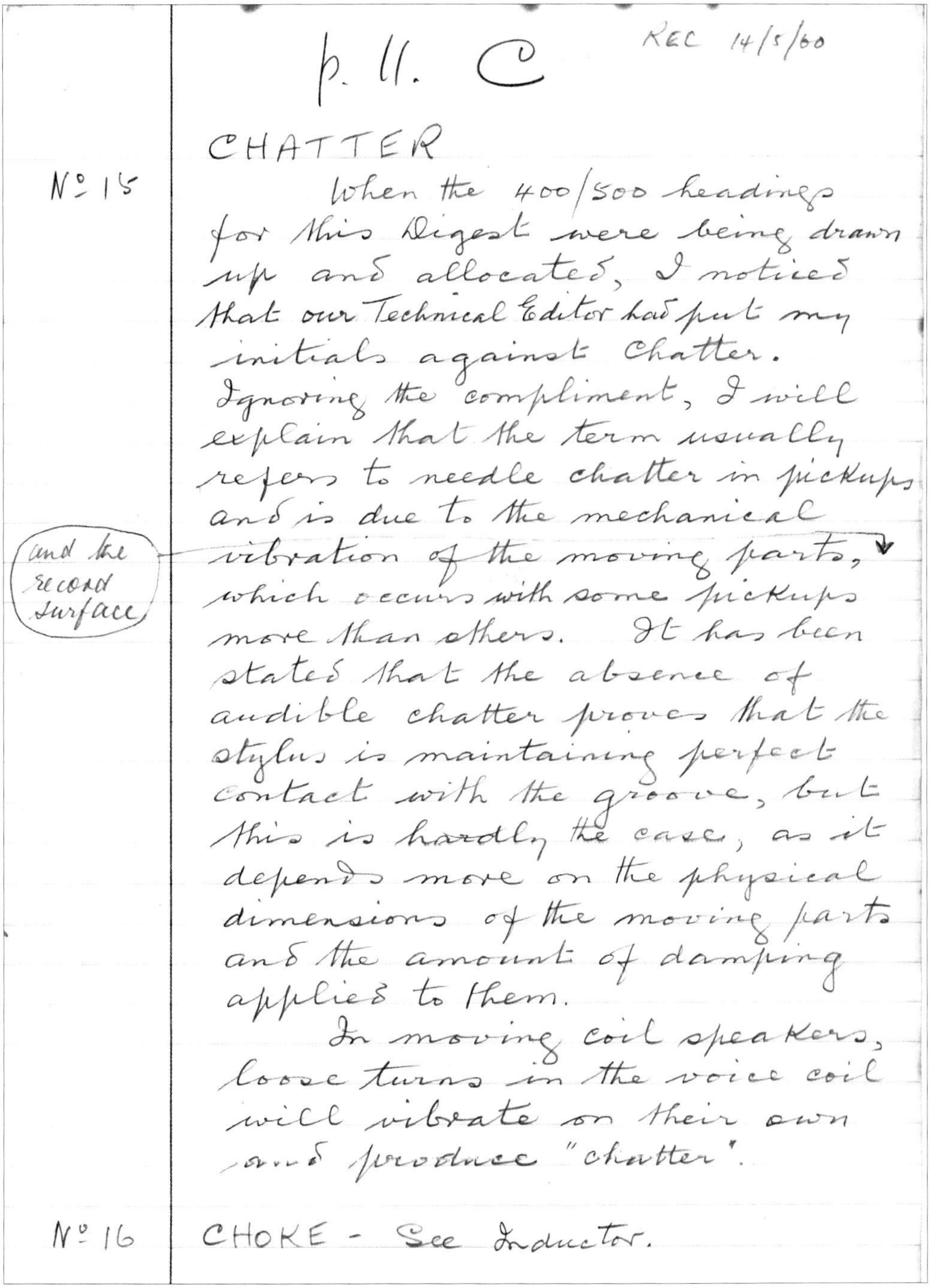

Chapter Twelve: Books

p. 11. C   REC 14/5/60

CHATTER

No. 15

When the 400/500 headings for this Digest were being drawn up and allocated, I noticed that our Technical Editor had put my initials against Chatter. Ignoring the compliment, I will explain that the term usually refers to needle chatter in pickups, and is due to the mechanical vibration of the moving parts, *(and the record surface)* which occurs with some pickups more than others. It has been stated that the absence of audible chatter proves that the stylus is maintaining perfect contact with the groove, but this is hardly the case, as it depends more on the physical dimensions of the moving parts and the amount of damping applied to them.

In moving coil speakers, loose turns in the voice coil will vibrate on their own and produce "chatter".

No. 16 CHOKE — See Inductor.

Figure 12.7 Specimen page from the manuscript of *A to Z in Audio*. (Scan courtesy of the Briggs family.)

'assisted as technical editor'—and, for the first time, got his name on the cover as such—he was much closer to being a true co-author of this book. Gilbert gave him the handwritten manuscript as a souvenir and Cooke displayed it at the 1974 AES meeting referred to before.[8] It still survives and a specimen page is reproduced in (12.7). The handwriting, in pencil, is Gilbert's and the addition is Cooke's. Who came up with the final title is unknown, but the book started out as *Audio Digest*. 10,000 copies were printed and they were all eventually sold, 7000 having been so by the end of 1964. An American edition was published by Gernsback Library Inc. in 1961.

Audio Biographies by G.A. Briggs
November 1961, 344 pages, price 25 s (£1.25)

The seed for this book was planted as early as 1953 when Gilbert was attending the New York Audio Fair in the company of Harold Leak. During a reminiscing session, in which Gilbert learned about Leak's early background in audio for the first time, it occurred to him that the personal stories of other contributors to the development of the field would make interesting material for a book, perhaps as a retirement project. He pursued the idea a bit further in his 1958 'Looking Back' chapter in *Loudspeakers 5* and the favourable reception, even amongst the technical reviewers, encouraged him. He got down to work, following the publication of *A to Z in Audio*, by inviting over 100 people he considered eligible (a fairly elastic definition), all of whom were known to him personally, to contribute. He provided a template of questions relating to personal details, education, positions held etc., requested a passport type photograph, and asked for 1000–2000 words on experiences in the audio field.

He received 10 direct refusals, including four from BBC engineers (on the grounds that it would count as publicity—a tragedy of officialdom) and 15 never responded. Another 15 failed to deliver the goods in time, including Gilbert's hero, Paul Voigt; he was unwell at the time and in his letter to Edna, following Gilbert's death (described in Chapter 10), he very much regretted that he had been unable to contribute. Three were omitted 'by mutual agreement' and the remaining 64 made up the list reproduced in (12.8). They comprised of many engineers, including those involved in the recording industries, professional technical writers, entrepreneurs and sundry others who had an interesting tale to tell, including his wife, Edna, and younger daughter Valerie! Gilbert himself contributed the first entry and called his reminiscence 'Looking back Again', which included additional information on the early days of Wharfedale and the 'live versus recorded' concert-demonstrations.

How many copies were printed is not known—Gilbert was unusually reticent in providing the information for this book. It was well received within the audio industry which it reflected, but did not sell well. Harold Leak, when finally cornered by Gilbert for his contribution, told him frankly that, in his opinion, there would be no demand for such a book. Not only did he not contribute, to Gilbert's disappointment, but he was proved correct. By the end of 1967 only 1653 copies had been sold and Gilbert noted in the books leaflet for 1968:

> 'Although this book cannot be counted as a best-seller, I still think it is one of our most interesting publications which increases in historic value as the years roll on.'

Chapter Twelve: Books

In this assessment he was undoubtedly correct. The book was still selling at the rate of two per month when Gilbert died in 1978. When asked, in a 1975 radio interview,[9] which was the personal favourite of his books, he replied without hesitation *Audio Biographies*, because it is the most interesting to read'.

CONTRIBUTORS

Name	Page	Name	Page
E. Aisberg	23	T. M. Henslow	183
D. W. Aldous	27	W. P. Hickson	187
R. Arbib	31	H. E. Holman	193
J. N. Borwick	41	G. D. L. Horsburgh	201
R. W. Bradford	44	W. A. Jamieson	209
D. E. Briggs (Mrs.)	49	S. Kelly	214
L. Carduner	52	P. W. Klipsch	218
J. M. Carson	57	M. D. Kramer	223
R. V. de Carvalho	62	R. W. Lowden	228
W. A. Chislett	67	C. G. McProud	234
H. A. M. Clark	72	R. W. Merrick	237
P. D. Collings-Wells	77	E. D. Nunn	240
J. D. Collinson	79	V. E. Pitchford (Mrs.)	246
J. M. Conly	84	A. M. Poniatoff	255
N. H. Crowhurst	87	W. L. Procter	262
F. K. Dawson	92	C. Rex-Hassan	266
F. L. Devereux	96	J. D. Rogers	270
J. F. Doust	100	A. J. Schmitt	273
H. A. Dunn	108	H. H. Scott	277
A. C. Farnell	113	J. Sieger	280
C. Fowler	117	F. W. Tetley	286
E. Fowler	120	F. Thistlethwaite	290
Free Grid	132	G. W. Tillett	295
J. C. G. Gilbert	139	E. M. Villchur	300
T. R. C. Goff	144	P. J. Walker	303
L. A. Guest	150	J. Walton (Miss)	309
B. M. Guest (Mrs.)	154	C. E. Watts	313
A. C. Haddy	157	R. L. West	323
D. Hall	164	P. Wilson	326
P. W. Harris	168	P. O. Wymer	335
H. A. Hartley	172	J. Young	338
A. Hatton	178		
W. Hatton (Mrs.)	181		

Figure 12.8 Contributors to *Audio Biographies*, reproduced from the book itself.

Wharfedale

CABINET CONSTRUCTION SHEET (September 1957)

LEAFLET A

ISSUE 4

INTRODUCTION
by G. A. BRIGGS

The following diagrams show the essential features of enclosures which give optimum results (in relation to size) with Wharfedale foam surround speakers. Other units can be used in these cabinets provided the cone resonance is not higher than the equivalent Wharfedale speaker; but no advice on such alternatives can be given. It should also be remembered that the performance of a Wharfedale foam surround speaker can be ruined by mounting in an unsuitable enclosure. The BJ and Rogers corner horns have been tested and found satisfactory; it is regretted that opinions on other designs cannot be given by letter.

Size
The larger enclosures always give more output at low frequencies, but the smaller designs described give a satisfactory response consistent with their size. Outside dimensions are given, together with constructional specifications and recommended materials. Plywood thinner than that specified should not be used. The share of any model may be changed without affecting the low frequency performance, provided the total volume does not vary by more than 10 per cent.

It is usually satisfactory to mount a speaker in an enclosure designed for a larger unit. It is, however, quite wrong to reverse the procedure and fit, say, a 10" unit in a cabinet designed for an 8" model.

Openings
Loudspeaker and port apertures should be left open or covered with an open mesh. For this purpose Wharfedale anodised aluminium mesh is available in gold or bronze finish.

The placing of such openings is not critical. In fact, the speaker and vent openings can be reversed in position, or the cabinet may be laid on its side, without upsetting the performance. Sound waves fortunately do not know when they are upside down or sideways on.

Lining
Where absorbent lining is specified this should consist of cellulose wadding, fibreglass or soft felt, not less than 1" thick. Papier-mâché egg-trays, acoustic tiles or foam plastics are NOT suitable absorbents in this application.

An absorbent lining is not recommended in the 9 cu. ft. corner enclosure, because results are better without it. Listening tests have been carried out in ordinary rooms and concert halls: the unlined enclosure was preferred on speech and music by all the judges. Even so, there is no real objection to lining such a model if an individual user prefers it that way.

Dust Exclusion
It is important that no foreign matter or dust from fibre glass should be allowed to enter the magnet gap; the cotton bags fitted to the open voice coil speakers should therefore be permanently retained.

Walls
The corner walls of a room, if brick built, form an ideal backing for a corner reflex enclosure, and are superior to any cabinet. The front and top panels should form an airtight fit to the walls; gaps are easily filled up by glueing layers of cloth or felt to the edge of panels.

Sand-filling
Next to concrete or bricks, a sand-filled panel gives the least resonance. Two sheets of plywood are spaced ½" or 1" apart—the larger the area the wider the spacing—and the cavity is filled with dry sand, which adds weight and absorbs vibration. Ordinary builder's sand is satisfactory.

Tweeters
The 8" and 3" units should not be placed in the bass enclosure because they are fitted with open chassis and the cones would be affected by the L.F. sound waves, thus offsetting one of the benefits of the crossover network.

Open mounting for middle and treble units is recommended to avoid enclosure resonances. Directional effects are avoided by facing the speakers upwards or towards a wall or corner at a suitable angle.

Users are free to put all the units into one cabinet and face them towards an armchair or reading lamp if they so desire; but it is rather optimistic to write to us and ask us to approve of arrangements which put appearance before performance, as this is a purely personal question.

N.B. The object of the above remarks and the following diagrams is to help the home constructor to obtain the best results he can, and at the same time to curtail unnecessary correspondence. In case of difficulty we are nevertheless always pleased to answer questions briefly and give further guidance by letter.

A brief reference to each design and its main purpose now follows:—

Fig. 1. This gives optimum results in a very small size when fitted with the Super 8/FS/AL. The new 8" Bronze/FS/AL is a reasonable substitute at a considerable saving. The acoustic filter is placed lower down than in the diagram in the previous leaflet because the foam surround unit requires a larger volume of air immediately behind the cone than the cloth surround type. The same

1

PRINTED IN ENGLAND

Figure 12.9 Pages from Cabinet Construction Sheet no. 4 issued in 1957. Scans courtesy of Alec Broers.

Fig. 5. 9 cu. ft. SAND-FILLED CORNER PANEL

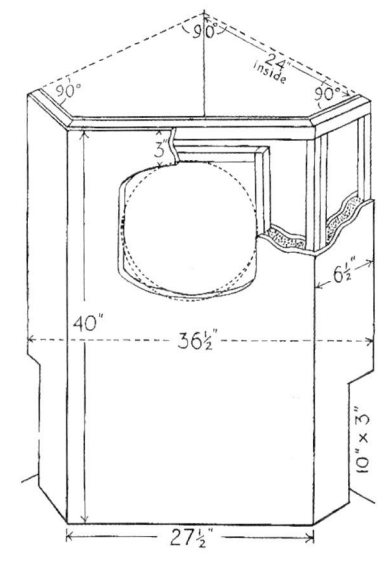

Recommended Units

W15/FS (bass only), Super 12/FS/AL (full range unit).

This enclosure gives excellent results with the W12/FS in a budget three-speaker system. 10" and 8" units also perform better than in any smaller cabinet.

Materials

Solid wood frame 1" thick, faced on both sides with sheets of $\frac{1}{2}$" plywood. Space between plywood films filled with tightly-packed dry sand. Top in 1" plywood or blockboard. For maximum bass response an air-tight fit to walls should be ensured by fitting strips of felt or Bostik white sealing strip.

Sub-baffle about 16" x 16" in $\frac{3}{8}$" plywood should be fitted up to the rear side of the front plywood skin, inside the frame shown in drawings.

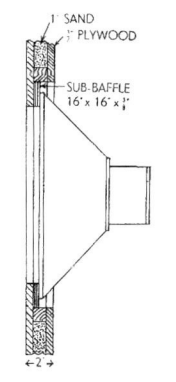

Distance along wall from corner to front of lid is 26¼". Weight of front panel 124 lbs.

Diagram to show method of fitting sub-baffle to sand-filled panel.

TWO-SPEAKER SYSTEM

Where two speakers are used with crossover network, the treble unit should be mounted on an open baffle placed above the reflex cabinet. Plywood $\frac{3}{8}$" thick is satisfactory and a crossover at 1,000 cycles is correct with the following sizes :

    8" Treble unit, baffle 14" × 12" approx.
   10"   ,,   ,,   ,,   16" × 14" ,,

Under these conditions, a quarter section series network gives excellent results.

If a 500 cycle crossover is used, the baffle should be 3 or 4 inches bigger.

With a 3" treble unit, the baffle size may be reduced to about 6" × 4" and the crossover frequency should not be lower than 4,000 c/s.

BEAM EFFECT.

To reduce directional effect and spread the H.F. beam, the treble unit in all the above cases may be mounted horizontally with the cone facing upwards, or at an angle of 45° facing into the corner of the room for good reflection effects.

5

Figure 12.9 (continued) Pages from Cabinet Construction Sheet no. 4 issued in 1957. Scans courtesy of Alec Broers.

Cabinet Handbook by G.A. Briggs
May 1962, 112 pages, price 7 s 6 d

Gilbert dedicated this book to 'Audiophiles and DIY stalwarts'. Wharfedale, along with other manufacturers in the UK and the USA, provided free Cabinet Construction Sheets. They had been doing this since about 1954 and by 1962 some 20,000 per year were being distributed through agencies, Wharfedale speaker stockists and direct from the factory. Two example pages from issue 4 of 1957 are shown in (12.9)

The aim of *Cabinet Handbook* was to supply all the information his dedicatees might wish for to supplement their construction sheets and Gilbert pulled together his usual array of facts, figures and insights from a wide variety of sources. He discussed materials in general and plywood in particular, machines used in materials preparation and mass production, cabinet assembly techniques, veneering and polishing, meshes, resonances, absorbents (contributed by James Moir), cabinet design principles, electric guitar issues and acoustics in rooms. Once again, when addressing his core audience, Gilbert hit the bulls-eye. In 18 months sales had passed 15,000 and by the end of 1967 over 23,000. By mid-1972 it was still selling well, with total sales having exceeded 33,000. Altogether, there were eight reprints between 1963 and 1976.

More about Loudspeakers by G.A. Briggs
March 1963, 136 pages, price 8 s 6 d

One of the reasons why Gilbert was able to include, or collate, information from such a variety of sources in his books was that, once he had embarked on a strategy of continuous publication, he kept scrap books in which he collected cuttings and related notes, from papers, magazines, journals and letters, as well as references to things he found in books. He was able to turn to these at a moment's notice when writing which gave his output such an up-to-date and wide-ranging feel. *More about Loudspeakers* seems to be more a collection of such items than anything else. Four years after *Loudspeakers 5* appeared he was probably wondering whether to bring out yet another revised edition, but instead he opted for an additional 'modest volume' on the subject. As he said in the Introduction, the title was wide and non-committal, which gave him the freedom to include anything he thought would interest his readers. Fred Tetley provided an update on magnet developments and James Moir vetted a couple of chapters, but otherwise Gilbert did this by himself.

Raymond Cooke, of course, had left before *Cabinet Handbook* but there is an interesting footnote about him which relates to this book. Whatever spat there may have been when Cooke resigned they were soon back in communication. When Gilbert was planning the contents in June 1962 he wrote to him as follows:

'In order to keep out of mischief I am starting on another little book, which will probably be entitled 'More about Loudspeakers', and I am thinking of including a section dealing with interesting speaker developments during the last few years. This would include your model [K1] with particular reference to your rectangular diaphragm and I shall be pleased to devote a page to the subject if you will let me have suitable photographs or drawings together with relevant details. In short, we propose to give you a bit of free publicity.'

Cooke replied that Gilbert's letter gave him a lot of pleasure, especially because it brought news of a new book. When the book appeared it was rather like the early editions of *Loudspeakers* and the section at the back on recent models from a range of makers, including the K1 from KEF, had the layout first used in 1948. In fact Gilbert included a copy of such a page with his letter to remind Cooke of the format.

The first printing was of 10,000 copies, which sold out in just over two years, with a reprint of 4500 which lasted until about 1969.

Audio and Acoustics by G.A. Briggs with J. Moir (as sub-editor)
November 1963, 168 pages, price 12 s 6 d

During 1962, as stocks of *Sound Reproduction 3* dwindled, Gilbert was again faced with the problem of reprint or revise. Since this third edition was nearly 10 years old he wanted to revise but, as can be judged from the increase in size through the three editions, any new edition would be a major project requiring two years to complete. By then, he quipped, he would 'probably be languishing in an old-folks home.' He decided instead to divide up this large subject into more manageable portions and was looking for someone to be his sub-editor. When James Moir, author of the book *High Quality Sound Reproduction*[10] and Acoustical Consultant, agreed to help him Gilbert capitalised on his expertise by making acoustics the major subject of the first 'portion'. He also decided not to include 'hi-fi' in the title because, by now, there were many books on this subject, but to use the general term 'audio' to cover the more general aspects of listening to sound.

After a chapter on the history of the subject, which included the marvellous illustration shown in (12.10), and the importance of the human ear, the book covered resonance, echo and reverberation, room acoustics, anechoic chambers, transients, stereo, schools, concert halls and different types of studio. The first printing of 5000 copies had been sold in a little over a year and a reprint of 4000 copies, in February 1965, lasted until about 1970.

Figure 12.10 'Live versus recorded', 1908. A VICTOR advertisement sent to Gilbert by Edgar Villchur. Reproduced from *Audio and Acoustics*.

Aerial Handbook by G.A. Briggs with R.S. Roberts
First edition October 1964, 144 pages, price 12 s 6 d
Second Edition January 1968, 176 pages, price 22 s 6 d

Whatever Gilbert's intention was for the contents of the next part of the 'Sound Reproduction in several books' series, it was derailed by his inclusion of aerials. In trying to put together a chapter on this subject, which he had not attempted previously, he soon realised it was too vast and really needed a separate book. Once again he was fortunate to find an expert to help him, probably through his friend Ralph West, in R.S. Roberts who also lectured (in telecommunications) at the Northern Polytechnic in London besides being a technical consultant to Antiference Ltd.

After going over the general principles of transmission and reception, the book considered all the different bands for radio and TV signals and their outdoor reception, indoor aerials, diplexers and multiplexers, boosters and attenuators, transmitters (written by Michael G Foster of the BBC Engineering Information Department), relay and communal systems and the situation abroad (information obtained from a questionnaire sent out to several countries). Whilst Roberts did much of the basic writing, Gilbert produced his usual array of illuminating facts, statistics, and illustrations, both current and historical, to enliven the contents. 5000 copies were printed and nearly 3000 were sold in the first two months. When the time for a reprint arrived in early 1966, new developments in colour television and multiplex stereo were underway, so Gilbert decided on a substantial revision when these came to fruition. Together with new material relating to the effects of UHF transmission on aerial requirements, this led to the second edition of January 1968 being 32 pages longer and with 50 new illustrations. Again the print run was 5000 and 4000 had been sold by the end of 1971.

Musical Instruments and Audio by G.A. Briggs
October 1965, 240 pages, price 32 s 6 d

Whether this book was ever considered to be a part of the 'Sound Reproduction in several books' series is unclear. Certainly, Gilbert did not mention the project again, after *Aerial Handbook* blew him off course in 1963/4. The behaviour of the orchestral instruments, in terms of their frequency range and harmonic composition, had been discussed several times in chapters of the earlier books, especially in relation to the ability of loudspeakers to reproduce them. In this book Gilbert went into the subject in great detail and, as usual, called upon numerous collaborators to provide specialist input and advice. Oscillograms, collected by his technical editor (and Technical Manager) Ken Russell and his assistant Bill Jamieson, featured once again. The book was in gestation for rather longer than normal. Gilbert started work on it in 1964 when he was Chairman at Wharfedale and practical help from the Technical Department was still available, but once he retired for good in February 1965 this was increasingly difficult to come by. One example of the analysis of musical instrument characteristics is shown in (12.11).

The book went into the basis of music, the nature and analysis of sounds, the description and characterisation of 63 instruments and the human voice, formants in speech and musical instruments, distortion in sounds and their reproduction, organs and electronic organs, pianos, tuning and music in schools. Donald Aldous, reviewing it in *Audio and Record Review* (December 1965), started as follows:

'Pressing work was forgotten, and the old Briggs' magic enticed me to read on—and on! You have been warned—don't start on this book unless you have the time to follow through, although dipping into the pages will almost certainly produce a few choice plums of audio wisdom.'

The print run was 5000. The topic, however, was one for specialists and sales were a bit disappointing. By the end of 1967 sales had reached 2242 and by mid-1972, 3450.

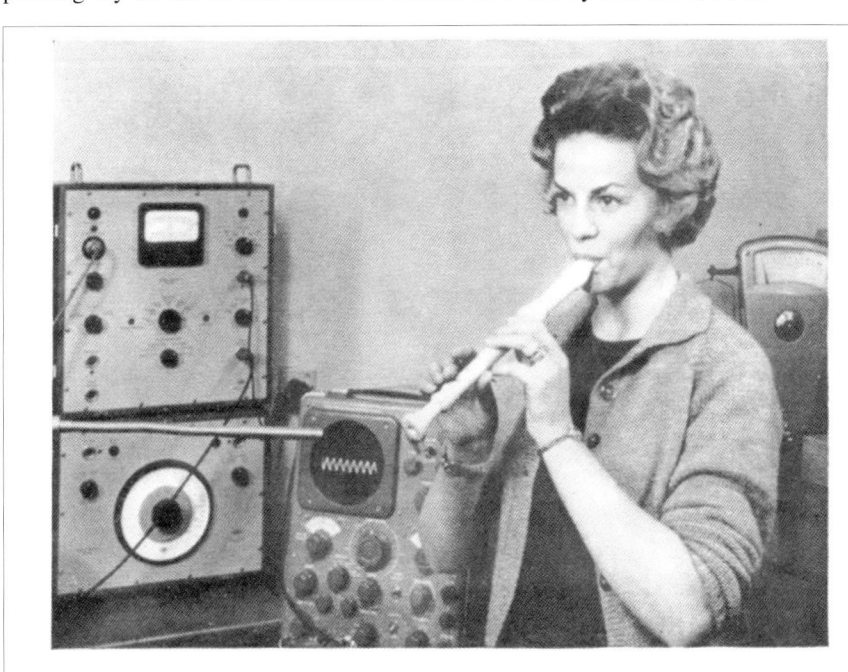

FIG. 4/21. *Lab. photograph showing waveform produced on 'scope by descant recorder.*

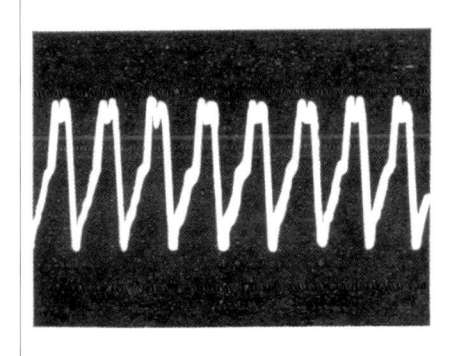

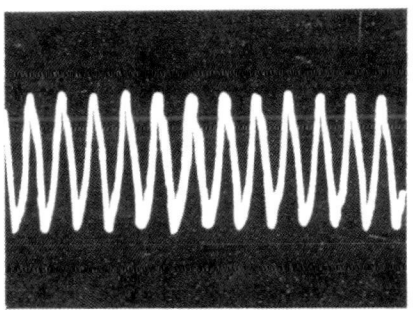

FIG. 4/22. *Low note and high note from descant recorder.*

Figure 12.11 Analysis of notes played, on the descant recorder, by Gilbert's secretary, Dorothy Dawson. Reproduced from *Musical Instruments and Audio*.

313

About Your Hearing by G.A. Briggs with J. Moir
May 1967, 132 pages, price 15 s 6 d (paperback) or 22 s 6 d (hardback)

With no technical support to call on, and following his lack of success with the self-published *Puzzle and Humour Book* of 1966, Gilbert returned to an audio-related subject which Rank Wharfedale were happy to publish, but which did not require experimental input from Ken Russell. As with musical instruments, the subject of hearing had cropped up in previous books, and this is the area he decided to try and write about. James Moir was happy once again to act as sub-editor and as usual collaborators old and new were charmed into making specialist contributions. Not the least of these was a Harley Street consultant in ear surgery, who had to be anonymous for professional reasons (and whose identity remains unknown).

Gilbert started with a short historical survey of devices used to increase hearing levels before discussing the relationship between sound and hearing. Then followed chapters on how the ear works, hearing tests, forms of deafness, hearing aids, noise, listening to reproduced sound with hearing impairment, surgical treatment of deafness and issues relating particularly to children. In a final chapter the 'questions and answers' technique which Gilbert had found useful in earlier books, was invoked for the last time, since this was to be his final book. He admitted it was the most difficult, but it kept him occupied and his persistence in tracking down unusual statistics and interesting illustrations was undiminished. The print run is not known, but it was probably his usual, by this time, 5000. The subject was, by some distance, the furthest from the core established 20 years earlier, and sales may have suffered for exactly the same reason that his Voluphone suffered in 1937/8 (see Chapter 4). Five years after publication sales had reached 2400.

Conclusion

The helpful sales assistant at Webb's Radio, who suggested that Gilbert write a much-needed book about loudspeakers, launched a minor publishing phenomenon. There was no competition in the UK and the available textbooks from the USA were expensive and difficult to come by in the post-war period. Either by luck or intuition, Gilbert's formula for the layman's guide found an immediate resonance, not just in the UK but in the USA, where there was competition in principle, and around the world. His style developed quickly and became highly idiosyncratic, soon being turned into a selling point. His following, built up through *Loudspeakers* and *Sound Reproduction*, was loyal and most of his books were snapped up in their thousands within months of being printed. His particular appeal to the DIY audio enthusiast was amply demonstrated by the spectacular success of *Cabinet Handbook*. Though not published until 1962, 14 years after his first offering, it was still selling about 1000 a year in 1972 when sales had passed 33,000.

There are many testimonies to the fact that these books not only introduced many young people to audio and hi-fi, and passed on the author's passion for music, but led many of them to a career in the industry. Thanks to Gilbert's love of statistics and tabulations the information on how sales were going was constantly laid before the public through advertisements, sales leaflets and book covers, and is the source of all my data. Whilst 260,000 copies sold is a remarkable statistic, the number of readers must be many times greater, through library and personal lending. It is therefore safe to conclude that Gilbert's books were not only hugely influential in terms of education, but also of incalculable benefit to the Wharfedale brand.

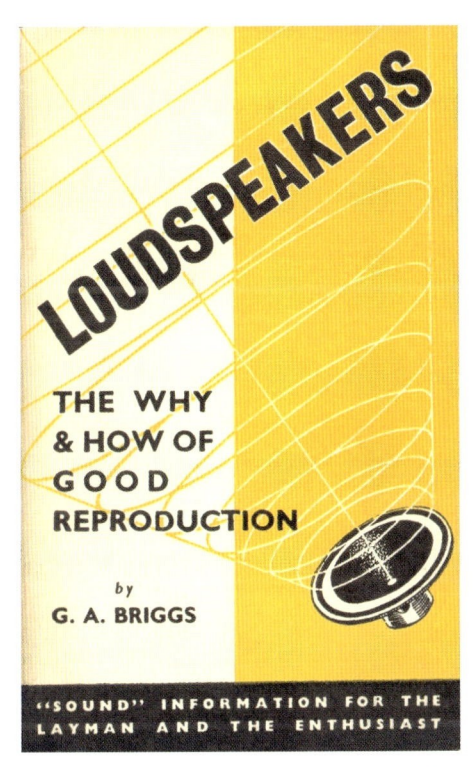

Chapter Twelve: Books

LS1 1948

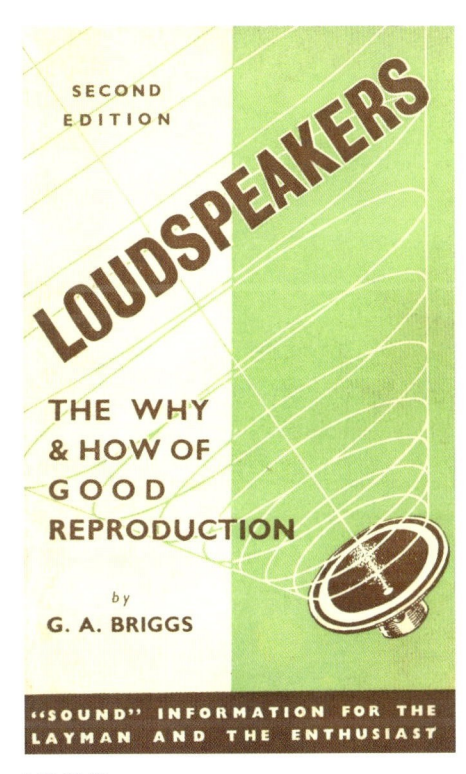

LS2 1948

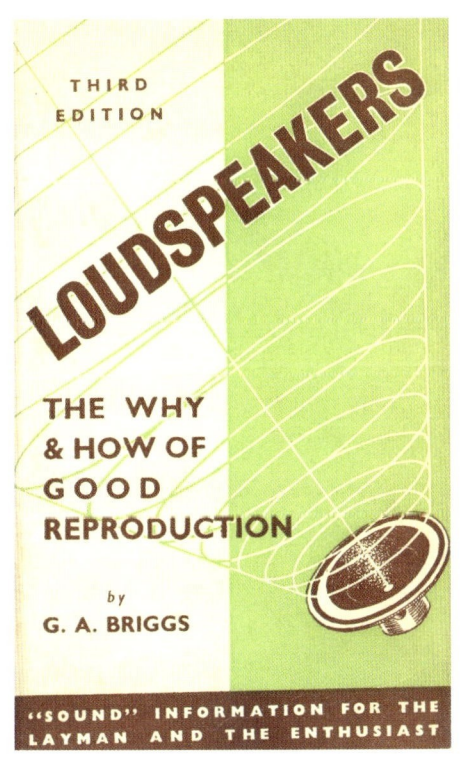

LS3 1949

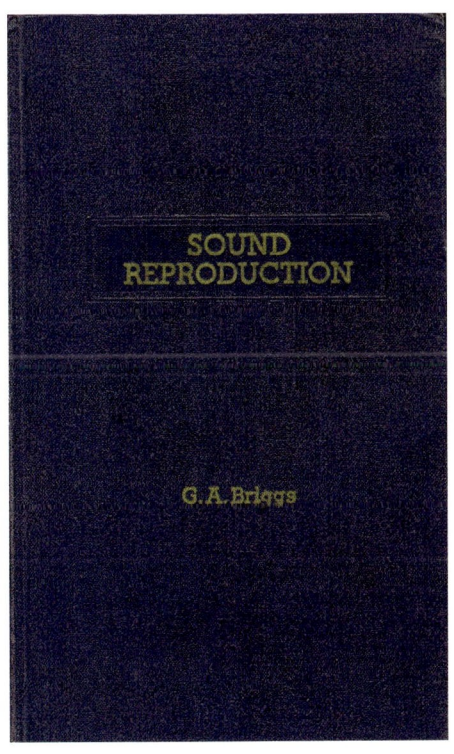

SR1 1949

315

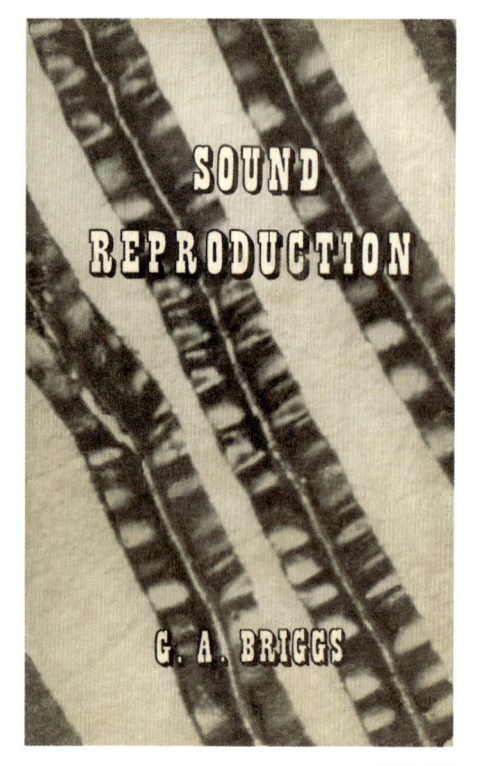

A Pair of Wharfedales: The Story of Gilbert Briggs and his Loudspeakers

SR2 1950

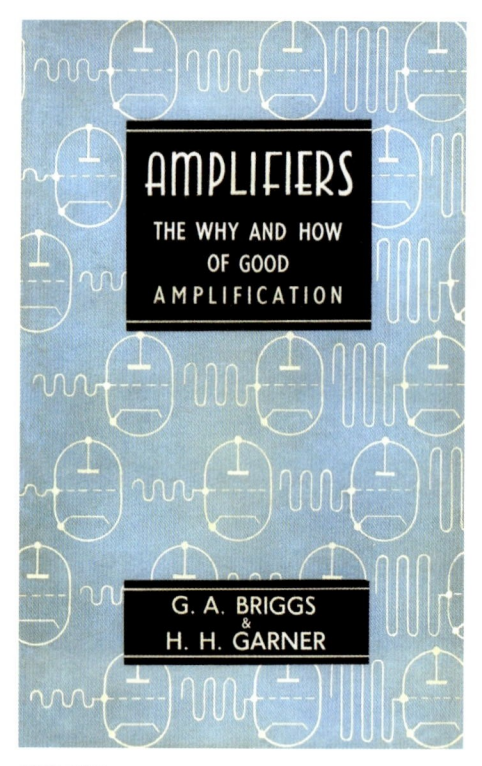

AMP 1952

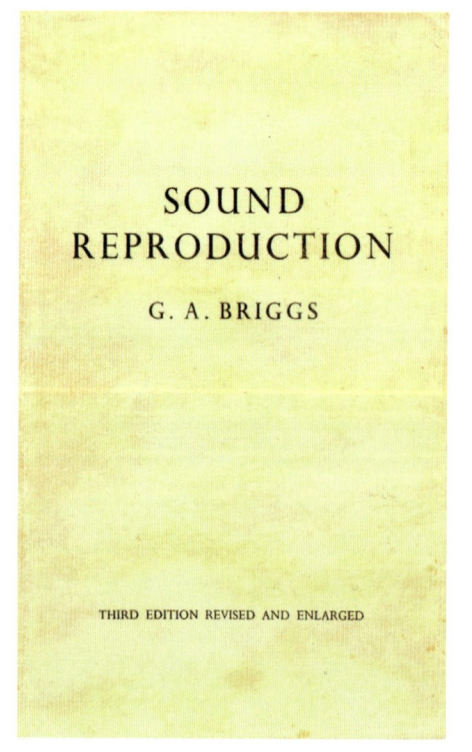

SR3 1953

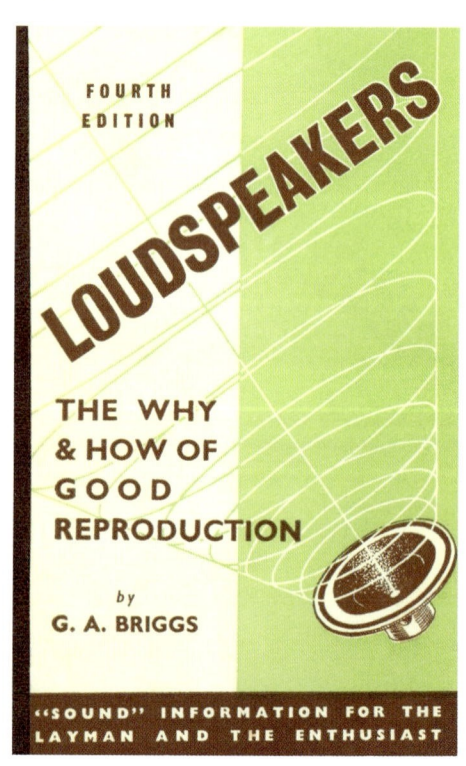

LS4 1955

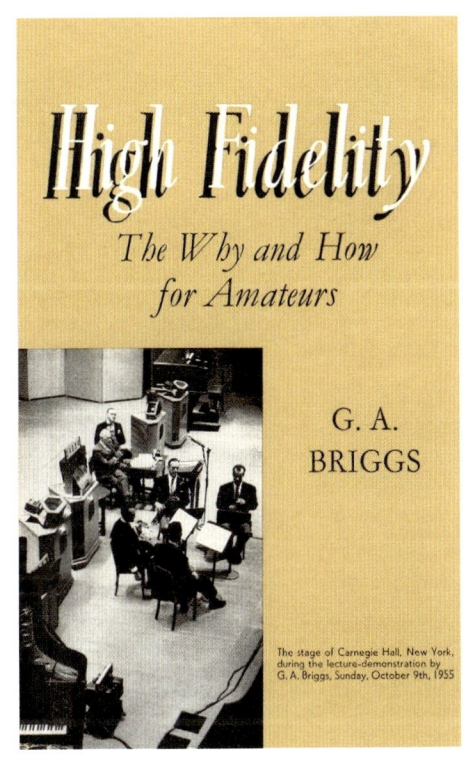

Chapter Twelve: Books

HF 1956

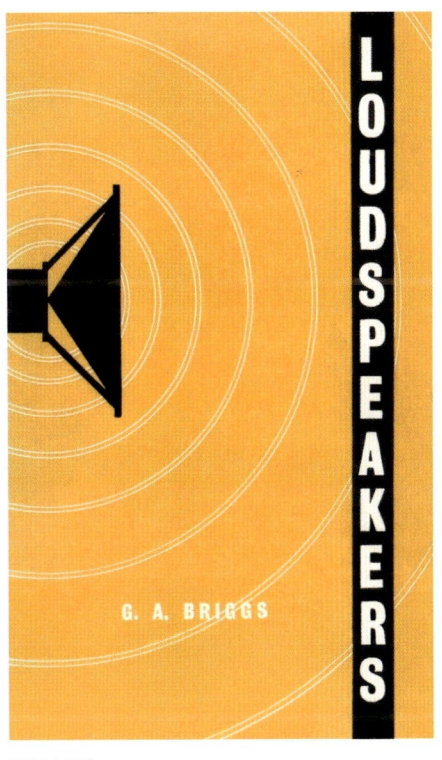

LS5 1958

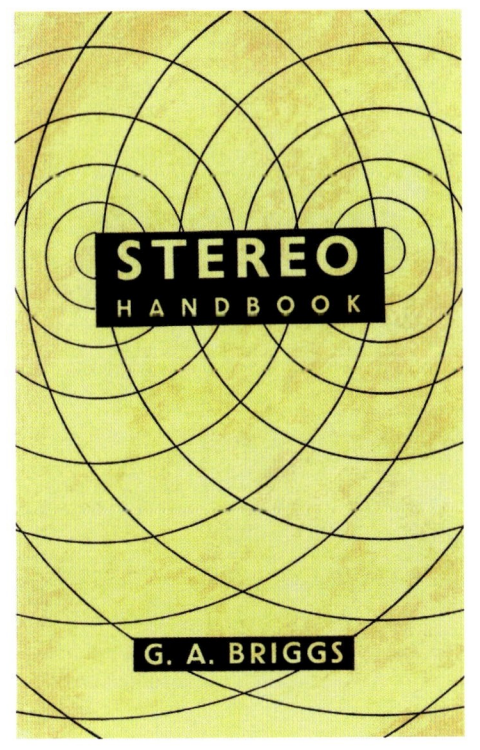

SH 1959

AZ 1960

AB 1961   CH 1962

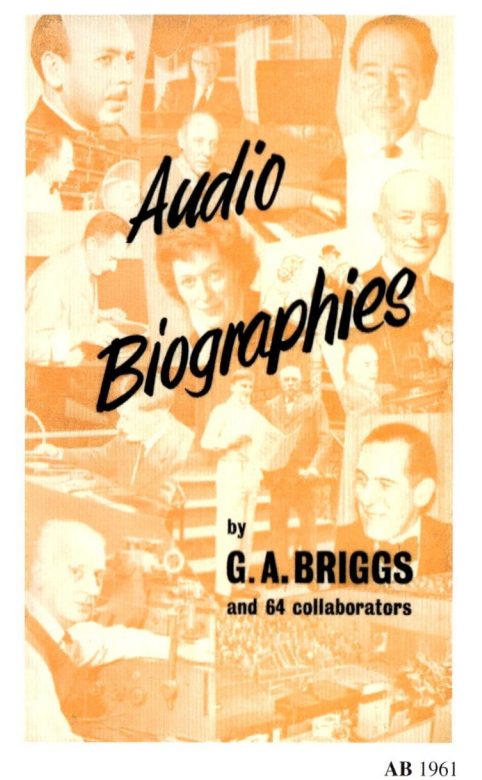

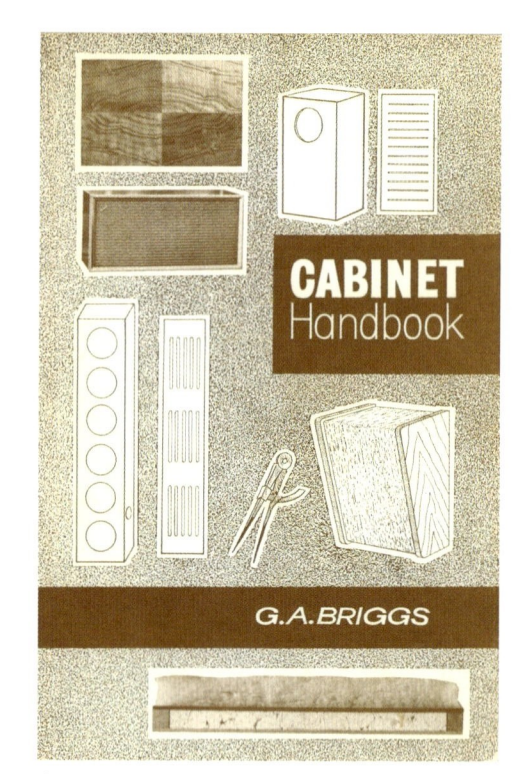

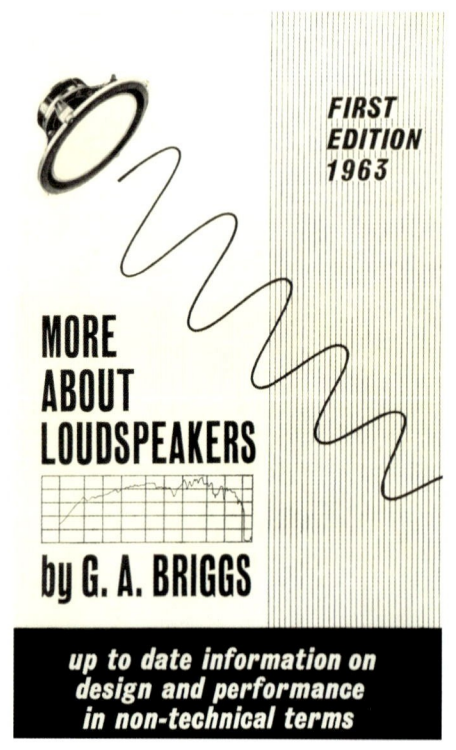

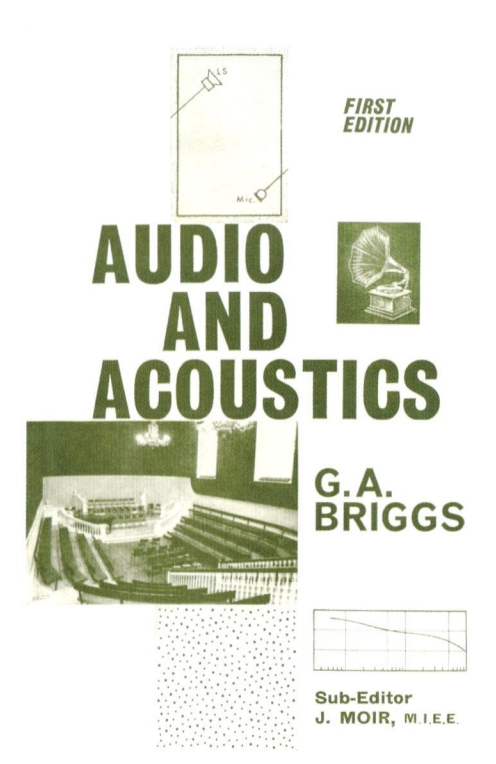

MLS 1963   AA 1963

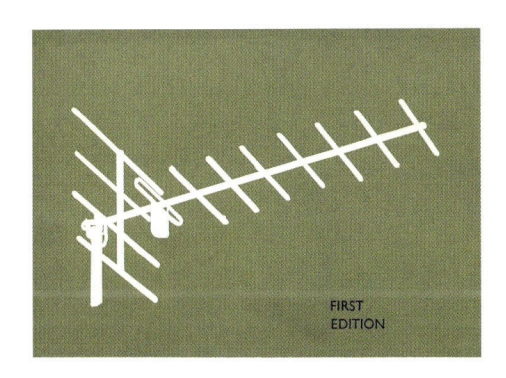

Chapter Twelve: Books

AH1 1964 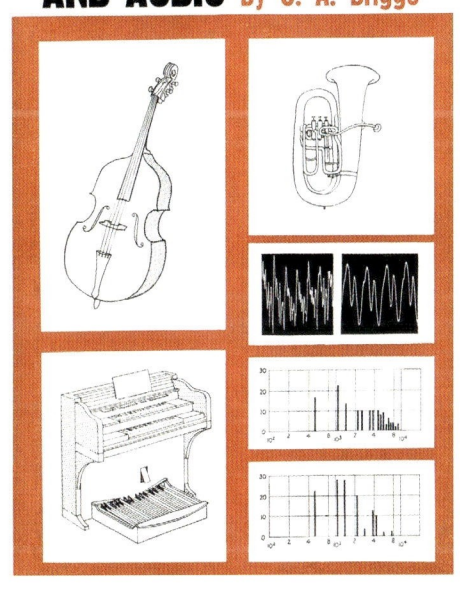 **MIA** 1965

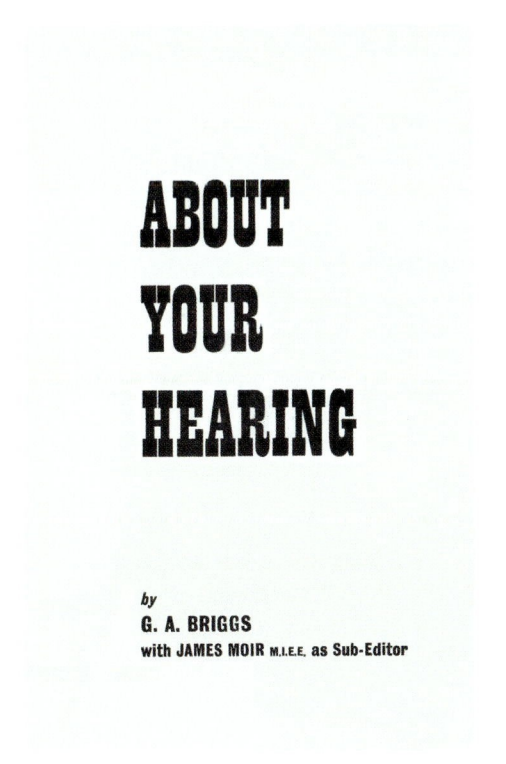

AYH 1967  **AH2** 1968

319

Epilogue

The fate of the Wharfedale brand

In late 1979 Rank gave up the unequal struggle with electronics from South Asia and abandoned the Leak brand. All electronics production ceased immediately and the sandwich speakers soon followed. The Wharfedale operation was re-evaluated in 1980 because of intensifying competition, rising production costs and the recession, leading to a reduction in the workforce to around 500.

In the Spring of 1982, as the 50th anniversary of the founding of Wharfedale Wireless Works approached, the Rank-Wharfedale internal newsletter, *The World of Wharfedale*, had two contrasting items on its front page. The first was an acknowledgement by the Export Product Manager that the year was going to be difficult for the hi-fi industry as a whole. The second was an article entitled '50 Years of Wharfedale' written by Ken Russell, who was still Technical Manager, which gave an excellent summary of the history of the company and ended:

> *'With the resources of the Rank Organisation behind them, Wharfedale can afford to look forward to a second 50 years as a respected name all over the world.'*

Sadly, the prediction of the first article was the more accurate and Rank decided to sell Wharfedale that summer. The buyer was Peter Newman, and Englishman who owned a loudspeaker manufacturer, Audio Industrias, in Spain. He closed down the cabinet making operation and outsourced this activity, but otherwise left things unchanged. At this point Ken Russell left and Gareth Millward, who had been Chief Engineer since 1977, effectively took over his role (he went on to serve as Technical Director from 1987 to 1993).

Around two years later Newman sold out to the Vallance family who owned a large white-goods retail operation in Yorkshire. In February 1985 they decided that the Highfield Road site was too large so they left Idle and moved into a new factory at Crossgates in East Leeds, with everything in one building. All the workers at Idle were offered jobs there, with bus transport organised from key collection points around Bradford, and most accepted. This move marked the end of the process of withdrawing from volume production and the return to a more specialised product range.

Not long after the move the Crossgates factory was visited by Denis Healey, previously Chancellor of the Exchequer in the Labour Government of the mid 1970s, in whose Leeds constituency it resided. He is shown below talking to Phil Escott, Service Department Manager, who provided so much material for this book.

In 1987 the Vallences sold the company to a consortium of investors, which included The Yorkshire Enterprise Board, 3i and Acumen Partners, and the company became Wharfedale plc. Within a couple of years there were mergers with Fane Acoustics and Cambridge Audio and this new enterprise was named the Verity Group plc. Another competitor, Mission, was incorporated around 1992. At some point the retailer Argos bought the rights to use the Wharfedale brand name on consumer electronics, such as televisions and mp3 players, but not loudspeakers.

Epilogue

Around 1993 Stan Curtis became MD with a mission 'to turn Wharfedale around' and two years later he moved to Quad to do a similar job when that company was rescued by the Verity Group. By 1997 Verity had moved into NXT technology and, in a bid to raise capital to concentrate on this development, they sold Wharfedale and Quad. Stan Curtis headed a management buyout which ultimately led to the brands becoming part of International Audio Group (IAG).

Edna Briggs

After Gilbert's death in 1978, Edna left Ilkley and went to live at her farm near Blubberhouses. She had been building up the activity there since she bought it in 1966 and had become a successful breeder of Charolais cattle and Arab horses, winning many prizes for both of them. In 1982, aged 82, she was forced to give up the farm because of ill-health and went to live close to Ninetta. Her spirit was indomitable and, despite being in a wheelchair, in order to work outdoors she took up gardening with long-handled tools. She died in 1997.

Denis Healey, MP (right) visiting the Crossgates factory during the late 1980s and talking to Phil Escott. (Picture courtesy of Philip S Escott.)

Appendix 1: Letters in the BBC Written Archive

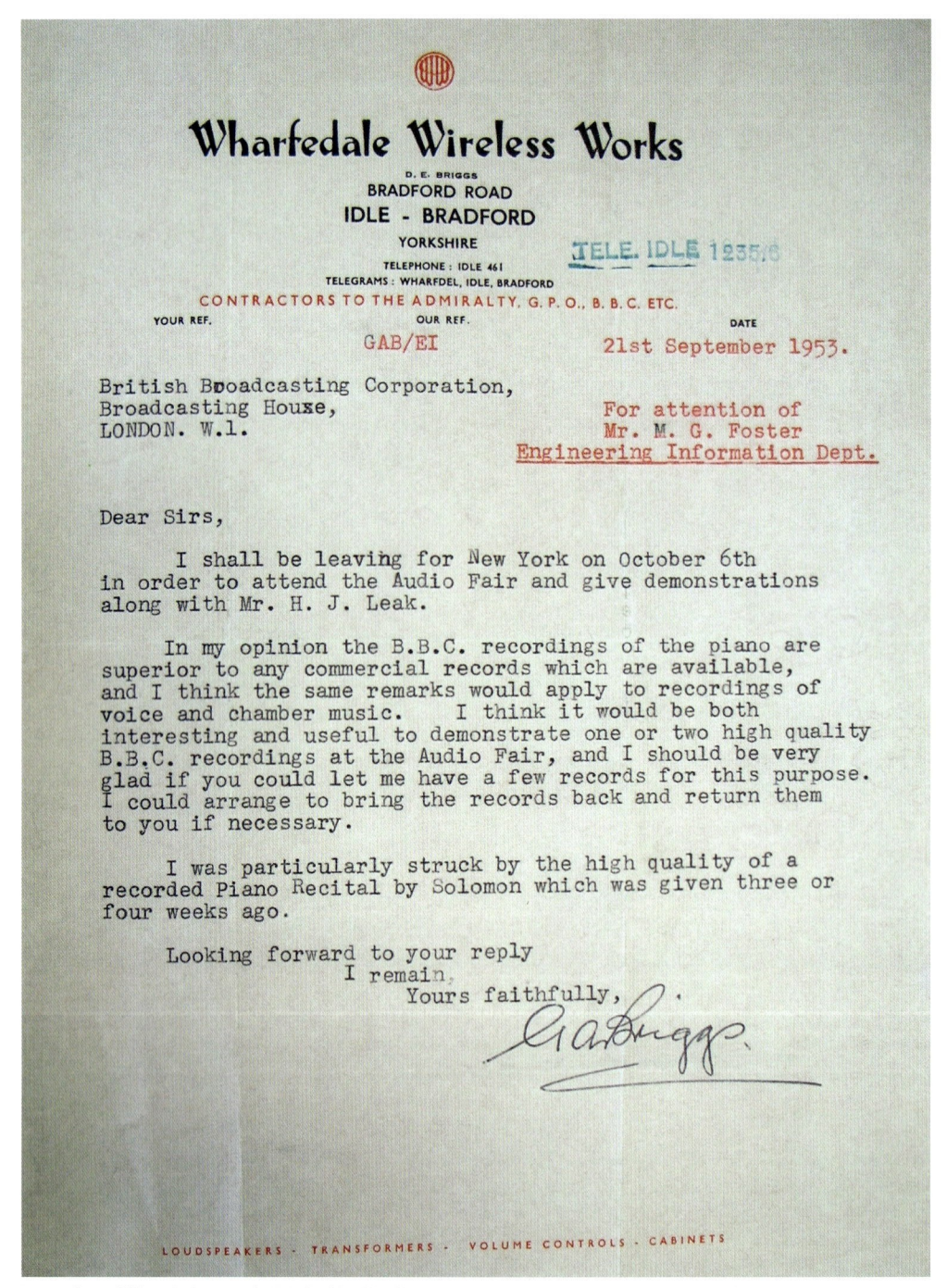

Letter from Gilbert Briggs to the BBC of 21.09.53. Reproduced by permission of the Briggs family, source BBC Written Archive Centre, file R46/552: Wharfedale Wireless Works File 1 (1953–1955).

Appendices

Wharfedale Wireless Works

D. E. BRIGGS
BRADFORD ROAD
IDLE - BRADFORD
YORKSHIRE
TELEPHONE : IDLE 461
TELEGRAMS : WHARFDEL, IDLE, BRADFORD
CONTRACTORS TO THE ADMIRALTY, G. P. O., B. B. C. ETC.

YOUR REF.

OUR REF. GAB/EI

DATE 30th April 1954.

British Broadcasting Corporation
Broadcasting House
LONDON, W.1.

For the attention of Mr. F. Miles Coventry.

Dear Sirs,

EXPERIMENTS IN SOUND.

We have booked the Royal Festival Hall for a Lecture-Demonstration on this subject on November 1st next. We intend to engage several artists, so that live performances can be compared with records, and we are assured of the full support of E.M.I. and Decca.

I note that it is against the policy of the B.B.C. to lend records for reproduction outside the circle of the Corporation; but in view of the importance of a Demonstration in the Festival Hall - which is something which has not previously been attempted in public - I think the Corporation could make an exception and lend me one of your finest musical recordings in order to demonstrate the high quality attained by the B.B.C. engineers. I should be glad if you would give this suggestion your serious consideration.

I enclose a programme of a similar Demonstration which we recently gave in Bradford. You will observe that there was no advertising.

Yours faithfully,
per pro WHARFEDALE WIRELESS WORKS LTD.
MANAGING DIRECTOR.

LOUDSPEAKERS · TRANSFORMERS · VOLUME CONTROLS · CABINETS

Letter from Gilbert Briggs to the BBC of 30.04.54. Reproduced by permission of the Briggs family, source BBC Written Archive Centre, file R46/552: Wharfedale Wireless Works File 1 (1953–1955).

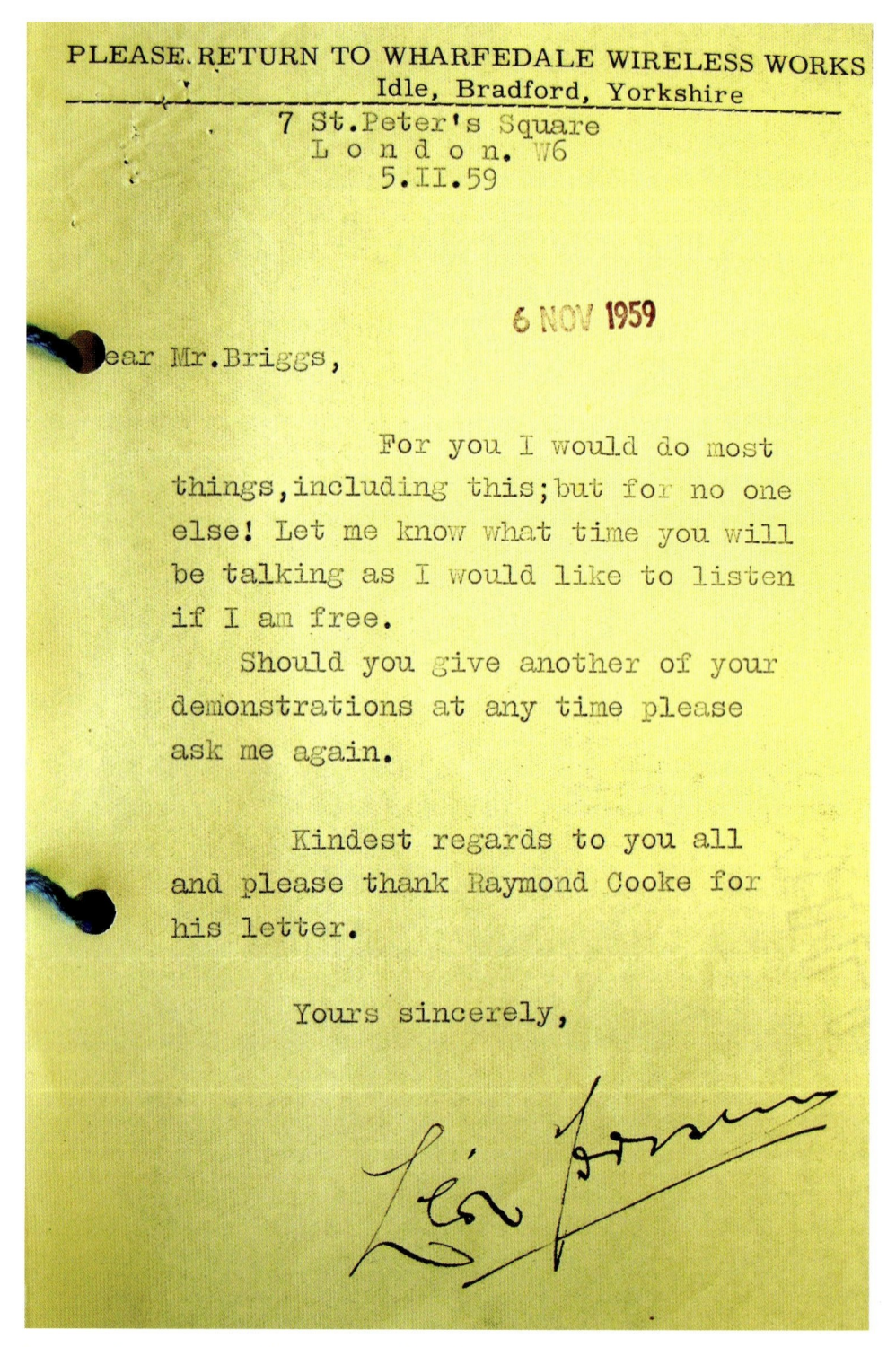

> **PLEASE. RETURN TO WHARFEDALE WIRELESS WORKS**
> **Idle, Bradford, Yorkshire**
>
> 7 St. Peter's Square
> London. W6
> 5.II.59
>
> 6 NOV 1959
>
> Dear Mr. Briggs,
>
> For you I would do most things, including this; but for no one else! Let me know what time you will be talking as I would like to listen if I am free.
>
> Should you give another of your demonstrations at any time please ask me again.
>
> Kindest regards to you all and please thank Raymond Cooke for his letter.
>
> Yours sincerely,
>
> Leon Goossens

Letter from Leon Goossens to Gilbert Briggs of 06.11.59. Reproduced by permission of the Goossens family, source BBC Written Archive Centre, file R46/772/1: Wharfedale Wireless Works File 2 (1956–1964).

Appendix 2: Details of major concert-demonstrations including live versus recorded items

Unless otherwise stated, recordings are mono and discs are 33.33 revolutions per minute (rpm), microgroove (m). Alternatives are 78 rpm and the earlier standard groove (eg 78 s). Tape speeds are in inches per second (ips).

Programme Front Covers

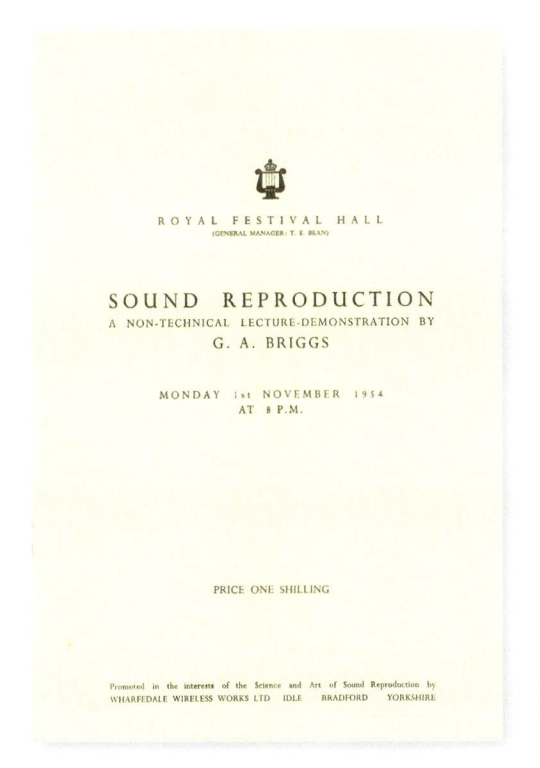

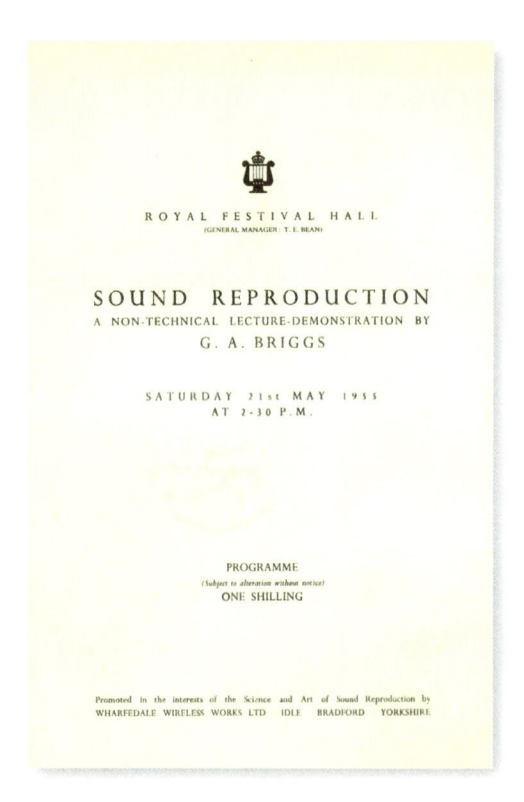

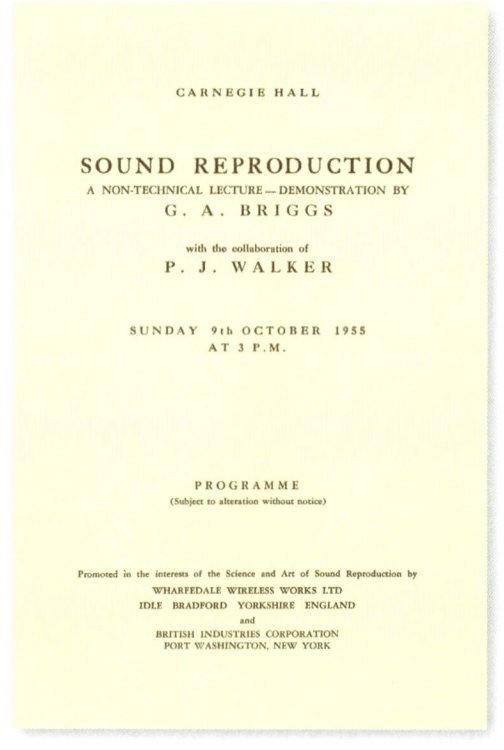

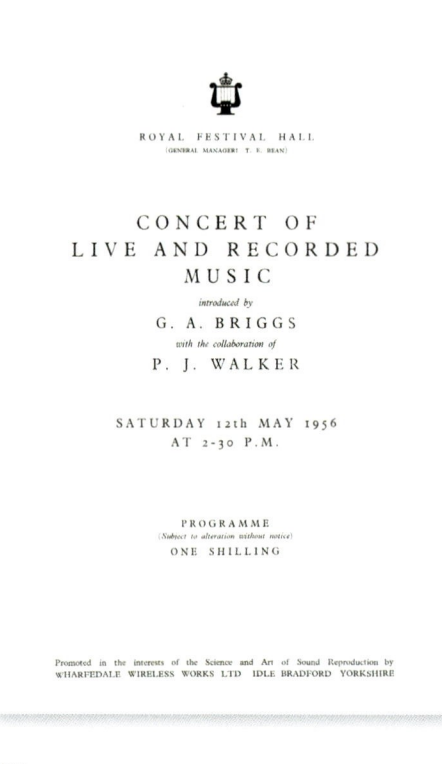

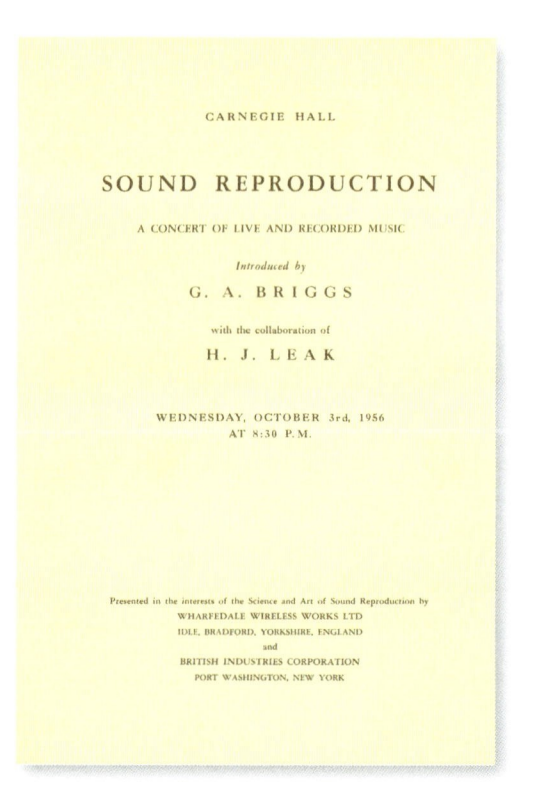

A Pair of Wharfedales: The Story of Gilbert Briggs and his Loudspeakers

ROYAL FESTIVAL HALL
(GENERAL MANAGER: T. E. BEAN)

SOUND REPRODUCTION
A NON-TECHNICAL LECTURE-DEMONSTRATION BY
G. A. BRIGGS

SATURDAY 21st MAY 1955
AT 2-30 P.M.

PROGRAMME
(Subject to alteration without notice)
ONE SHILLING

Promoted in the interests of the Science and Art of Sound Reproduction by
WHARFEDALE WIRELESS WORKS LTD   IDLE   BRADFORD   YORKSHIRE

CARNEGIE HALL

SOUND REPRODUCTION
A NON-TECHNICAL LECTURE — DEMONSTRATION BY
G. A. BRIGGS

with the collaboration of
P. J. WALKER

SUNDAY 9th OCTOBER 1955
AT 3 P.M.

PROGRAMME
(Subject to alteration without notice)

Promoted in the interests of the Science and Art of Sound Reproduction by
WHARFEDALE WIRELESS WORKS LTD
IDLE   BRADFORD   YORKSHIRE   ENGLAND
and
BRITISH INDUSTRIES CORPORATION
PORT WASHINGTON, NEW YORK

ROYAL FESTIVAL HALL
(GENERAL MANAGER: T. E. BEAN)

CONCERT OF
LIVE AND RECORDED
MUSIC

introduced by
G. A. BRIGGS

with the collaboration of
P. J. WALKER

SATURDAY 12th MAY 1956
AT 2-30 P.M.

PROGRAMME
(Subject to alteration without notice)
ONE SHILLING

Promoted in the interests of the Science and Art of Sound Reproduction by
WHARFEDALE WIRELESS WORKS LTD   IDLE   BRADFORD   YORKSHIRE

CARNEGIE HALL

SOUND REPRODUCTION

A CONCERT OF LIVE AND RECORDED MUSIC

Introduced by
G. A. BRIGGS

with the collaboration of
H. J. LEAK

WEDNESDAY, OCTOBER 3rd, 1956
AT 8:30 P.M.

Presented in the interests of the Science and Art of Sound Reproduction by
WHARFEDALE WIRELESS WORKS LTD
IDLE, BRADFORD, YORKSHIRE, ENGLAND
and
BRITISH INDUSTRIES CORPORATION
PORT WASHINGTON, NEW YORK

Appendices

Programme 1

PHILHARMONIC HALL · LIVERPOOL

A CONCERT OF LIVE AND RECORDED MUSIC

Introduced by
G. A. BRIGGS

with the collaboration of
P. J. WALKER

TUESDAY 2nd JULY 1957
Commencing at 7-30 p.m.

LEON GOOSSENS	Oboe
KENNETH PAGE	Violin
KENNETH CLARE	Accompanist
GERALD GOVER	Piano
Dr. CALEB JARVIS	Organ

STEINWAY PIANOFORTE   WHARFEDALE LOUDSPEAKERS
GARRARD VARIABLE SPEED TRANSCRIPTION MOTOR (301)
FERRANTI PICKUP   ACOUSTICAL QUAD II AMPLIFIERS
FERROGRAPH TAPE RECORDERS

PROGRAMME . . . . PRICE SIXPENCE
(Subject to alteration without notice)

Promoted in the interest of the Science and Art of Sound Reproduction by
WHARFEDALE WIRELESS WORKS LTD. · IDLE · BRADFORD · YORKSHIRE

Programme 2

TEMPEST ANDERSON HALL · YORK

A CONCERT OF LIVE AND RECORDED MUSIC

Introduced by
G. A. BRIGGS

with the collaboration of
P. J. WALKER

WEDNESDAY 30th APRIL 1958
Commencing 7-30 p.m.

LEON GOOSSENS	Oboe
EDGAR KNIGHT	Piano
BARBARA ELSY	Soprano

FERROGRAPH TAPE RECORDERS   WHARFEDALE LOUDSPEAKERS
GARRARD VARIABLE SPEED TRANSCRIPTION MOTOR (301)
ORTOFON PICKUP   ACOUSTICAL QUAD II AMPLIFIERS
STEINWAY PIANOFORTE (by kind permission of The British Music Society of York)

PROGRAMME . . . . PRICE SIXPENCE
(Subject to alteration without notice)

Promoted by GUSSINS & LIGHT LTD., King's Square, York,
and WHARFEDALE WIRELESS WORKS LTD., Idle, Bradford,
in aid of THE CHESHIRE HOMES, York Committee Appeal

Programme 3

ROYAL FESTIVAL HALL
(General Manager : T. E. Bean, C.B.E.)

CONCERT OF LIVE AND RECORDED MUSIC

introduced by
G. A. BRIGGS

with the collaboration of
P. J. WALKER

SATURDAY 9th MAY 1959 at 3 p.m.
(Concluded at 5-10 p.m.)

LEON GOOSSENS	Oboe
DENIS MATTHEWS	Piano
RALPH DOWNES	Organ
*HAROLD BLACKBURN	Bass
*GERALD GOVER	Accompanist

* By kind permission of Sadler's Wells Trust Ltd.

STEINWAY PIANOFORTE   WHARFEDALE LOUDSPEAKERS
GARRARD VARIABLE SPEED TRANSCRIPTION MOTOR (301)
ACOUSTICAL QUAD II AMPLIFIERS AND STEREO PRE-AMP 22
DECCA STEREO PICKUP   FERRANTI & ORTOFON TYPE C MONO PICKUPS
FERROGRAPH TAPE RECORDERS

PROGRAMME . . . . PRICE SIXPENCE
(Subject to alteration without notice)

Promoted in the interests of the Science and Art of Sound Reproduction by
WHARFEDALE WIRELESS WORKS LTD IDLE BRADFORD YORKSHIRE

Programme 4

COLSTON HALL BRISTOL

CONCERT OF LIVE AND RECORDED MUSIC

introduced by
G. A. BRIGGS

with the collaboration of
R. E. COOKE and R. L. WEST

FRIDAY, 9th OCTOBER 1959 at 7-30 p.m.
(Concluding at 9-45 p.m.)

LEON GOOSSENS	Oboe		
*HAROLD BLACKBURN	Bass		
EDWARD FRY	Organ		
	Piano Trio		
GERALD GOVER	Piano	*KENNETH POPPLEWELL	Violin
	TERENCE WEIL	'Cello	

*Harold Blackburn appears by kind permission of Sadlers Wells Trust Ltd.,
and Kenneth Popplewell by kind permission of the B.B.C.

STEINWAY PIANOFORTE   WHARFEDALE LOUDSPEAKERS
GARRARD VARIABLE SPEED TRANSCRIPTION MOTOR (301)
and CONNOISSEUR TYPE 'B' MOTOR
ACOUSTICAL QUAD II AMPLIFIERS AND STEREO PRE-AMP 22
DECCA STEREO PICKUP   ORTOFON TYPE 'C' MONO PICKUP
AMPEX 351/2P AND FERROGRAPH TAPE RECORDERS

PROGRAMME . . . . PRICE SIXPENCE
(Subject to alteration without notice)

Promoted in the interests of the Science and Art of Sound Reproduction by
WHARFEDALE WIRELESS WORKS LTD IDLE BRADFORD YORKSHIRE

Programme Items

Royal Festival Hall, London, 1 November 1954

Barrell Symphony: (a) snare drums, (b) cymbals, (c) maracas, (d) triangle, (e) castanets, (f) tambourine. Each section recorded low and wide frequency range. EMI JGS 74 (78 s).

Glass Breaking (open air recording), R. Bradford (78 s).

Memories of You, Loring Nichols and his Band. Audiophile AP-7 (78 m).

Tugboat Noises (recorded on the Thames by G.A. Elliott). Mercury Sound Recordings Ltd (78 s).

Handel, Sonata in G (harpsichord). Record of Thurston Dart by C.E. Watts (78 s), live performance by Stanislav Heller.

Wasner, Lay down your staffs. Lutheran Choir of Milwaukee. Audiophile 1-A (78 m).

(a) Bach, Toccata in D min, (b) Stanley, Allegro –Voluntary in D, live v recorded organ playing by Ralph Downes. EMI tape (30 ips).

The High and the Mighty, Light Brigade Orchestra. Nixa XLPY 151.

Haydn, Symphony No. 100 in G, London Philharmonic Orchestra/Solti. Decca LXT 2984.

Beethoven, Sonata in D Minor. Live v recorded piano playing by Denis Matthews. EMI tape (30 ips).

Vaughan Williams, A Song of Thanksgiving (chorus and orchestra). Parlophone PMB 1003.

Echo Effects (recorded in the Hamilton Mausoleum, Lanarkshire). BBC (78 s).

Bartok, Viola Concerto, William Primrose/New Symphony Orchestra of London/Serly. Bartok Records BRS 309.

Bach, Qui sedes – Mass in B Minor, Kathleen Ferrier. Decca LXT 2757.

Beethoven, Symphony No. 6 (Pastoral), Concertgebouw Orchestra/Kleiber. Decca LXT 2872.

Vaughan Williams, A Sea Symphony, London Philharmonic Choir and Orchestra/ Boult. Decca LXT 2907.

St George's Hall, Bradford, 1 April 1955

Handel, Sonata in G (harpsichord), Thurston Dart. C.E. Watts (78 s).

Echo Effects (recorded in the Hamilton Mausoleum, Lanarkshire). BBC (78 s).

Wasner, Lay down your staffs. Lutheran Choir of Milwaukee. Audiophile 1-A (78 m).

Tugboat Noises (recorded on the Thames by G.A. Elliott). Mercury Sound Recordings Ltd (78 s).

Bolero from 8 Danses Modernes, M. Philippe-Gerard Ensemble. Nixa LPY 122.

Bach, Toccata in D Minor – Decca LW 5095 and Chorale Prelude – Decca LXT 2915. Live organ playing by G. Hankin v recordings by Jeanne Demessieux.

Schubert, Symphony No. 9, Hallé Orchestra/ Barbirolli. HMV ALP 1178.

Rachmaninoff, Preludes in G Minor and G Major. Columbia SCB 117 (45 m). Live piano playing by Edgar Knight v recordings by Geza Anda.

Vaughan Williams, This Day (Hodie), BBC Chorus and Choral Society/Watford Grammar School Boys' Choir/BBC Symphony Orchestra/Sargent. EMI recording of BBC broadcast from RFH.

(a) Quilter (arr), Drink to me only, (b) Rameau, Gavotte. Live v recorded by Leon Goossens (oboe) and Edgar Knight (piano). A.R. Sugden stereo tape (15 ips).

Schubert, Variations in B Flat, piano duet by Badura-Skoda and Joerg Demus. Nixa WLP 5147.

Bach, Qui sedes – Mass in B Minor, Kathleen Ferrier. Decca LXT 2757.

Mozart, Piano Concerto in A (K488), Denis Matthews/Philharmonia Orchestra/Schwarz. Columbia 33s 1039.

Britten, Diversions for Piano (left hand) and Orchestra, Julius Katchen/London Symphony Orchestra/Britten. Decca LXT 2981.

Vaughan Williams, A Sea Symphony, London Philharmonic Choir and Orchestra/ Boult. Decca LXT 2907. Live organ accompaniment by G. Hankin.

Royal Festival Hall, London, 21 May 1955

(a) Memories of You, Loring Nichols and his Band. Audiophile AP-7 (78 m). (b) Bolero from 8 Danses Modernes, M. Philippe-Gerard Ensemble. Nixa LPY 122.

(a) Marcello-Bach, Adagio in D Minor, (b) Teleman, Rondeau in G. Live v recorded harpsichord playing by Thurston Dart. C.E. Watts (78 s).

Tugboat Noises (recorded on the Thames by G.A. Elliott). Mercury Sound Recordings Ltd (78 s).

Schubert, Variations in B Flat, piano duet by Badura-Skoda and Joerg Demus. Nixa WLP 5147.

(a) Handel, Zadoc the Priest, (b) Noble, Souls of the Righteous. Live v recorded performance by section of Goldsmiths' Choral Union/Haggis with Ralph Downes playing organ in (a).

Beethoven, Symphony No. 6 (Pastoral), Concertgebouw Orchestra/Kleiber. Decca LXT 2872.

(a) Bach, Toccata in D min, (b) Stanley, Allegro –Voluntary in D, live v recorded organ playing by Ralph Downes. EMI tape (30 ips).

Vaughan Williams, This Day (Hodie), BBC Chorus and Choral Society/Watford Grammar School Boys' Choir/BBC Symphony Orchestra/Sargent. EMI recording of BBC broadcast from RFH.

Chopin, (a) Berceuse, (b) Waltz in A Flat, Live v recorded piano playing by Phyllis Sellick. EMI tape (30 ips).

Rachmaninoff, Prelude in G Major, Geza Anda. Columbia SCB 117 (45 m).

Bach, Qui sedes – Mass in B Minor, Kathleen Ferrier. Decca LXT 2757.

Mozart, Piano Concerto in A (K488), Denis Matthews/Philharmonia Orchestra/Schwarz. Columbia 33s 1039.

Schubert, Symphony No. 9, Hallé Orchestra/ Barbirolli. HMV ALP 1178.

Britten, Young person's Guide to the Orchestra, Concertgebouw Orchestra/Van Beinum. Decca LXT 2886.

Vaughan Williams, A Sea Symphony, London Philharmonic Choir and Orchestra/ Boult. Decca LXT 2907. Live organ accompaniment by Ralph Downes.

Carnegie Hall, New York, 9 October 1955

Handel, Sonata in G (harpsichord), Thurston Dart. C.E. Watts (78 s).

Memories of You, Loring Nichols and his Band. Audiophile AP-7 (78 m).

Tugboat Noises (recorded on the Thames by G.A. Elliott). Mercury Sound Recordings Ltd (78 s).

Handel, Concerto Grosso (violins, 'cello and dulcimer), Bamberger Symphony Orchestra. Archive 13010 AP.

(a) Chopin, Scherzo in B Minor, (b) Theme and Variations – Anchors Aweigh. Live v recorded piano playing by Leonid Hambro. Columbia tape (15 ips)

Wasner, Lay down your staffs. Lutheran Choir of Milwaukee. Audiophile 1-A (78 m).

(a) Bach, Toccata in D Minor, (b) Daquin, Noel Grand Jeu et Duo. Live v recorded playing by E. Power Biggs. Columbia tape (15 ips).

Thingamagig, Mel Powell (jazz) Trio. Vanguard VRS 8502.

Handel, Sonata in C Minor. Live v recorded playing by John De Lancie (oboe) and Leonid Hambro (piano). Columbia tape (15 ips).

Dvorak, Symphony No. 4, Philharmonia Orchestra/Sawallisch. Angel ANG 35214.

Schubert, Variations in B flat, Badura-Skoda and Joerg Demus. Westminster WL 5147.

Mozart, Wind Quartet, Live v recorded playing by members of Philadelphia Wind Quintet. Columbia tape (15 ips).

Beethoven, Quintet in E Flat. John De Lancie (oboe), Sol Schoenbach (bassoon), Anthony Gigliotti (clarinet), Mason Jones (french horn), Leonid Hambro (piano) playing live and recorded (except Rudolf Serkin on piano). Columbia ML 4834.

Sibelius, Symphony No. 4, Philadelphia Symphony Orchestra/Ormandy. Columbia ML 5045.

Mozart, Piano Concerto in A (K488), Denis Matthews/Philharmonia Orchestra/Schwarz. Columbia 33s 1039.

Bach, Qui sedes – Mass in B Minor, Kathleen Ferrier. London ffrr LL 688.

Britten, Young person's Guide to the Orchestra, Minneapolis Symphony Orchestra/Dorati. Mercury MG 50047.

Vaughan Williams, A Sea Symphony, London Philharmonic Choir and Orchestra/ Boult. London ffrr LL 972-3. Live organ accompaniment by E. Power Biggs.

Royal Festival Hall, London, 12 May 1956

Musical Instruments: kettle drum, trumpet, piccolo, saxophone, double bass, contra bassoon, xylophone, harp. BBC tape (15 ips).

Doc Evans jazz. Audiophile tape (15 ips) (also AP 31).

Wasner, Lay down your staffs. Lutheran Choir of Milwaukee. Audiophile 1-A (78 m).

Haydn, Symphony No. 103, live v recorded playing by London Mozart Players/Harry Blech. EMI tape (30 ips).

Bach, Organ Fantasia in G Minor, Karl Richter. Decca LXT 5029.

(a) Kronke, Sarabande, (b) Rameau, Gavotte. Live v recorded oboe playing by Leon Goossens accompanied by Mabel Lovering. EMI tape (30 ips).

Menotti, Sebastian – Ballet Suite, NBC Symphony Orchestra/Stokowski. RCA Victor LM-1858.

Beethoven, Sonata in D, live v recorded piano playing by Denis Matthews. EMI tape (30 ips).

Verdi, Requiem Mass, RIAS Symphony Orchestra, Berlin/Fricsay. DGG, DGM 18155.

Mozart, Overture: Marriage of Figaro, live v recorded playing by London Mozart Players/ Harry Blech. EMI tape (30 ips).

Sound Effects, recordings by BBC and Ewing D. Nunn of Audiophile Records. Tape (15 ips).

Joseph Strauss, Waltz from Rita Streich (soprano) Recital. DGG DG 17052.

(a) Handel, Sonata No. 1, (b) Hubay, Zephyr, live v recorded violin playing by Campoli accompanied by Eric Gritton. EMI tape (30 ips).

'Adios', Philip Green and His Orchestra. Columbia Stereosonic BTD 701 tape (7.5 ips).

Prokofiev, Classical Symphony, Philharmonia Orchestra/Malko. HMV Stereosonic STD 1750 tape (7.5 ips).

Mozart, Piano Concerto K466, live v recorded playing by Denis Matthews with London

Mozart Players/Harry Blech. EMI tape (30 ips).

Dvorak, Symphony No. 4, Philharmonia Orchestra/Sawallisch. Columbia 33 sx 1034.

Handel, Hallelujah Corus, London Philharmonic Choir and Orchestra/ Boult. Decca LXT 2924. Live accompaniment by Ralph Downes (organ) and the London Mozart Players.

Carnegie Hall, New York, 3 October 1956

Beethoven, Variation on a theme from Mozart's 'The Magic Flute'. Guitar solo, Lauring Almeida. Capitol P8341.

Brahms, Variations on a theme by Haydn for two pianos, live v recorded playing by Teicher and Ferrante. Columbia tape.

Jazz, (a) Doc Evans, Audiophile tape (15 ips) (also AP 31), (b) Sweet Georgia Brown, Audiophile AP 38.

Joseph Strauss, Waltz from Rita Streich (soprano) Recital. DGG DG 17052.

Shostakovich, Symphony No. 5, Philharmonic Symphony Orchestra of London/Rodzinski. Westminster WN 18001.

African Echoes from 'Soundproof'. Live v recorded playing by Teicher and Ferrante. Columbia tape (also Westminster WN 6014).

Tchaikovsky, Chant Sans Paroles for solo violin, Erica Morini. Westminster WN 18087.

(a) Soler, Concerto in G for two organs (first part recorded, second part live), (b) Purcell/Clarke, Trumpet Volountary. Live v recorded playing by E. Power Biggs. Columbia tape.

Bruch, Violin Concerto in G Minor, Isaac Stern, Philadelphia Orchestra/Ormandy. Columbia ML 5097.

Mozart (a) Organ Fantasia in F Minor, (b) Festival Sonata in C. Columbia K 3L-231.

Sound Effects, recordings by BBC and Ewing D. Nunn of Audiophile Records. Tape (15 ips).

Beethoven, Overture Leonore No. 1, Philharmonia Orchestra/Klemperer. Angel 35258.

Debussy, Prelude to 'Afternoon of a Faun', Boston Symphony Orchestra/Munch. RCA Victor LM-1984.

Haydn, Trumpet Concerto in E Flat, George Eskdale, Vienna State Opera Orchestra/Litschauer. Vanguard VRS-454.

Gould: Bell Carol, Percussion Piece, Parade. Live v recorded playing by percussion ensemble conducted by Morton Gould. Columbia tape.

Beethoven, Symphony No. 7, Chicago Symphony Orchestra/Reiner. RCA Victor LM 1991.

Gould, Challenge (for tap dancer and percussion). Live v recorded playing by percussion ensemble conducted by Morton Gould with Danny Daniels, tap dancer. Columbia tape.

Beethoven, Piano Concerto No. 3, Wilhelm Kempff/ Berlin Philharmonic/van Kempen. DGM 18130

Handel, Hallelujah Corus, London Philharmonic Choir and Orchestra/ Boult. London LL 983. Live accompaniment by E. Power Biggs (organ).

Philharmonic Hall, Liverpool, 2 July 1957

Rossini, Il Signor Bruschino: Overture, London Symphony Orchestra/Gamba. Decca LXT 5137.

Old Fashioned Love, Doc Evans and his band. Audiophile tape (15 ips).

Handel, Concerto Grosso in G, Bamberg Symphony Orchestra/Lehmann. Archive AP 13010.

Verdi, Otello: Era piu calmo, Victoria de los Angeles (soprano)/Orchestra of the Opera House, Rome/Morelli. HMV ALP 1284.

Chopin, Sonata No. 2 in B Flat Minor, live piano playing by Gerald Glover v recorded by Victor Schioler (HMV ALP 1243).

Rossini, Stabat Mater, Kim Borg (bass)/ St Hedwig's Cathedrale Choir/RIAS Symphony Orchestra, Berlin/Fricsay. DGG DGM 18203.

Bach, (a) Chorale Prelude: Kommst du nun, Jesu, (b) Fantasia in G Minor, Live organ playing by Dr Caleb Jarvis v recorded by Karl Richter (Decca LXT 5029).

Sound Effects, recordings by BBC and Ewing D. Nunn of Audiophile Records. Tape (15 ips).

Mozart, Trio and Quartet, Philadelphia Woodwind Quartet. Columbia tape (15 ips).

(a) Pugnani-Kriesler, Minuet, (b) Fiocco, Allegro, live v recorded violin playing by Kenneth Page accompanied by Kenneth Clare. Hollick and Taylor tape (15 ips).

Beethoven, Symphony No. 3 (Eroica), Philharmonia Orchestra/Klemperer. Columbia 33CX 1346.

Mozart, Piano Concerto No. 23, Denis Matthews/ Philharmonia Orchestra/Schwarz. Columbia 33S 1039.

(a) Kronke, Sarabande, (b) Rameau, Gavotte. Live v recorded oboe playing by Leon Goossens accompanied by Kenneth Clare. Hollick and Taylor tape (15 ips).

Haydn, The Seasons: The Hunt, Beecham Choral Society and Royal Philharmonic Orchestra/ Beecham. EMI Stereosonic tape (7.5 ips).

Saint-Saens, Cello Concerto No. 1, Mtislav Rostropovitch/Philharmonia Orchestra/Sargent, HMV Stereosonic tape SBT 1251 (7.5 ips).

Vaughan Williams, A Sea Symphony, London Philharmonic Choir and Orchestra/ Boult. Decca LXT 2907. Live organ accompaniment by Dr Caleb Jarvis.

Tempest Anderson Hall, York, 30 April 1958

Bach, Adagio in D Minor (harpsichord), Marcello. C.E. Watts (78 s).

Hushabye, Monte Sunshine (clarinet)/Chris Barber's band. Nixa NJT 502.

Handel, Concerto Grosso No. 2, Bamberg Symphony Orchestra/Lehmann. Archive AP 13010.

Verdi, Otello: Era piu calmo, Victoria de los Angeles (soprano)/Orchestra of the Opera House, Rome/Morelli. HMV ALP 1284.

Chopin, Sonata No. 2 in B Flat Minor, live piano playing by Edgar Knight v recorded by Victor Schioler (HMV ALP 1243).

Bach, Fantasia in G Minor, Karl Richter. Decca LXT 5029.

(a) Kronke, Sarabande, (b) Rameau, Gavotte. Live v recorded oboe playing by Leon Goossens accompanied by Edgar Knight. Hollick and Taylor tape (15 ips).

Beethoven, Symphony No. 3 (Eroica), Philharmonia Orchestra/Klemperer. Columbia 33CX 1346.

Sound Effects, recordings by BBC and Ewing D. Nunn of Audiophile Records. Tape (15 ips).

(a) Purcell, The blessed Virgin's Expostulation, (b) Fauré, Notre Amour, live v recorded solos by Barbara Elsy (soprano) accompanied by Stanley E. Toms. IBC tape (15 ips).

Rossini, Stabat Mater, Kim Borg (bass)/ St Hedwig's Cathedrale Choir/RIAS Symphony Orchestra, Berlin/Fricsay. DGG DGM 18203.

Mozart, Piano Concerto No. 23, Denis Matthews/ Philharmonia Orchestra/Schwarz. Columbia 33S 1039.

Vaughan Williams, A Sea Symphony, London Philharmonic Choir and Orchestra/ Boult. Decca LXT 2907.

Royal Festival Hall, London, 9 May, 1959

Beethoven, Variation on a theme from Mozart's 'The Magic Flute'. Guitar solo, Lauring Almeida. Capitol P8341.

Hushabye, Monte Sunshine (clarinet)/Chris Barber's band. Nixa NJT 502.

Verdi, Ernani, Victoria de los Angeles (soprano)/Orchestra of the Opera House, Rome/Morelli. HMV ALP 1284.

Greig, Peer Gynt: Solveig's song, Ilse Hollweg (soprano). HMV ASD 258 (stereo).

(a) Chopin, Nocturne in F, (b) Debussy, Prélude pour le piano, live v recorded piano playing by Denis Matthews. EMI tape (15 ips).

Schubert, Symphony No. 9, London Symphony Orchestra/Krips. Decca SXL 2045 (stereo).

Handel, Te Deum, Geraint Jones Singers and Orchestra/Jones. Archive SAMP-198008 (stereo).

Bach, (a) Fugue a la Giga, (b) Variations on 'Sei Gegrüsset, Jesu Gütig', live v recorded organ playing by Ralph Downes. Pye CSCL 70006 (stereo).

Ros on Broadway, Edmundo Ros and his Orcherstra. Decca SKL 4004 (stereo)

Handel, Concerto Grosso No. 2, Bamberg Symphony Orchestra/Lehmann. Archive AP 13010.

(a) Verdi, Aria from Simon Boccanegra, (b) Gover, Craigie Burnwood, live v recorded singing by Harold Blackburn (bass) accompanied by Gerald Gover. IBC tape (15 ips).

Dvorak, Dumky Trio, Suk Trio. DGG SLPE 133003.

Adam, Giselle, Paris Conservatoire Orchestra/Martinson. London CS 6098.

(a) Cesar Franck, Andantino in G, (b) F.S. Kelly, Jig, live v recorded playing by Leon Goossens (oboe) and Denis Matthews (piano). EMI tape (15 ips).

Vaughan Williams, A Sea Symphony, London Philharmonic Choir and Orchestra/ Boult. Decca LXT 2907. Live organ accompaniment by Ralph Downes.

Colston Hall, Bristol, 9 October 1959

Handel, Sonata in G (harpsichord), Thurston Dart. C.E. Watts (78 s).

Verdi, Otello: Era piu calmo, Victoria de los Angeles (soprano)/Orchestra of the Opera House, Rome/Morelli. HMV ALP 1284.

Greig, Peer Gynt: Solveig's song, Ilse Hollweg (soprano). HMV ASD 258 (stereo).

(a) Dvorak, Dumky Trio, (b) Mendelssohn, Trio in D Minor, live v recorded playing by Gerald Gover (piano), Kenneth Popplewell (violin) and Terence Weil (cello). EMI tape (15 ips, mono and stereo).

Adam, Giselle, Paris Conservatoire Orchestra/Martinson. Decca SXL 2128 (stereo).

(a) Bach, Trio Sonata No. 1, (b) Widor, Toccata from Symphony No. 5, live v recorded playing by Edward Fry. Bristol and West Recording Service tape (15 ips).

Ros on Broadway, Edmundo Ros and his Orcherstra. Decca SKL 4004 (stereo).

Handel, Concerto Grosso No. 2, Bamberg Symphony Orchestra/Lehmann. Archive AP 13010.

(a) Verdi, Aria from Simon Boccanegra, (b) Water Boy, live v recorded singing by Harold Blackburn (bass) accompanied by Gerald Gover. EMI tape (15 ips, mono and stereo).

Handel, Te Deum, Geraint Jones Singers and Orchestra/Jones. Archive APM 14124 (mono) and SAPM 198008 (stereo).

Chopin, Sonata No. 3 in B Minor, Victor Schioler (HMV ALP 1243).

(a) Cesar Franck, Andantino in G, (b) F S Kelly, Jig, live v recorded playing by Leon Goossens (oboe) and Gerald Gover (piano). EMI tape (15 ips).

Vaughan Williams, A Sea Symphony, London Philharmonic Choir and Orchestra/ Boult. Decca LXT 2907. Live organ accompaniment by Edward Fry.

Appendix 3: Published Articles by G.A. Briggs

Gramophone Needles and Grooves, *Industrial Diamond Review*, 10 (110), Jan 1950, 11–18

The Loudspeaker and the Ear, *High Fidelity*, 1(3) (Winter 1951) 17–21

Response Curves, *High Fidelity*, 1(4) (Spring 1952) 66–74

Room Acoustics, *High Fidelity*, 2(1) (Summer 1952) 69–74

The Loudspeaker, *High Fidelity*, 2(2) (Sep–Oct 1952) 39–43

Enclosures for Loudspeakers, *High Fidelity*, 3(4) (Sep–Oct 1953) 98–102, 126, 129

Enclosures for Loudspeakers, Part II, *High Fidelity*, 3(5) (Nov–Dec 1953) 97–100

Enclosures for Loudspeakers, Part III, *High Fidelity*, 3(6) (Jan–Feb 1954) 89–92

Enclosures for Loudspeakers, Part IV, *High Fidelity*, 4(1) (Mar 1954) 86–88

Sound Reproduction in the Royal Festival Hall, *The Gramophone*, 1955 (Feb), 464

Sound Reproduction in the Royal Festival Hall (with P. Wilson), *The Gramophone*, 1955 (Apr), 509

Mesures sur les Baffles (Measurements on Baffles), *Toute la Radio*, 22 (195 and 197), May 1955, 169–173 and July/August 1955, 255–259

Electrostatics, Watts, Realism and Concert Halls, *Audio*, 41(2), Feb 1957, 26, 28, 58–59.

All about Audio and Hi-Fi, *Radio and TV News*, May 1957 – Mar 1958:

Part 1: The Listening Ear, May 1957, 41–42

Part 2: Room Effects, Jun 1957, 38–39

Part 3: Room Resonance and Stereo, Jul 1957, 34–35

Part 4: Testing Loudspeakers, Aug 1957, 40–41, 95–97

Part 5: Checking Speaker Performance, Sep 1957, 66–68, 104

Part 6: Electrostatic Speakers and Transient Response, Oct 1957, 63–65, 106–107

Part 7: Transient and Directional Effects, Dec 1957, 60–61, 192

Part 8: Speaker Power and Efficiency, Feb 1958, 56–57, 146–147

Part 9: Speaker Mounting, Mar 1958, 64–65, 120

Fourth Festival Hall Concert (with P. Wilson), *The Gramophone*, July 1959, 81

Polystyrene Diaphragms: Absorbing Resonances in Shallow Enclosures, *Wireless World*, Jan 1962, 44–45

Appendix 4: Wharfedale Products 1932–1978

This listing is derived from a database assembled during the late 1970s by Les Halliwell of the Rank Wharfedale service department. The year that a product first appeared usually came from historical catalogue information, which may mean that a product launched towards the end of a year was actually given a 'start' date for the following year. Similarly 'finish' dates may be in error, but this is less likely. The prices before 1973 include any purchase tax. VAT replaced purchase tax in 1973 and for the items listed between * and ** it is not clear whether the price included tax. Prices following ** do include VAT. All prices have been converted to the current decimalised system.

	Type	Model	Start	Finish	Price (£)
1	Chassis (Driver)	Bronze Wharfedale	1932	1934	1.98
	Chassis	Blue Wharfedale	1933	1934	1.63
	Chassis	Cadmium Wharfedale	1933	1934	1.33
	Chassis	Golden Wharfedale	1933	1934	2.93
	Chassis	Bronze (8") relay	1934	?	not retail
	Chassis	Golden (10") relay	1934	?	not retail
	Chassis	Junior (8")	1934	1936	1.63
	Extension speaker	Bijou	1934	1956	1.98
	Extension speaker	De Luxe	1934	1936	2.62
10	Extension speaker	Nubian	1934	1936	2.02
	Extension speaker	Rexine Junior	1934	1936	1.93
	Extension speaker	Type E Bronzian	1934	1936	1.67
	Misc	Class B unit	1934	1935	2.85 + valve
	Transformer	Standard (output)	1934	1946	0.38
	Transformer	Universal (output)	1934	1953	0.48
	Chassis	MR relay (8")	1935	1936	not retail
	Extension speaker	Grecian	1935	1936	1.30
	Extension speaker	MR relay	1935	1936	not retail
	Misc	Hand Microphone	1935	1939	1.38
20	Misc	Truqual Volume Control	1935	1966	0.15
	Transformer	De Luxe (output)	1935	1953	0.75
	Chassis	Standard (8")	1936	1950	1.18
	Chassis	Twin Cone Auditorium	1936	1937	4.50
	Extension speaker	Bronzian	1936	1956	3.28
	Extension speaker	Console	1936	1940	7.00
	Extension speaker	Coronet	1936	1941	2.48
	Extension speaker	Oval Type Relay	1936	1937	0.98
	Extension speaker	Ring Type Relay	1936	1937	1.25
	Misc	Voluphone	1936	1939	1.98

Type	Model	Start	Finish	Price (£)
Extension speaker	Corner	1937	1940	2.75
Extension speaker	De Luxe	1937	1950	4.75
Extension speaker	Meritor	1937	1947	1.55
Transformer	Service	1937	1940	0.30
Chassis	6"	1938	1945	0.98
Chassis	Portland (10")	1938	1947	5.25
Extension speaker	Console Junior	1938	1940	3.28
Extension speaker	Gem	1938	1947	1.28
Extension speaker	Moderne	1938	1941	1.35
Extension speaker	W66	1938	1939	3.38-6.75
Misc	Remote Volume Control	1938	?	0.45
Chassis	5"	1939	1945	1.00
Chassis	Coronet (8")	1939	1941	1.40
Extension speaker	Langham	1939	1940	8.25
Misc	Speaker Switch	1939	1964	0.18
Chassis	10" Bronze	1940	1962	1.90
Extension speaker	Louvre	1940	1941	3.68
Factory	Factory Speaker, 8 or 10"	1940	1946	1.83, 1.93
Transformer	Class B Driver	1940	1946	0.38
Transformer	CT3 (output)	1940	1963	0.21
Transformer	Multi-ratio Service	1940	1942	0.33
Transformer	OP3 (output)	1940	1963	0.21
Transformer	Permalloy LF	1940	1946	0.31
Transformer	Permalloy QPP	1940	1946	0.33
Transformer	QPP Service	1940	1942	0.38
Transformer	Service LF	1940	1942	0.35
Transformer	Type P (output)	1940	1963	0.27
Chassis	Midget (3½")	1941	1955	1.43
Extension speaker	Rexdale	1941	1942	1.98
Extension speaker	Rexine Gem	1941	1942	1.45
Factory	Flare loudspeaker	1941	?	6.75
Chassis	W12	1942	1956	7.50
Non-Audio	Polishing Outfit	1942	?	0.25
Non-Audio	W35 Table	1942	?	1.38
Non-Audio	W35N Table	1942	?	1.25
Non-Audio	W36 Table	1942	?	1.63
Non-Audio	W36C Table	1942	?	1.63
Non-Audio	W37 Table	1942	?	1.63
Non-Audio	W40 Table	1942	?	1.98
Non-Audio	W45 Table	1942	?	2.38
Non-Audio	W50 Table	1942	?	1.98
Extension speaker	W50S Table speaker	1942	?	4.13
Non-Audio	W55 Table	1942	?	2.38

Appendices

	Type	Model	Start	Finish	Price (£)
	Non-Audio	W56 Table	1942	?	2.98
	Transformer	GP8 (output)	1942	1965	0.48
	Transformer	W12 (output)	1945	1965	0.88
	Chassis	W15 (15")	1946	1956	12.00
	Chassis	W5	1946	1953	1.38
	Chassis	W6	1946	1949	1.55
	Extension speaker	Tyny	1946	1950	2.38
80	Factory	Factory/Bronze	1946	1956	From 2.75
	Factory	Factory/Golden	1946	1956	From 3.75
	Factory	Factory/W12	1946	1956	From 6.75
	Chassis	W10/CS	1947	1951	7.00
	Chassis	W12/CS	1947	1956	7.25
	Cabinet system	2-way corner unit	1947	1950	48.50
	Misc	¼ section crossover	1947	1963	3.75
	Misc	Choke-type V.C.	1948	1953	1.38
	Misc	Switch Box	1948	?	2.98
	Cabinet system	Varitone	1949	1949	From 9.50
90	Cabinet system	Super 8/CS Reflex	1949	1949	16.25
	Chassis	Super 8	1949	1956	3.25
	Chassis	Super 8/CS	1949	1956	3.75
	Chassis	Super 12	1949	1949	12.25
	Chassis	Super 12/CS	1949	1949	12.75
	Chassis	W15/CS	1949	1956	11.50
	Chassis	Golden/CS	1949	1950	5.30
	Extension speaker	Bantam	1949	1950	2.25
	Extension speaker	De-Luxe Baffle	1949	1949	9.25
	Extension speaker	Gem Baffle	1949	1950	3.44
100	Extension speaker	Sylvan Baffle	1949	1951	5.00
	Extension speaker	W10/CS Baffle	1949	1953	11.00
	School	School Baffle	1949	1962	From 8.00
	School	W12/Reflex	1949	1959	19.50
	School	WM10	1949	1950	7.63
	School	WM8	1949	1959	7.13
	Transformer	W15 (output)	1949	1966	3.00
	Cabinet system	Sand Filled Corner Baffle	1950	1956	53.75
	Chassis	Super 8/CS/AL	1950	1956	4.25
	Chassis	Golden CSB	1950	1956	4.50
110	Chassis	Super 12/CS/AL	1950	1956	12.00
	Extension speaker	Super 8/CS Baffle	1950	1953	7.10
	Misc	Microgroove Equaliser	1950	1955	2.50
	School	Essex Grey	1950	1966	6.88
	Chassis	Super 5	1951	1956	4.63
	Extension speaker	Super 5 Baffle	1951	1952	7.13

A Pair of Wharfedales: The Story of Gilbert Briggs and his Loudspeakers

Type	Model	Start	Finish	Price (£)
Cabinet system	3-Speaker System	1953	1963	73.50
Chassis	8" Bronze/AL	1953	1956	3.53
Extension speaker	Reflex Baffle	1953	1966	3.00
Extension speaker	Bronze Baffle	1953	1966	6.33
Misc	HS/CR3/2 Crossover	1953	1962	From 8.50
Misc	400/CR3/2 Crossover	1953	1963	12.50
Cabinet	Twin Treble Cabinet	1954	1963	8.50
Cabinet	R-J Cabinet	1954	1957	9.50
Cabinet	RJ-2 Cabinet	1955	1957	10.50
Cabinet system	Golden /CSB Corner Panel	1955	1969	12.25
Cabinet system	FS/3-way Cabinet	1955	1956	90.00
Chassis	10" Bronze/CSB	1955	1957	5.57
Cabinet	Super 3 Cabinet	1956	1960	From 3.50
Cabinet	Bronze Reflex Cabinet	1956	1957	14.00
Cabinet	FS/10 Reflex Cabinet	1956	1957	26.00
Cabinet system	SFB/3	1956	1962	39.50
Chassis	Super 3	1956	1960	7.00
Cabinet	AF10 Reflex Cabinet	1957	1960	15.75
Cabinet system	R-J8	1957	1958	11.50
Chassis	8" Bronze/FS/AL	1957	1963	4.55
Chassis	Super 8/FS	1957	1963	7.00
Chassis	Super 8/FS/AL	1957	1962	7.35
Chassis	10" Bronze/FSB	1957	1963	5.56
Chassis	Golden/FSB	1957	1962	8.75
Chassis	W10/FSB	1957	1961	12.12
Chassis	W12/FS	1957	1960	10.25
Chassis	Super 12/FS/AL	1957	1960	17.50
Chassis	W15/FS	1957	1960	15.50
Transformer	WMT1 (matching)	1957	1969	0.68
Cabinet	AF12 Reflex Cabinet	1959	1963	24.50
Cabinet system	W2, W2-Dovedale in 65	1959	1968	29.50
Cabinet system	W3	1959	1966	39.50
Cabinet system	W4	1959	1966	49.50
Chassis	Coaxial 12	1959	1963	25.00
School	P8	1959	1962	6.30
School	LS/7	1959	1962	40.00
Cabinet	PST/8	1959	1966	10.50
Cabinet	Column 8	1958	1960	21.75
Chassis	8/145	1958	1966	7.00
Transformer	SM1 (stereo mixing)	1960	1963	1.50
Cabinet system	Airedale	1961	1969	65.00
Chassis	W12/RS	1961	1962	10.50
Chassis	RS12/DD	1961	1969	11.50

	Type	Model	Start	Finish	Price (£)
	Chassis	Super 12/RS/DD	1961	1969	17.50
160	Chassis	W15/RS	1961	1969	17.50
	Cabinet system	Slimline 2	1962	1969	22.50
	School	LS/6B	1962	1962	58.00
	Chassis	Super 8/RS/DD	1962	1973	7.10
	Chassis	W12/EG	1962	1966	10.50
	Chassis	W15/EG	1962	1965	17.50
	Cabinet	EG12	1962	1966	From 8.75
	Cabinet	EG15	1962	1965	From 12.50
	Kit	9" pipe column	1962	1968	3.75
	Kit	11" pipe column	1962	1968	4.90
170	Chassis	PST/4	1962	1966	3.58
	Chassis	8" Bronze/RS	1962	1963	4.23
	Chassis	10" Bronze/RS	1962	1963	5.16
	Chassis	Golden 10/RS/DD	1962	1969	8.78
	Chassis	Super 10/RS/DD	1962	1973	12.18
	Chassis	W12/RS/PST	1962	1965	10.75
	Chassis	W12	1963	1963	?
	Cabinet system	Bookshelf 2	1963	1965	16.50
	Transformer	WMT2 (matching)	1963	1969	0.68
	Chassis	8" Bronze/RS/DD	1964	1973	3.79
180	Chassis	10" Bronze/RS/DD	1964	1969	4.64
	Misc	HS/400/3 Crossover	1964	1969	6.50
	Misc	QS/800 Crossover	1964	1969	3.25
	Cabinet system	Linton	1965	1967	18.63
	Cabinet system	Dalesman	1965	1967	25.50
	Misc	QS/3000 Crossover	1966	1969	3.25
	Public Address	PA30 portable system	1966	1971	69.50
	Extension speaker	XT30 (for PA30)	1966	1968	27.50
	Transformer	PA5 (matching)	1966	1968	0.75
	Transformer	PA10 (matching)	1966	1968	1.75
190	Amplifier	GP20	1966	1968	39.50
	Amplifier	GP50	1966	1968	68.50
	Public Address	GP503 speaker	1966	1968	6.88
	Public Address	GP505 speaker	1966	1968	7.88
	Public Address	GP545 speaker	1966	1968	8.25
	Public Address	GP575 speaker	1966	1968	32.50
	Public Address	GP585 speaker	1966	1968	65.00
	Public Address	GP600/B speaker combo	1966	1968	64.00
	Public Address	GP510 cine speaker	1966	1967	19.00
	Chassis	W12/FRS	1966	1969	11.75
200	Cabinet system	Teesdale	1966	1968	From 52.50
	Cabinet system	Super Linton	1967	1972	41.20 pr

Type	Model	Start	Finish	Price (£)
Cabinet system	Denton	1967	1971	32.53 pr
Electronics	WHF20 amplifier	1967	1968	75.00
Electronics	WFM-1 tuner	1967	1968	53.38/56.68
Electronics	WTT-1 turntable (Thorens)	1967	1968	?
Kit	Unit 3	1968	1973	10.50
Cabinet system	Melton	1969	1971	29.50
Cabinet system	Rosedale	1969	1973	55.00
Cabinet system	Dovedale III	1969	1971	39.50
Cabinet system	Triton	1970	1971	55.00
Cabinet system	Aston	1970	1971	43.00
Kit	Unit 4	1970	1973	16.00
Kit	Unit 5	1970	1973	23.50
Electronics	100.1 Receiver (tuner-amp)	1970	1972	131.25
Electronics	DC-9 cassette recorder	1971	1973	110.00
Electronics	Linton amplifier	1971	1973	60.00
Electronics	Linton tunrtable (W30)	1971	1975	33.00
Cabinet system	Denton 2	1971	1975	39.90 pr
Headphones	DD1	1971	1977	11.00
Cabinet system	Linton 2	1971	1975	49.90 pr
Cabinet system	Triton 3	1971	1974	65.00 pr
Cabinet system	Melton 2	1971	1974	35.00
Cabinet system	Dovedale 3	1971	1975	45.00
Electronics	Linton tuner-amplifier	1973 *	1974	107.66 *
Electronics	Denton amplifier	1974	1975	52.80
Electronics	Linton amplifier Mk2	1974	1975	69.39
Electronics	Denton tuner-amplifier	1974	1975	99.50
Electronics	Linton tuner-amplifier Mk2	1974	1976	115.60
Electronics	WHD-20D cassette deck	1974	1975	105.60
Cabinet system	Denton 1	1974	1976	32.00 pr
Cabinet system	Glendale 3	1974	1975	79.55 pr
Cabinet system	Kingsdale 3	1974	1975	81.96
Kit	Linton 2 kit	1974 **	1976	33.45 pr**
Kit	Glendale 3 kit	1974	1976	56.24 pr
Kit	Dovedale 3 kit	1974	1977	95.49 pr
Headphones	Isodynamic	1974	1979	27.33
Cabinet system	Denton 2XP	1975	1978	57.29 pr
Cabinet system	Linton 3XP	1975	1978	79.69 pr
Cabinet system	Glendale 3XP	1975	1978	107.00 pr
Cabinet system	Dovedale SP	1975	1977	182.71 pr
Cabinet system	Airdale SP	1975	1977	274.10 pr
Cabinet system	Chevin XP	1975	1978	41.50 pr
Electronics	SXP tuner-amplifier	1975	?	229.85
Electronics	XP casette recorder	1976	?	

Type	Model	Start	Finish
Kit	Denton 2XP kit	1976	1978
Kit	Linton 3XP kit	1976	1978
Kit	Glendale 3XP kit	1976	1978
Cabinet system	E50	1977	1981
Cabinet system	E70	1977	1981
Cabinet system	Dovedale SP2	1977	1979
Cabinet system	Teesdale SP2	1977	1980
Cabinet system	Chevin XP2	1978	1980
Cabinet system	Denton XP2	1978	1980
Cabinet system	Shelton XP2	1978	1980
Cabinet system	Linton XP2	1978	1980
Cabinet system	Glendale XP2	1978	1980

Notes and References

References to books by G.A. Briggs use the following abbreviations:

AA: *Audio and Acoustics*
AB: *Audio Biographies*
AH1, AH2: *Aerial Handbook, Editions 1 and 2*
AMP: *Amplifiers*
AYH: *About Your Hearing*
AZ: *A to Z in Audio*
CH: *Cabinet Handbook*
HF: *High Fidelity*
LS1-5, *Loudspeakers, Editions 1-5*
MIA: *Musical Instruments and Audio*
MLS: *More about Loudspeakers*
PPS: *Pianos, Pianists and Sonics*
SH : *Stereo Handbook*
SR1-3: *Sound Reproduction, Editions 1-3*

Chapter 1

1. Briggs was professor of geometry at Gresham College, London when Napier published his theory of logarithms in 1614. He realised its significance for easing the difficult calculations involved in his work, which included the construction of astronomical tables for use in navigation. He decided, however, that using the base 10 rather than Napier's base e in the construction of log tables would be of greater practical use and, with Napier's support, he spent several years calculating the logs of thousands of natural numbers and all the trigonometric functions, often to fourteen decimal places. These became known as 'common' or Briggsian logarithms.

2. AB 134

Chapter 2

1. LS5 9

2. G. Dalby, *BODKIN* (Journal of the Bradford Family History Society), **40**, 20 (1995)

3. *Lynn News,* February 1894

4. AB 11 and 68

5. All the information about the school in this section is taken from R. Taylor, A. Kafel and R. Smith, *Crossley Heath School*, Tempus Publishing, Stroud (2006)

6. Facsimile, recreated from a Crossley and Porter's archive document dated 1904. Scan of original

courtesy of Crossley Heath School.

7. AB 68,   8. AB 12,   9. PPS 13

Chapter 3

1. LS5 9

2. Bill Lang, 'Wharfedale' Rules the Radio Waves, *The Dalesman*, 1966 (July), 290

3. Background information about the Bradford textiles industry in this chapter is largely derived from: M. Keighley, *Wool City*, G. Whitaker and Co Ltd, Ilkley, 2007

4. AB 12

5. *Yorkshire Observer*, 4 July 1951

6. PPS 145

7. *Bradford Telegraph and Argus*, 1 February 1941

8. Saltaire village, together with the Salt's Mill complex, was designated a UNESCO World eritage Site in 2001.

9. For a detailed history of early recording see A. Millard, *America on Record: A History of Recorded Sound*, Second Edition, Cambridge University Press, 2005

10. For the early days of radio see K. Geddes and G. Bussey, *The Setmakers*, BREMA, 1991

11. AB 121,   12. AB 169,   13. LS5 11,   14. LS5 10

Chapter 4

1. LS5 10,   2. LS5 11,   3. AB 50,   4. LS5 13,   5. AB 238

6. *Wireless World*, 19 August 1932

7. Bill Lang, 'Wharfedale' Rules the Radio Waves, *The Dalesman*, 1966 (July), 291

8. The Hall, built in 1904, and capable of seating over 2000 people, was almost destroyed by a fire in 1996, but the impressive façade, with its stained glass window, has been restored and now fronts a development of luxury apartments.

9. AB 14

10. Later Price became well known through his textbooks, written with Harold Buckingham: *Principles of Electronics* (Cleaver-Hume, 1953), *Principles of Electrical Measurements* (English Universities Press, 1955) and *Electro-technology for National Certificate Courses Vol III* (English Universities Press, 1959) which, with later editions, were all current until the 1970's.

11. LS5 12, 12. AB 93, 13. AB 178, 14. LS5 14, 15. AB 13, 16. LS5 18, 17. AB 173

18. AH1 123, 19. LS5 17, 20. AB 99, 21. LS5 16, 22. PPS 13

Chapter 5

1. In a contribution, from the floor, to a meeting of the British section of the Audio Engineering Society held as 'G.A. Briggs, An Appreciation' at the Royal Institution, London, 29 May 1974 (tape recorded for Gilbert).

2. LS5 18, 3. LS5 19, 4. LS5 20, 5. LS5 17, 6. AB 292, 7. AB 15,16

8. Quoted in *Microbes and Men*, R. Reid, BBC Publications, London, 1974

Chapter 6

1. LS5 46, 2. LS5 12, 3. LS5 62, 4. LS5 24, 5. LS5 22, 6. LS5 27, 7. LS1 7, 8. LS5 28

9. Wharfedale catalogue, September 1949

10. As retold by Raymond Cooke in his invited contribution to a meeting of the British section of the Audio Engineering Society held as 'G.A. Briggs, An Appreciation' at the Royal Institution, London, 29 May 1974 (tape recorded for Gilbert).

11. HF 156

12. From an interview with John Borwick, *The Gramophone*, 1982 (March), 1311

13. www.kef.com/history/1970/emipresentation.pdf, accessed 03.11.

14. BBC Written Archive file R57/9 –Technical Recording: Amplifiers and Loudspeakers 1944-55

15. AB 248

16. Agnes H. Watts, *Cecil E. Watts: Pioneer of Direct Disc Recording*, 74 (1972) (self-published)

17. AB 118

Notes and References

Chapter 7

1. LS5 297, 2. LS5 298, 3. AB 85

4. W. Joseph and F. Robbins, The R-J Speaker Enclosure, *Audio Engineering*, Dec 1951

5. W. Joseph and F. Robbins, Practical Aspects of the R-J Speaker Enclosure, *ibid*, January 1953

6. Wharfedale catalogue dated August 1954

7. AB 17

8. This trumpet version reached number 1 in the singles chart on 5 January 1954 and stayed there for nine weeks.

9. AB 11

10. In an interview with John Borwick, *The Gramophone*, 1982 (March), 1311

11. LS5 24, 12. LS5 235, 13. HF 188, 14. LS5 171

15. Percy Wilson, *The Gramophone*, May 1957, 476

16. AB 22

17. In his invited contribution to a meeting of the British section of the Audio Engineering Society held as 'G.A. Briggs, An Appreciation' at the Royal Institution, London, 29 May 1974 (tape recorded for Gilbert).

18. Column Eight leaflet, 1959

19. LS5 14, 20. AB 51, 21. AB 64, 22. AB 264

23. *Radio and TV News*, May 1957, 40

Chapter 8

1. *Billboard Magazine*, 9 Feb 1959

2. AB 301, 3. LS5 191

4. Wharfedale leaflet, April 1959

5. AB 209, 6. LS5 47 and ML 22

7. In an interview with John Borwick, *The Gramophone*, 1982 (March), 1311

8. S. Chalke and D. Hodgson, *No Coward Soul: The Remarkable story of Bob Appleyard*, Fairfield Books, Bath, 2003

9. K. Kessler and A. Watson, *KEF: 50 Years of Innovation in Sound*, GP Acoustics International, 2011

10. www.kef.com/history/en/1960/k1.asp, accessed 05.11

11. The bass unit had a flat radiused rectangular diaphragm (10x14") moulded from expanded polystyrene, with a rear surface tapered in the form of a shallow pyramid, coated on both sides with a 0.001" layer of aluminium. The mid-range unit had an elliptical cone shaped diaphragm of similar sandwich construction. The tweeter was a Melinex (polyester) spherical dome radiator. [MLS, 122]

12. In 1961 Gilbert wrote the following: *We all know him* [Donald Aldous] *as Technical Editor of Record Review and I was astonished a few months ago to see his face in the audience at a Saturday TV Juke Box Jury event, which used to fascinate me by the revolting records which topped the poll. When I wrote and chided him on the way technical writers appeared to spend their spare time in London, he replied neatly that he in turn was astonished, by the letters he had received, how many so-called highbrows watched the programme.* [AB, 28]

13. *Audio and Record Review*, September 1962

14. CH 108

15. *Bradford Telegraph and Argus*, 10 January 1964

16. AB 235,   17. AZ 115

18. In his contribution, from the floor, to a meeting of the British section of the Audio Engineering Society held as 'G.A. Briggs, An Appreciation' at the Royal Institution, London, 29 May 1974 (tape recorded for Gilbert).

Chapter 9

1. S. Spicer, *Firsts in Hi-Fi, The History and Products of H J Leak & Co Ltd*, Audio Amateur Press, USA, 2000

2. Invented by Don Barlow in 1958 whilst working for Alcan, this light but very stiff cone was made from a thick section of expanded polystyrene sandwiched between outer skins of aluminium. Having published his research, Barlow was induced to move to Leak and develop the speaker, which was launched in 1961. At the time of the Rank takeover there were two Leak loudspeakers

produced at Downham Market: the Sandwich Two-way Mk 2 (bass driver and tweeter both sandwich construction) and the Mini-Sandwich. Leak had always sub-contracted cabinet manufacture so, following the takeover, the assembled sandwich units were shipped to Idle for incorporation into cabinets made there. Don Barlow, who had parted company with Leak in 1968, was re-engaged by Rank-Leak at Bradford to ensure that sandwich cone expertise would be available when the manufacture of drivers was also transferred to Idle.[1]

3. Geoffrey Horn, *Audio Record Review*, August 1969

4. 'Audio Talk', *The Gramophone*, December 1969, 1062

5. Castle soon established their own high reputation in the loudspeaker market and continued to produce hand-crafted cabinets, in the old Wharfedale tradition, when others moved to highly automated production methods. They ended up supplying bespoke cabinets to other firms. They also took over the repair and renovation of Wharfedale speakers over 6 years old, which Gilbert had always undertaken as part of his customer-relations focus, when the Wharfedale owners stopped this in about 1984 (Gilbert's policy had been to keep spares for 10 years after a model ceased production). Castle Acoustics folded in 2007 but the brand was acquired by IAG.

Chapter 10

1. Bill Lang, *The Dalesman*, July 1966, 290

2. AB 16

3. Adrian Hope, *Hi-Fi for Pleasure*, December 1975, 45

4. In the short account, written entirely from memory at least 30 years after the event, Sargent gives the date as 'winter 1972'. But he also mentions filming an anechoic chamber at Idle, which was not built until about 1976. Discussions with the author failed to resolve the date question.

5. Robert Parker found fame for his digital re-mastering of early jazz recordings dating back to the 1920s. Commissioned by the Australian Broadcasting Corporation, his long-running radio series *Jazz Classics in Digital Stereo* started in May 1982 and was later aired by the BBC on Radio 2.

6. Stefan Sargent, *Digital Video*, December 2007

7. The term was coined in 1927 by H.A. Hartley, who later regretted the fact, see Chapter 4.

8. Bert Whyte, *Audio*, January 1978, 66

9. Tribute at Gilbert's funeral, 16 January 1978

10, 11. In his invited contribution to a meeting of the British section of the Audio Engineering

Society held as 'G.A. Briggs, An Appreciation' at the Royal Institution, London, 29 May 1974 (tape recorded for Gilbert).

Chapter 11

1. AB 179,   2. HF 138,   3. AZ 113

4. G.A. Briggs, contribution to Agnes H. Watts, *Cecil E Watts: Pioneer of Direct Disc Recording*, 105 (1972) (self-published)

5. HF 142,   6. AZ 114,   7. AB 85

8. In Peter Walker's invited contribution to a meeting of the British section of the Audio Engineering Society held as 'G.A. Briggs, An Appreciation' at the Royal Institution, London, 29 May 1974 (tape recorded for Gilbert).

9. AB 82,   10. AA 154,   11. AB 80,   12. AB 83

13. *Wireless World*, Dec 1954

14. Sound Reproduction in the Royal Festival Hall, Part 1, G.A. Briggs, *The Gramophone*, February 1955, 424

15. Sound Reproduction in the Royal Festival Hall, Part 2, P.J. Walker, *ibid*, March 1955, 467

16. Sound Reproduction in the Royal Festival Hall, Summary, G.A. Briggs and P. Wilson, *ibid*, April 1955, 509

17. AB 304

18. *The Times*, 23 May 1955

19. *High Fidelity*, August 1955

20. *Harpers Magazine*, December 1955

21. Draft copy of an article by G.A. Briggs entitled 'Carnegie Hall – And After', dated 27 October 1955 and intended for, if not submitted to, *High Fidelity* magazine.

22. *Hi-Fi News*, June 1956

23. From announcement of second Carnegie Hall concert, *High Fidelity*, September 1956.

24. LS5 302,   25. AZ 114,   26. AA 159

27. Gilbert frequently railed about the overpowering volumes used in cinema sound.

28. AB 63,   29. LS5 202,   30. AB 229,   31. SH 129

32. *The Times*, 11 May 1959

33. SH 131,   34. LB5 19,   35. AZ 115,   36. HF 157

37. In a contribution, from the floor, to a meeting of the British section of the Audio Engineering Society held as 'G.A. Briggs, An Appreciation' at the Royal Institution, London, 29 May 1974 (tape recorded for Gilbert).

Chapter 12

1. Agnes H. Watts, *Cecil E Watts: Pioneer of Direct Disc Recording*, 105 (1972) (self-published)

2. AB 320,   3. SR3 76,   4. SR3 361,   5. LS5 31,   6. AB 184,   7. LS5 32

8. In Raymond Cooke's invited contribution to a meeting of the British section of the Audio Engineering Society held as 'G.A. Briggs, An Appreciation' at the Royal Institution, London, 29 May 1974 (tape recorded for Gilbert).

9. Radio Leeds, *Oasis*, 21 September, 1975

10. J. Moir, *High Quality Sound Reproduction*, Second Edition, Chapman and Hall (1961)

Index

A

About Your Hearing 242, 314
Acoustic Research Inc. 171
Acoustical Manufacturing Co. see Quad
Admiralty 80
Aerial Handbook
  editions 1–2 312
    first edition 208, 241
    second edition 242
Airedale Works (Apperley Bridge) 222
Aldous, Donald 312, 348
Altech 291
Ambassador Radio 67, 72
Ampex 291
Ampleforth College 290
Amplifiers 129, 130, 178, 299–302
Appleyard, Bob 187
Ashley, Ted 222
A to Z in Audio 183, 211, 304–306
Audio 250
Audio and Acoustics 194, 208, 212, 311
Audio and Record Review 197
Audio Biographies 189, 194, 306
Audio Engineering 143–145
Audio Engineering Society 242–247, 291
Audio Fair
  London 152
  New York 143, 152, 160, 172, 211, 253, 274
  Washington D.C. 171
Audiophile recording 259, 263
Automatic Telephone Manufacturing Co. 190

B

Bakers Selhurst 36
Baldock, Rex 242, 243, 250
Bale, Harry 215, 248
Balls, I. John 161, 174, 182–188
Bank, Graham 232
Barber, William C. 13
Barlow, Don 348
Barlow, Gordon 237
Barrell, W.S. 262

Bartlett's *Familiar Quotations* 293
BBC 64, 70, 71, 80, 130, 143, 149, 150, 160, 211, 254–269, 288, 295, 312
Beardsley, Roger 246, 248
Beaumont, Frank H. 72, 112, 298, 301
Bell Labs 34, 35
Berliner, Emile 33
Biggs, E. Power 273, 281
Birmingham and Midland Institute 282
Black Dyke Mills 6
Blackburn, Harold 285, 288
Blech, Harry 277
Blubberhouses 91, 213
Blue Spot 36
Blumlein, Alan 162
Books by G.A. Briggs 293
    see also individual titles
Borwick, John 211
Boyen, Father Bernard 290
Bradford
  Dyers Association 105
  Mechanics Institute 19
  Radio Society 44
  St. George's Hall 146, 151, 159, 254–258, 265–270
  Technical College 17
  Telegraph and Argus 201, 256, 257
  Textile industry 17–25, 46
  Bradford Observer 5
Bradford Road site 97, 137, 227, 229
Bradford, Robert W. 263
Brans, P.H. Ltd 294
Brick-built corner unit 168
Briggs, Asa 6
Briggs, Bernard J. 5, 17, 18, 23, 37
Briggs, Brigg 6
Briggs, Claris 73, 78, 91, 95–98, 123, 167
Briggs, D. Edna 41–46, 63–69, 81, 89, 113, 133–137, 146–148, 167, 241, 250, 306
  farm 212
  goat breeding 95
Briggs, Emma 20
Briggs, G.A. Ltd 241

352

Index

Briggs, Gilbert A.
  AES award 243
  AES recognition 242, 291, 306
  ancestors 3
  betting 212
  birth 3
  books 52, 65, 194, 208
  concert-demonstrations 141–146, 151–157, 168, 208, 248, 253
  death 250
  family life 63, 89
  financial crisis 31
  first book 112
  funeral 250
  grandchildren 135, 167, 189, 210, 211, 242, 303
  harmonium 210
  health 89, 149, 298
  house addresses 9, 95
  India trips 133
  lecture on the East 20
  lecture-demonstrations 59, 143, 253
  music 63, 91, 113, 131
  musical lectures 23
  piano lessons 63, 132
  pianos 15, 65, 113, 132
  radio broadcasts 211, 246, 307
  reading 131, 192, 241
  religion 9, 63
  retirement 201, 208, 237
  romance 27
  schooling 9
  sense of humour 44, 118, 167, 185, 192
  textiles export merchant 17
  USA trips 113, 143, 152, 157–160, 169, 274
  wedding 27
  writing method 303
Briggs, Henry 3
Briggs House 249
Briggs, Jane Anne 167
Briggs, Mabel 13, 73, 91, 95, 167
Briggs, Ninetta M. 43, 63–65, 89, 90–95, 131–135, 167, 210, 212
Briggs, Peter R. 39, 43, 63–65, 89–93, 189, 250
  tragic death 93
Briggs, Phineas 11, 91

Briggs, Valerie E. 43, 63, 91, 114, 131, 135, 147, 174, 209, 210, 237, 250, 265, 277, 282, 286, 306
Briggs-Wharfedale Studio 237
Brighouse 67
British Industries Corporation 115, 143–152, 157–160, 171–173, 180–183, 208–216, 220, 223, 258, 274, 281, 282, 293, 294
  Achromatic Systems 208, 215
British Sound Recording Association 151, 194, 253, 259, 270, 277, 284
  exhibition 129, 154, 253
Broadley, Ezra R. 47, 52, 56, 62, 71, 97, 104, 105, 122, 124, 139, 164, 165, 191, 192, 232, 248
Byrne, Peter 246

C

Cabinet Construction Sheet 157, 310
Cabinet Handbook 40, 51, 194, 201, 310, 314
Campoli 277
Canby, Edward T. 276
Carduner, Bill 116
Carduner, Leonard 115, 116, 144, 158, 161, 164, 171–173, 182, 281
Carnegie Hall 151–157, 211, 265, 273–276, 290
  concert October 1955 273
  concert October 1956 281
Carvalho, R.V. de 168
Castle Acoustics Ltd 228, 349
Celestion 69
Chapman, Leonard 187, 201
Churchill, Winston S. 89, 303
Clare, Kenneth 282, 283
Clayton 3
Claytonian entertainers 20
Clifford Briggs Ltd (printers) 112
Clifton, David 250
Collinson, John D. 219, 223–228, 232, 259, 260–266, 274, 279, 282–288
Colston Hall, Bristol 288
Columbia 122, 256, 263, 273–277, 282
Concert-demonstrations
  cost 290
  stereo 266, 282, 285

353

Conly, John M. 144
Cooke, Raymond E. 126, 129, 149–201, 213, 232, 243, 247, 251, 280–290, 310
  books collaboration 297, 301–304
Cosens, C.R. 298
Countess of Harewood 116
Crabtree, Boyd 190
Craggs, Geoffrey 249
Crossley and Porter's School 11, 73, 243
Crowhurst, Norman H. 298, 303
Crystal set 34
Cunnington, Alan 228
Curzon, Clifford 92
Cutforth, Marguerite 211

D

Daily Telegraph 212, 241
Daniels, Danny 281
Dart, Thurstan 259, 271
Darwins Ltd 100
Davis, John H. 161, 174, 222, 238, 240, 250
Dawson, Dorothy P. 190–194, 237–241, 250, 313
Dawson, F. Keir 49–56, 112, 113, 164, 298, 301
Dawson, Julian 290
Decca 122, 256–260, 295
Devereux, Frederick L. 61, 288, 301
Direct disc recording 106, 122
Disc recording development 122, 159
Dolby Laboratories 291
Dolby, Ray 243, 291
Downes, Ralph 260, 264, 271–280, 285
Dublin Gramophone Society 290
Dye, Peter 208
Dyer, John H. 271

E

Eaton Audio Fitments 198, 222
Edison, Thomas 33
Electric guitars 196
Electrovoice 172
Elsey, Barbara 284
EMI 168, 183, 257–272, 277, 285, 288, 295
Emsley, Arthur 9
Emsley, Emma 9
Emsley, Lloyd 113
Emsley, Luther B. 161, 192, 201
Emsley, Mary Ann 5, 9, 18
Emsley, Phineas 9
Escott, W.S. (Bill) 97, 98, 122–124, 130, 137, 158, 173, 175, 182, 184, 187–201, 228, 238, 240, 255
Essex Education Committee 178, 301

F

Fine Arts Quartet 172
Fitton, R. Noel 67, 72
Foster, Michael G. 312
Fountain, Guy 69
Fowler, Charles 134, 144
French, Cyril 69
Fryer, Peter 230, 232

G

Garner, Alex 226, 232, 233, 243–245
Garner, H.H. 178, 301
Garrard Ltd 116, 147, 157, 260
Gasman, Arthur 144, 145, 281
Gernsback Library Inc. 306
Gigliotti, Anthony 273
Gilels, Emil 288
Globe Mills 228
Goldsmiths' Choral Union 271
Goodmans 42, 102, 167
Goodmans Industries 260
Goossens, Leon 190, 211, 243, 266, 270, 277, 282–288, 290
Gott, Arthur W. 92
Gould, Morton 281
Gover, Gerald 283, 285, 288
GPO 70, 71, 73, 80
Graham, Ann 237
Gramophone 33, 35
Gramophone Company 33, 35
Graphophone 33
Groll, Kees 26, 55
Gutmann, Robert 215–223

H

Haggis, Frederick 271
Haigh, John H. 10

Halifax 11, 91
Hambro, Leonid 273
Harper's Magazine 276
Harris, Percy W. 34
Harrison, Henry C. 34, 35
Hartley, H.A. 59
Hatton, Arnold 52, 62, 97, 124, 152–158, 248–55, 272–281, 290
Hatton, Winifred 97, 157
Hayward, Richard 290
Heller, Stanislav 259, 264
Henrich, Dr J. Charles 25
Henslow, Miles 298
Hess, Myra 92
Hi-Fi for Pleasure 248
Hi-Fi Sound 222
High Fidelity
  bibliography 169
  book 150, 154, 301, 303
  magazine 134, 144, 157, 158, 263, 273, 277
Highfield Road site 215, 220–229, 234, 248
Holdsworth, Lund & Co. 17, 21–25
Holland, Frank W. 301
Hollick and Taylor 282–284
Holl, Tim 226, 232
Hope, Adrian 248
Horne, Stewart 165
Horn, Geoffrey 223
Horsburgh, George D.L. 100, 113, 114

I

Idle 88
Ilkley 91, 150
  Briggs-Wharfedale Studio 237
  Concert Club 92
  Gramophone Society 91
International Broadcasting Company 282
International Harvesters 137
Ionophone 164
Isles, Edith 123, 124, 158, 186, 190, 303

J

Jackson, Peter 232
James Nelson Ltd 27, 31
Jamieson, W.A. (Bill) 183–190, 196, 212, 219, 241, 312

Jarvis, Caleb 283
Jensen, Peter L. 39
Johnson, Eldridge 33
Johnson, Samuel 169
Jones, Malcolm 189
Jones, Mason 273
Joseph, William 145
Jowett Cars Ltd 88, 97

K

KEF Electronics 126, 188–196, 348
Kellogg, Edward W. 39, 60
Kelly, Stanley 151, 161, 169, 215, 280, 298, 303
Kent Engineering & Foundary 187
Kentner, Louis 92
King's Lynn Technical School 10, 210
KLH 171, 172
Kloss, Henry 172
Knight, C. Edgar 63, 132, 255, 266, 270, 282, 284, 290
Kokinis, Alex 249
Kramer, Milton D. 144, 145, 172

L

Ladhams, E.E. 304
Lancie, John De 273
Lang, Bill 240
Leak, Harold J. 143–148, 222, 258, 259, 281, 282, 306
Leak, H.J. and Co. 157, 182, 222, 281
Leeds–Bradford airport 238
Leeds University 282, 290
Listening test 67
Liverpool Post 284
Live versus recorded 146, 151, 152, 159, 168, 169, 254, 291
Livingstone, T.B. (Stan) 69
Lodge, Oliver 39
London Mozart Players 277
Longden, John 250
Loudspeaker
  adhesives 191
  balanced armature 34
  cone 71, 102, 116, 119
  cross over network 47, 107
  early types 36

electrostatic 151–154, 164, 223
magnet 39, 46, 52, 59, 87, 100
magnet (ceramic) 185
moving coil 39, 152
moving-iron 60
surround 104, 116, 150, 155, 158, 176, 185
transformer 40, 46, 52, 55, 62
voice coil 39, 45, 47, 52, 62, 124, 182
Loudspeakers 135
editions 1–4 169, 294, 311, 314
fifth edition 164, 172, 185, 302, 306, 310
first edition 113, 122, 123, 129, 241
Lowden, Richard W. 284
Lund, Thomas B. 189
Lympany, Moura 92

M

Macintosh, Eric 69
Magnavox 36
Maltby, Peter 232
Manchester Radio Exhibition 49
Mann, Frank 52, 62, 97, 191
Marconi, Guglielmo 33
Mart, Louise 242
Mason, Fred 123, 124, 132, 137, 161
Matthews, Denis 92, 260, 277, 279, 285
Maxwell, Joseph P. 34, 35
McGraw-Hill 114
McProud, Charles G. 211
Millward, Gareth 232, 245
Moir, James 208, 242, 243, 302, 310–314
Moiseivitsch 92
More about Loudspeakers 194, 310
Mortimer, Edmund W. 147
M.S.S. Recording Co. 295
Musical Instruments and Audio 241, 312

N

Naylor Jennings & Co. Ltd 27, 31
Northern Audio Fair 248
Nunn, Ewing D. 259, 263

O

Oasis
see Radio Leeds 246

Oats, Nellie 123, 124
Olson, Harry F. 112, 113
Oscillogram 142, 298, 312
Owen, William B. 33

P

Pachmann, Vladimir de 19
Page, Kenneth 282, 283
Parker, Robert 248
Pearch, Bob 187, 188
Pearson, Brian 233
Permanent Magnet Association 100
Philbin, Philip J. 171
Philharmonic Hall, Liverpool 282
Phonograph 33
Pianos, Pianists and Sonics 65, 260, 262, 298
Pitchford, H. John 209, 286
Pollack, Anton 39
Polystyrene
in cabinets 163
in diaphragms 187, 195
Popplewell, Kenneth 288
Price, Ernest M. 47, 70, 108, 123, 142, 165, 260, 270, 294, 298, 301
Pridham, Edwin S. 39
Proctor, Wilfred L. 169, 279
Puzzle and Humour Book 241, 293, 314
Pyett, David 91

Q

Quad 151, 154, 168, 219, 223, 255, 259–270, 281, 288
Queen Mary (widow George V) 116
Queensbury 5, 6

R

Rackman, Ronald 69
Radio
development 32
relay 55
Radio and Television News 294
Radio and TV News 164, 169, 302
Radio Leeds 246
Radio London 250
Radiolympia 45–49, 55, 116–119

Rank, J. Arthur 161
Rank Organisation 160, 161, 171–175, 182, 187, 189, 192, 201, 208, 248
  Leak aquisition 222
  logo 175
  Rank Audio Visual division 224
  Rank Hi-Fi division 234, 248
  Rank Radio International division 226, 228, 234
Rank Wharfedale
  cabinet systems 216, 230–234
  cassette recorders 226, 230, 234
  disco speakers 234
  electronics 220, 223, 226
  employee numbers 220, 229
  isodynamic headphones 227
  kits 220, 222, 226, 230, 234
  logo 216
  marketing 223, 224, 232
  named as such 216
  output 228
  public address 218
  research 219, 223, 226–232, 348
  Wharfedale Book Department 237
Read, Herbert 13
Re-entrant horn 35
Response curve 67, 71, 73, 165, 227
Reynard, Harold 192, 241
Rice, Chester W. 39, 60
Ridehalgh & Sons Ltd 27–31
River Edge (cabinets) 157
R-J enclosure 145, 157, 257
RK 36
Robbins, Franklin 145
Roberts, R.S. 208, 312
Rogers, Jim 131
Rola 70
Rola–Celestion 167
Royal Aircraft Establishment, Farnborough 284
Royal Festival Hall 147, 151, 152, 168, 169, 183, 211, 246, 264, 265, 269, 273, 291
  concert May 1955 270
  concert May 1956 280
  concert May 1959 285
  concert November 1954 257
Russell, Kenneth F. 189–196, 208, 212, 219, 223–251, 290, 312

S

Salt, Titus 13
Saltaire 29
Sand-filled panel 125, 146, 155, 168, 185
Sargent, Sir Malcolm 277
Sargent, Stefan 248
Sarsby, C.H. (Harry) 42, 46, 47
Scaife, Steve 232
Scharrer, Irene 92
Schoenbach, Sol 273
Selleck, Phyllis 271
Sharp, Ronald 245
Shorter, D.E.L. 70, 71, 149, 160, 230, 254
Siemens, Werner von 39
Slade, Hector V. 147
Smith, Albert 55, 89, 158, 160, 192, 196, 216, 238, 273, 274, 290, 293
Smith, Cyril 92
Smyth, J.B. (Barney) 143, 183, 253–259, 274
Societé des Editions Radio 298, 303
Sound (BBC programme) 211
Sound Reproduction 135
  editions 1–3 169, 295, 314
  first edition 119, 122, 129, 134
  second edition 129, 301
  third edition 129, 142, 149, 165, 311
Sounds Good
  see Radio London 250
Spencer, K.J. 169
Spink, Arthur 133
Stagg, Alan E. 282
Stereo 152, 158, 162, 171–177, 185, 189
  disc 159, 162, 171, 277, 303
  in concert-demonstration 277, 285, 288
  mixing transformer 180
  pickup 163
  tape 159, 162, 183, 277
Stereo Handbook 180, 303
Stevens, Dorothy E. 78, 95–98, 123, 124, 161, 175, 192, 208, 210, 222, 240, 270
Stokowski, Leopold 250
Stroh, John 39
Sugden, A.R. and Co. Ltd 266
Sugden, Arnold R. 159–164, 266
Swift–Levick 42, 100, 113, 185, 302

T

Tainter and Bell 33
Tannoy 67, 243
Tape recorder 159, 199
Teicher and Ferrante 281
Tempest Anderson Hall, York 284
Tetley, Frederick W. 100, 113, 114, 302, 310
The Gramophone 150, 224, 243, 301, 302
  loudspeaker review 155
The Music Teacher 262, 299
Theobald, Leslie 135
The Pianomaker 298
Thistlethwaite, Frank 106–108
Thomas B. Lund and Co. 42, 43, 46, 47
Thomas Dyson Ltd 44
Thompson, Diamond and Butcher Ltd 161
Tillett, George W. 215
Tivoli Hall, Lisbon 284
Tobin, J. Raymond 262, 299
Toms, Stanley E. 284
Trianon Electric Ltd 163
Truqual (volume control) 55–60

U

Ultrasonic tinning 182
U.M. De Muiderkring 304

V

Valentim de Carvalho Ltd 168, 284
Van Dyk, Herman 15
Victor Talking Machines Co. 311
Villchur, Edgar M. 171, 172, 311
Voigt, Paul G.A.H. 60, 172, 185, 250, 254, 306
Voluphone (headphone) 60, 61

W

Walker, Peter J. 151–155, 211, 219, 243, 251, 259, 262–266, 269–274, 279–290
Wallis, Alfred 11
Wall, Peter 230, 232
Ward, Alton 20
Watt indicator 260, 261, 270
Watts, Agnes 134, 255, 295

Watts, Cecil E. 134, 255, 259, 264, 295
  photomicroscopy of records 295
Webb's Radio 112, 314
Weil, Terence 288
Wesley Guild 63
West, Ralph L. 160, 190, 288, 298, 303, 312
Wharfedale Wireless Works
  acoustic filter 152, 157–162, 176, 195
  as Limited Company 137
  brick-built corner unit 118, 121, 168, 248
  cabinet design 52, 215
  cabinet making 62, 72
  cabinet systems 152
  catalogue design 125, 216
  chassis 102, 116, 159
  column speakers 162
  crossover networks 108, 139, 144
  employee numbers 81, 124, 192
  established 41, 44
  exports 55, 130
  extension speakers 51–55, 61, 64, 67–73, 78, 116
  factory speakers 78
  first three-way system 139
  first two-way system 106
  Gem extension speaker 69
  headphones 60
  in Bradford 39
  in Brighouse 67
  in Idle 97
  kits 198
  logo 53, 216
  loudspeaker components see Loudspeaker
  Microgroove Equaliser 125
  microphone 59, 61
  output 53, 72, 80, 98, 158
  overseas agents 56, 154, 174, 304
  Polishing Outfit 80
  public address 200
  relay cabinets 56–60
  R-J cabinet 145, 150, 275
  Sand Filled Baffle (SFB/3) 158
  school speakers 178, 199
  sell-out to Rank 160, 174, 187
  slimline cabinet 195, 196
  speaker/amplifier 52
  speaker switch 73

    tables 80
    trade mark 98
    transformers 59, 72, 78, 105, 155, 180, 185, 190, 200
    twin-diaphragms 67, 178, 185
    volume controls 56, 72, 108, 111, 124
    Voluphone (headphone) 71, 73
    works outing 124
White, Ralph 192, 194
Whyte, Bert 250
Wilkinson, Clarence 24, 27
Wilkinson, Maurice 20
Wilson, Percy 155, 243, 265
Winkles, Kenneth 161, 174, 175
Wireless Constructor 34
Wireless World 40, 55, 89, 195, 257–264, 271, 285, 288
    book review 301
    loudspeaker review 53, 59, 61, 70, 71
    term 'high fidelity' first used 59
    Wharfedale advertisement 39, 49, 58, 141, 146, 167, 276, 299
Wood, C.H. (photographers) 126, 290
World War 1 13, 17, 21–23, 34, 91
World War 2 55, 73, 159

Y

Yeadon airport
    see Leeds–Bradford airport 238